Advanced Materials and Nano Systems: Theory and Experiment-Part 1

Edited By

Dibya Prakash Rai
Department of Physics
Pachhunga University College, Mizoram University
Aizawl, 796001
India

Advanced Materials and Nano Systems: Theory and Experiment

(Part 1)

Editor: Dibya Prakash Rai

ISBN (Online): 978-981-5050-74-5

ISBN (Print): 978-981-5050-75-2

ISBN (Paperback): 978-981-5050-76-9

First published in 2022.

need for a court order if at any point you breach any terms of this License Agreement. In no event will any delay or failure by Bentham Science Publishers in enforcing your compliance with this License Agreement constitute a waiver of any of its rights.

3. You acknowledge that you have read this License Agreement, and agree to be bound by its terms and conditions. To the extent that any other terms and conditions presented on any website of Bentham Science Publishers conflict with, or are inconsistent with, the terms and conditions set out in this License Agreement, you acknowledge that the terms and conditions set out in this License Agreement shall prevail.

Bentham Science Publishers Pte. Ltd.
80 Robinson Road #02-00
Singapore 068898
Singapore
Email: subscriptions@benthamscience.net

CONTENTS

FOREWORD

The editor **Dr. Dibya Prakash Rai** has requested me to read the first edition of the book *"Advanced Materials and Nanosystems: Theory and Experiment"*. I am fortunate and delighted to have the opportunity to read and review this book before it is officially published. I'm particularly pleased to learn that this book will be published by one of the reputed publishers, **"Bentham Science"** which has published over 150 articles, books, and other works in multidisciplinary fields. Not only that, it has a large number of followers, readers, and subscribers worldwide. The book's title looks very attractive, broad, trendy, and interdisciplinary in the rapidly growing research areas. All the chapters and their titles are very much within the scope of the book as proposed.

This book's collection provides brief and comprehensive information on current advanced materials and nanosystems research. All the contributing authors are experts in their respective fields, such as in Physical Sciences, Chemical Sciences, Material Sciences, Nano Sciences, BioSciences, and Engineering Sciences. Awareness of Nanotechnology is a strategic aim in today's situation to meet the significant objectives of energy harvesting, generation, usage, and storage, along with water treatment, waste management, medical sciences and more.

This book is written and visualized as an instrumental material that intends to offer information for technology progress to students, scholars, scientists, engineers, and professors. The present phases of nanotechnology and the creation of novel energy materials are highlighted in this book, defined by the integration of fundamental knowledge, the fabrication of nanostructure samples, and analysis using standard theory and computing. Nanotechnology is crucial both in terms of experiment and theory to tame and utilize material energy in devices. There has been a worldwide scientific revolution concerning the preparation and integration for the development and uses of new technologies. The current novel discoveries, findings, and results from diverse experimental methodologies and theoretical calculations were discussed in several chapters of this book. Many new topics are covered, including wastewater treatment using nanomaterials, nanoparticles in medical research (targeted drugs, pharmaceutics, dental implants), nanomaterial for air purification, nanomaterials for food preservation, nano-sensors, cosmetics, *etc*. This book contributes to the present trend by bringing together numerous current and foreseen scopes on various themes. It also presents a broad range of viewpoints and opinions on the current state of the art and future prospectus. Whether the readers are novices or experts in the subject, they will start a thought-provoking exploration of contemporary advances in Material Sciences.

My best wishes to **Dr. Dibya Prakash Rai**, for collecting beautiful chapters, compilation, and publication of the first edition of this edited book in **Bentham Science**. I know Dr. Rai for the last ten years as a dynamic and hardworking researcher. I believe he will continue his leadership role in advancing computational material sciences and advanced materials with this book.

P.K. Patra
Department of Physics
North-Eastern Hill University (NEHU)
Shillong, India

PREFACE

In the present scenario, the research in materials sciences are devoted to enslaving the solid-state materials and utilising their properties in devices to perform desired applications from nanotechnology. The introduction of new materials and manipulation of their exotic properties in device fabrication for advancing technology is crucial, which has urged me to think for this title, "**Advanced Materials and Nanosystems: Theory and Experiment**". Nanomaterials and Nanoscience have taken this development in science and engineering to a new height for the well-being of humankind. This topic covers all sectors of research field like Physics, Chemistry, Engineering, Biosciences, *etc*. This will only be the possible book covering maximum relevant and latest topics both from a theory and the experimental point of view. The present compilation would help many students, scholars, teachers, scientists, *etc.*, in their curriculum development and research work. This book highlighted the latest development and the significant role of different new materials in future technology.

In **Chapter 1**, Thakur *et al.*, elucidated the theoretical progress of the emergent materials like Carbon and boron nitride nanostructures for hydrogen storage applications from the first-principles simulations. They discussed phenomenological models and explore the essential chemical and physical properties for integration into the applied sciences. The direct combinations of theory and model provide concise pictures in understanding the diverse phenomena of 2D nanomaterials for potential H_2-storage capability, they can thoroughly clarify the critical mechanisms in pristine subsystems and even the composite systems. Interestingly, there exist a lot of theoretical predictions on the stationary ion transport. However, how to develop a unified theoretical framework suitable for the very complicated and active chemical environments in ion-related batteries remains to be urgently solved during the near-future studies. The close relations between the optimal current density and the vanishing/slight/great asymmetries of crystal structures are expected to be a studying focus.

Chapter 2, discusses the prospective future nanomaterials for retinal implant technique. From a medical science point of view, this work would be of great interest in which the authors have discussed the artificial vision for blind patients suffering from retinal diseases with the help of advancement in CMOS technology. The materials such as titanium nitride (TiN), iridium oxide (IrOx), platinum grey, and carbon nanotube (CNT) were employed in recent years in many retinal prosthetic projects. This chapter discusses the important and desired physical properties of nanomaterials viz. conductivity, tensile strength, absorption of photons, and adsorption of water molecules for the subretinal implant technique.

Chapter 3, this chapter talks about the advancement of electrode materials for rechargeable batteries. The demand for high-performance batteries has exploded like never before. To meet such a prospect massive amount of research endeavours in the design and development of high-performance rechargeable batteries are being taken. Starting with such critical analysis, they discuss the fundamental principle that governs the performance of electrochemical devices. They reviewed the state-of-the-art advancement of various types of materials used in the fabrication of electrodes including a description of their structures and storage mechanisms.

Chapter 4, Kumar *et al.* reported the production of ammonia from Nitrogen Reduction Reaction (NRR) adopting an eco-friendly approach. They design cost-effective catalysts holding on a substrate for the nitrogen reduction reaction. As an alternative, the direct conversion of nitrogen has been carried out by photocatalysis and electrocatalysis. This chapter discusses the challenges faced by researchers to formulate righteous catalysts for the sustainable reduction of nitrogen by studying each of these types with a few examples.

Chapter 5 discusses the nanoparticles in environmental remediation. Nanocomposites offer an exclusive advantage over bulk materials in terms of efficiency on account of their greater surface area, higher reactivity, ease of modification, good dispersion and hence, multi-faceted applications. The various forms of nanocomposites derived from low-cost resources, especially carbon-based materials are of unique interest. Activated carbons offer a unique advantage as the matrix for nanocomposites synthesis, large surface area and porosity offer vivid applications in various fields such as environmental remediation as adsorbents, suitable sorbents in analytical determination of organics, targeted drug delivery, diagnostic agents, fuel cells and sensors, to name a few. The role of nanocomposites as sensors and environmental remediation tools includes adsorption, nano-catalysis, membrane filtration, *etc.*, for pollutants ranging from inorganic ions, heavy metals, pesticides, dyes, anti-bacterial, oil spills and many more.

Chapter 6 reveals the concentration-dependent optical properties of aqueous, ethanol and toluene binary solutions, the refractometry method was used. The direct relation between the chemical bonds of the molecules and their chemical structure is discussed.

Chapter 7 overviews a description of the Nanotechnology-Based Nanomaterials focusing on the developments and challenges of Metal Oxide Nanoparticles such as chromium oxide (Cr_2O_3) nanoparticles, indium oxide nanoparticles (In_2O_3), and magnesium oxide (MgO) nanoparticles. These materials are considered novel materials for biological and smart applications such as antimicrobial, drug delivery systems and cancer therapy. The mechanism of anti-microbial activities of metal oxide nanoparticles is discussed here in detail.

Chapter 8 describes the analysis of the effect of load direction on the stress distribution in orthopaedic implants. Characteristics of implant materials such as rigidity, corrosion, biocompatibility, surface morphology, tissue receptivity, and stability are the key factors that influence the choice of the implant material. The mechanical properties of the implants are one of the significant factors for bone substitution. In this study, 3-dimensional modelling of implant and simulation using the finite element analysis software were incorporated to investigate the effect of load direction on the stress distribution in different orthopaedic implant materials.

Chapter 9 accentuates the latest breakthrough in the catalysis, sensing and wastewater treatment applications of advanced and smart materials. The number of catalytic and sensing operations of advanced and smart materials is discussed in detail. Catalysis and sensing phenomena involve the conversion of obtained signals into a readable format and advanced materials with their exemplary optical, semiconducting or physical properties are studied widely. With the advancement in the latest synthesis and functionalization methods, these advanced materials are becoming nanohybrid systems. It covers the implementation of these nanohybrid systems for catalysis, wastewater treatment and sensing.

Chapter 10 discusses the recent advancement of topological materials. Topological materials are characterized by a unique band topology that is prominently distinct from ordinary metals and insulators. This new type of quantum material exhibits insulating bulk and conducting surface states that are robust against time-reversal invariant perturbations. Bi_2Se_3, Sb_2Te_3 and Bi_2Te_3 were predicted as 3D Topological insulators (TIs) with a single Dirac cone at the surface state. For application purposes, however, bulk conductivity due to Se vacancy in Bi_2Se_3 or anti-site defects in Bi_2Te_3 has been a challenging issue. To achieve an enhanced surface conductivity over the bulk, nanomaterials are irreplaceable. Nanostructures' high surface to volume ratio provides a good platform for investigating the topological existence of surface states.

Chapter 11 is a final chapter that provides information about the hollow nanostructure materials, the most studied topics in current nanoscience research. These hollow structures can be in the form of nanospheres, nanocages, nanorods, nano boxes, single-layered, multi-layered, *etc*. All these variations in hollow structures like carbon buckyball, nanotubes, *etc*. open up several applications in various fields of research from biomedicines to optoelectronics. The observed properties of a material in a hollow shape, like better conductivity, trapping capacity, and catalytic effect, *etc*. This chapter covers the basic information about different kinds of hollow structures like carbon buckyball, variations in their properties along with recent developments, and their applications. Also, includes detailed research about buckyball structures of ZnO, ZnS, and Al-doped ZnO using simulations, with their comparative study and future applications.

Guest Editor:

Dibya Prakash Rai
Department of Physics
Pachhunga University College, Mizoram University
Aizawl, 796001
India

List of Contributors

A. Banik	Department of Electrical Engineering, National Institute of Technical Teachers' Training & Research (NITTTR), Kolkata, India
A.K. Shrivastav	Department of Physic, National Institute of Technology, Raipur, India
Abdulmutallib M. Kokhkharov	Institute of Ion-plasma and laser technologies, Tashkent, Uzbekistan
Abhishek Kumar Gupta	Department of Physics and Material Science, Madan Mohan Malaviya University of Technology, Gorakhpur (U. P.), 273010, India
Anjali Oudhia	Department of Physic, Government Nagarjuna P.G. Science College, Raipur, India
Ashish Tiwari	Computational Nanomaterial Lab, Department of Applied Physics and Department of Electronics and Telecommunication, Shri Shankaracharya Technical Campus, Bhilai, Chhattisgarh, India
Asma Almontasser	Department of Applied Physics, Z. H. College of Engineering & Technolog, Aligarh Muslim University, Aligarh-202002, India
Azra Parveen	Department of Applied Physics, Z. H. College of Engineering & Technolog, Aligarh Muslim University, Aligarh-202002, India
B. Chettri	Physical Science Research Center (PSRC), Department of Physics, Pachhunga University College, Mizoram University, Aizawl 796001, India Department of Physics, North-Eastern Hill University, Shillong793022, India
D.P. Rai	Physical Science Research Center (PSRC), Department of Physics, Pachhunga University College, Mizoram University, Aizawl 796001, India
Debarati Pal	Department of Physics, Himachal Pradesh University, Shimla, India
Deeksha R.	Department of Chemistry, Faculty of Mathematical and Physical Sciences, M.S. Ramaiah University of Applied Sciences, Bengaluru-560058, India
Deepak Kumar	Department of Chemistry, Faculty of Mathematical and Physical Sciences, M.S. Ramaiah University of Applied Sciences, Bengaluru-560058, India
Dilbar T. Bozorova	Institute of Ion-plasma and laser technologies, Tashkent, Uzbekistan
Feruza T. Umarova	Institute of Ion-plasma and laser technologies, Tashkent, Uzbekistan
Ishika Tulsian	Department of Physics, Aditya Birla World Academy, Mumbai, India
K.O. Obodo	HySA Infrastructure Centre of Competence, Faculty of Engineering, North-WestUniversity (NWU), P. Bag X6001, Potchefstroom, 2520, South Africa
Mahavir Singh	Department of Physics, Himachal Pradesh University, Shimla, India
Manit Gosalia	Department of Physics, Aditya Birla World Academy, Mumbai, India
Mavlonbek A. Ziyayev	Institute of Bioorganic Chemistry, Tashkent, Uzbekistan
Mohan L.Verma	Computational Nanomaterial Lab, Department of Applied Physics and Department of Electronics and Telecommunication, Shri Shankaracharya Technical Campus, Bhilai, Chhattisgarh, India
Nihar R. Mahapatra	Department of Physics, Aditya Birla World Academy, Mumbai, India

Oksana B. Ismailova   Institute of Bioorganic Chemistry, Tashkent, Uzbekistan
Turin Polytechnic University in Tashkent, Uzbekistan
Uzbekistan-Japan Innovation Center of Youth, Tashkent, Uzbekistan

Premanjali Rai   Trace Organics Laboratory, Central Pollution Control Board, Parivesh Bhavan, East Arjun Nagar, Delhi-110032, India

Sakshi Sharma   Department of Physic, National Institute of Technology, Raipur, India

Sarvesh Kumar Gupta   Department of Physics and Material Science, Madan Mohan Malaviya University of Technology, Gorakhpur (U. P.), 273010, India

Shikha Rana   Department of Physics, Himachal Pradesh University, Shimla, India

Shivani Gupta   Department of Physics and Material Science, Madan Mohan Malaviya University of Technology, Gorakhpur (U. P.), 273010, India

Shukur P. Gofurov   Tsukuba University, Ibaraki-shi, Japan
Uzbekistan-Japan Innovation Center of Youth, Tashkent, Uzbekistan

Swapnil Patil   Department of Physics, Himachal Pradesh University, Shimla, India

Y.T. Singh   Physical Science Research Center (PSRC), Department of Physics, Pachhunga University College, Mizoram University, Aizawl 796001, India
Department of Physics, North-Eastern Hill University, Shillong793022, India

CHAPTER 1

Carbon and Boron Nitride Nanostructures for Hydrogen Storage Applications; A Theoretical Perspective

Y.T. Singh[1,2]**, B. Chettri**[1,2]**, A. Banik**[3]**, K. O. Obodo**[4] **and D.P. Rai**[1,*]

[1] *Physical Science Research Center (PSRC), Department of Physics, Pachhunga University College, Mizoram University, Aizawl796001, India*

[2] *Department of Physics, North-Eastern Hill University, Shillong793022, India*

[3] *Department of Electrical Engineering, National Institute of Technical Teachers' Training & Research (NITTTR), Kolkata, India*

[4] *HySA Infrastructure Centre of Competence, Faculty of Engineering, North-WestUniversity (NWU), P. Bag X6001, Potchefstroom, 2520, South Africa*

Abstract: We present the recent progress in hydrogen storage in carbon and boron nitride nanostructures. Carbon and boron nitride nanostructures are considered advantageous in this prospect due to their lightweight and high surface area. Many researchers highlight the demerits of pristine structures to hold hydrogen molecules for mobile applications. In such cases, weak van der Waals interaction comes into account. Hence, the hydrogen molecules make weak bonds with the host materials and, therefore, weak adsorption energy and low hydrogen molecules uptake. So, to tune the adsorption energy and overall kinetics, methods such as doping, light alkali-alkaline earth metals decoration, vacancy, functionalization, pressure variation, application of external electric field, and biaxial strain have been adopted by many researchers. Physisorption with atoms decoration is promising for hydrogen storage applications. Under this condition, the host materials have high storage capacity, average adsorption energy and feasible adsorption/desorption kinetics.

Keywords: Adsorption energy, Boron nitride, Carbon nanotube, Chemisorption, Density Functional Theory, Desorption temperature, Graphene, Hydrogen storage, Physisorption, Pressure, Temperature.

INTRODUCTION

Population increase and rapid population surge in different parts of the world, accompanied by the need for a sustainable and better quality of life, resulted in

* **Corresponding author D. P. Rai**: Physical Science Research Center (PSRC), Department of Physics, Pachhunga University College, Mizoram University, Aizawl 796001, India; Tel: +918132832252; E-mail: dibya@pucollege.edu.in

a significant increase in energy demands [1]. Currently, fossil fuels are the primary and dominant energy source due to their established infrastructure, ease of delivery, and cost competitiveness compared to other energy sources. Fossil fuels are known to cause serious harm to the environment due to the emission of harmful pollutants by their use in different industries. The need for renewable energy sources to mitigate these environmental challenges and the secure energy future due to the limited availability of fossil fuels is excellent. The transition to a clean, renewable energy source is essential due to the various drawbacks of fossil fuels.

Lithium-ion (Li-ion) batteries in Unmanned Aerial Vehicles (UAVs) have limitations in their operating range due to low energy density. As hydrogen has a very high mass and volume-specific energy value, it can provide a significant range improvement over Li-ion batteries. Hence, efforts are ongoing to investigate the potential of hydrogen-fueled power plants for small UAVs [2]. The use of hydrogen fuel cell electric vehicles (HFCEVs) produces zero tailpipe pollutant emissions and is more traditional than gasoline-based internal combustion engine vehicles (ICEVs). Study shows that an HFCEV, even fueled by hydrogen from a fossil-based production pathway, uses 5%–33% less WTW fossil energy and has 15%–45% lower WTW greenhouse gas emissions than a conventional gasoline ICEV [3].

Hydrogen is abundant in nature but not in a free state. It can be utilized as an energy carrier because it does not have any harmful by-products during combustion (by-product is water) and has a high energy density compared to other elements [4, 5]. Even though hydrogen has many advantages over other energy sources, the problem lies in its storage. As per the United States Department of Energy, the benchmark hydrogen uptake capacity should be above 6.5wt%. Different means of storing hydrogen have been explored, such as compressed gas, liquid organic hydrogen carriers, inorganic systems, *etc* [6 - 8]. The average adsorption energy benchmark is set to 0.2-0.8 eV per H_2 molecule at ambient conditions. However, considering the recent progress made in hydrogen storage materials, they lack at achieving all the benchmark criteria. Also, the current experimental storage methods are not cost-effective and have safety concerns.

High-pressure storage is the most accomplished and easiest method to store hydrogen. But the main obstacle to this method of storing hydrogen is the high manufacturing and development costs. For vehicle application, hydrogen storage needs to be at extremely high pressure of 700-1000 bar. The consumed power is 10% of the gas energy content at the mentioned hydrogen storage pressure range [9 - 11]. Liquid hydrogen storage is another method of hydrogen storage that shows significant storage density and safety benefits compared with pressure storage. The total power consumption in this method is about 35% of the energy content of the stored hydrogen, which is relatively higher than other hydrogen storage techniques. The liquid hydrogen storage method is quite popular for space and flight applications as high volumetric and gravimetric energy storage density is required regardless of its high energy consumption [12, 13]. In the cryo-compress method, optimized temperature and pressure values for hydrogen storage enhance the storage density compared to the pressure storage method. The power consumption value reduces to 25% compared to the liquid hydrogen storage method [13].

Recently, many researchers have focused on material-based hydrogen storage techniques. The materials used for storing hydrogen in such techniques should have specific characteristics such as good reversibility, affordable price, high storage capacity at operating temperature, and pressure. The possibility of hydrogen storage on nanomaterials like silicene, TMDs, carbon nitrides, silicon carbides, boron carbides, metal hydrides, magnesium-based hydrides, metal nitrides/amides/imides, polymers, clathrate hydrates, zeolites, metal-organic frameworks (MOFs) has been studied. Most of these materials have a storage capacity in the desired range, but the hydrogen molecules are mostly chemically absorbed by the host materials. In such cases, the materials are not considered fit for onboard applications. At present, researcher's interest is in MOFs, which under doping show a favorable response with better adsorption/desorption kinetics towards hydrogen storage [7, 8, 9 - 16, 17, 18]. Due to the high porosity, one-dimensional nanostructures have become an attractive prospect for hydrogen storage applications. Kumar *et al.* theoretically investigated magnesium functionalized on the boron clusters for hydrogen energy storage application. Boron clusters (six boron atoms) with two magnesium atoms decoration have a gravimetric density of 8.10 wt% [26]. Magnesium oxide($Mg_{12}O_{12}$) nanotube showed a higher possibility of surface adsorption of 12 - 24 hydrogen atoms. The adsorption of hydrogen on Mg and O sites decreased the work function of the nanotube [27]. When functionalized with transition metals, silicon carbide nanocages and nanotubes shows favorable responses towards hydrogen molecule adsorption [28, 29].

Carbon nitride nanostructures where the bonding between carbon and nitrogen atoms is sp^2-type and similar to graphene have also garnered interest from researchers. The hydrogen adsorption possibility on such nanostructures has garnered interest due to its lightweight and high surface area. Wang *et al.* also highlighted the possibility of lithium and calcium co-doped g-CN nanostructures for hydrogen storage applications. They reported that the hydrogen molecules adsorb on the host material with an adsorption energy of 0.26 eV/H_2 and a gravimetric density of 9.17 wt%. The adsorption mechanism is through polarization of the H_2 molecules by the transfer of charge from lithium and calcium atoms [30]. B_2S monolayer, which also shares graphene-like properties, shows a promising response towards hydrogen storage. Liu *et al.* reported the gravimetric density of 9.7 wt% on lithium decorated B2S nanosheet with a binding energy of 0.14 eV/H2 using dispersion corrected density functional theory [31]. GeC(germanium carbide) monolayer, which shows semiconducting characteristics, is another interesting 2D nanomaterial that is highlighted by Arellano *et al.* for its hydrogen adsorption and storage ability. They decorated the GeC nanosheet with alkali and alkaline earth metals. Weakly adsorption of alkaline earth metals on the GeC nanosheets discarded the possibility of hydrogen adsorption. Alkali metals, when chemically adsorbed on the GeC nanosheet, tune the hydrogen adsorption ability. Potassium adsorbed GeC nanosheet shows better kinetic towards hydrogen adsorption [32]. Light alkali metals decorated borophene with a gravimetric density of 13.96 wt% was reported recently by Wang *et al.* [33]. Calcium decorated MoS_2 under the application of external field stores six hydrogen molecules by Kubas-interactions with adsorption energy ~ 0.14 eV per H_2. Hydrogen molecules are found to be strongly bonded on such bulky systems but are unable to meet the gravimetric density criteria set by USDOE [34].

Carbon nanotubes and graphene were first experimentally synthesized by Ijima *et al.* Novoselov *et al.* in 1991 and 2004, respectively [35, 36]. Since then, the scientific community has highly explored graphene and carbon nanotubes (single and multi-walled) for their applications in device fabrications, sensors, and drug delivery due to their unique chemical, electronics, thermal, and mechanical properties. Bolotin *et al.* highlighted the high electron transport and mobility in suspended graphene. Later, the scope of graphene for electronics and photonics device applications owing to its high charge mobility and electron transport was reported by Avouris *et al.* [37, 38]. Functionalized graphene and CNTs are proving to be useful in biomedical applications. Their in-vitro and in-vivo toxicity upon drug delivery analysis is more likely to be utilized in many biomedical applications [39 - 44]. Due to its superior thermal property, graphene suits well for application in thermal interface materials. The thermal conductivity improves by a factor of ~20 when graphene is present as one of the thermal interface materials. The addition of graphene as thermal interface material proves promising to build thermoelectric devices in the future [45 - 47].

Experimental growth of boron nitride nanotubes using arc-discharge was first made possible in 1995 by Chopra *et al.* [48]. The tight binding theoretical model predicted the possibility of BN nanotubes in the year 1993 by Rubio *et al.* and a year later by Blase *et al.* using the ab initio pseudopotential method and predicting the single and multi-walled BNNTs(Boron nitride nanotubes) [49, 50]. Boron nitride nanotubes are an exciting prospect due to their high young modulus ~ 1.2 TPa, high thermal stability, and chemical inertness. Defects like vacancies, Stone-wales are common during the growth of nanostructures. Such defects tune the nanomaterial properties for device applications [51 - 53]. Boron nitride nanosheets, also known as 'white graphene' due to the similarity in structural properties, were experimentally realized by Novoselov *et al.* in 2005 [54]. As the boron nitride nanosheets are chemically inert, Vatanparast *et al.* investigated the possibilities of anti-cancer drug delivery using density functional theory and molecular dynamics. The adsorption and delivery of the anti-cancer drugs to the target cell prove feasible [55 - 57].

Recently, Avval *et al.* highlighted the BNNTs capability for drug delivery application for breast cancer therapy [58]. Due to their high porosity and unique partial ionic B-N bonding, boron nitride nanotubes are favorable for gas sensing applications. The transition metal-doped BNNTs are feasible for nitrogen monoxide and carbon monoxide sensors. Also, the application of an external electric field tunes the ammonia adsorption on the BNNTs [59 - 62]. SW and DW-BNNTs have shown a favorable response to water purification. They can capture a higher concentration of methylene blue particles present in the water [63]. Boron nitride nanotube incorporation into thermoplastic materials using solution blending was reported to tune the thermal conductivity. Thus, can be utilized as a filler in polymers [64]. Like graphene, BNNTs also have the potential for their application as thermal interface materials [65, 66]. Spin-splitting effects persist on the open BNNTs. Similarly, the local magnetic moment was introduced on BNNTs and BNNSs upon functionalization, doping by transition metals, or creating a vacancy at B/N sites. Such BNNTs are promising for spintronics devices [67 - 73].

Lightweight solid-state materials with i) high surface area and ii) more outstanding hydrogen molecule adsorption capabilities provide a better opportunity for hydrogen energy-based onboard application. In this regard, carbon and boron nitride nanostructures are considered efficient and promising for hydrogen storage. We have highlighted the recent progress and exciting findings on the carbon and boron nitride-based nanostructures for hydrogen storage applications by researchers and research communities through theoretical (DFT) insights.

APPLICABLE METHODS

Density functional theory (DFT) [74] implemented computational software such as VASP, Quantumwise ATK, CASTEP, GAMESS, Dmol^3, *etc.* were employed by the researchers to study the hydrogen adsorption properties of the boron nitride and carbon nanostructures [75 - 79]. In DFT, the Kohn-Sham equation is solved considering many-body electron-electron, electron-ion effects. Some of the standard approximation methods used are local density approximation (LDA) and generalized-gradient approximation (GGA) to deal with the energy exchange-correlation potential [80, 81]. However, as LDA and GGA are said to have some drawbacks in calculating interaction energies between host materials surface and hydrogen molecules, researchers also adopted an advanced method of van der Waals(vdW) dispersion correction methods such as DFT-D2, D3 [74, 75].

Different parameters like binding energy, adsorption energy, average adsorption energy, and desorption temperature are calculated to frame the H_2 storage capabilities of the solid-state nanomaterials.

An equation to calculate the adsorption energy is given below [84 - 86]

$$E_{ad} = E_{tot}(BN/C + H_2) - E_{tot}(BN/C) - nE_{free}(H_2) \tag{1}$$

Similarly, the average adsorption energy is calculated using the following relation [76 - 78],

$$E_{ad} = \{E_{tot}(BN/C + H_2) - E_{tot}(BN/C) - nE_{free}(H_2)\}/n \tag{2}$$

where the BN/C denotes the carbon and boron nitride host material, and $E_{tot}(BN/C+H_2)$ is the total energy of the hydrogen molecules adsorbed system, $E_{free}(H_2)$ is the total energy of a free H_2 molecule, $E_{tot}(BN/C)$ is the total energy of the host material (carbon and boron nitride nanostructures) and n denotes the number of adsorbed H_2 molecules on the host materials.

The hydrogen uptake capacity of the carbon and boron nitride nanomaterials is calculated using the following relations [84 - 86],

$$H_2(wt\%) = \frac{nM_{H_2}}{(nM_{H_2} + M_{Host})} *100 \tag{3}$$

where M_{H2}, M_{Host}, and n are the masses of H_2, host material (Boron Nitride and Carbon nanostructure), and the number of H_2 molecules, respectively.

To quantitatively analyze the desorption process, the desorption temperature (T_D (K)) is estimated using the van't Hoff's equation [87],

$$T_D = \left(\frac{E_{ads}}{K_B} \right) \left(\frac{\Delta S}{R} - \ln P \right)^{-1} \qquad (3)$$

where K_B is the Boltzmann Constant (1.38×10^{-23} J K^{-1}), S = 75.44 J K^{-1} Mol^{-1} is the H_2 entropy change from gas to the liquid phase at equilibrium pressure P=1atm and R(= 8.31 J K^{-1} Mol^{-1}) is the gas constant.

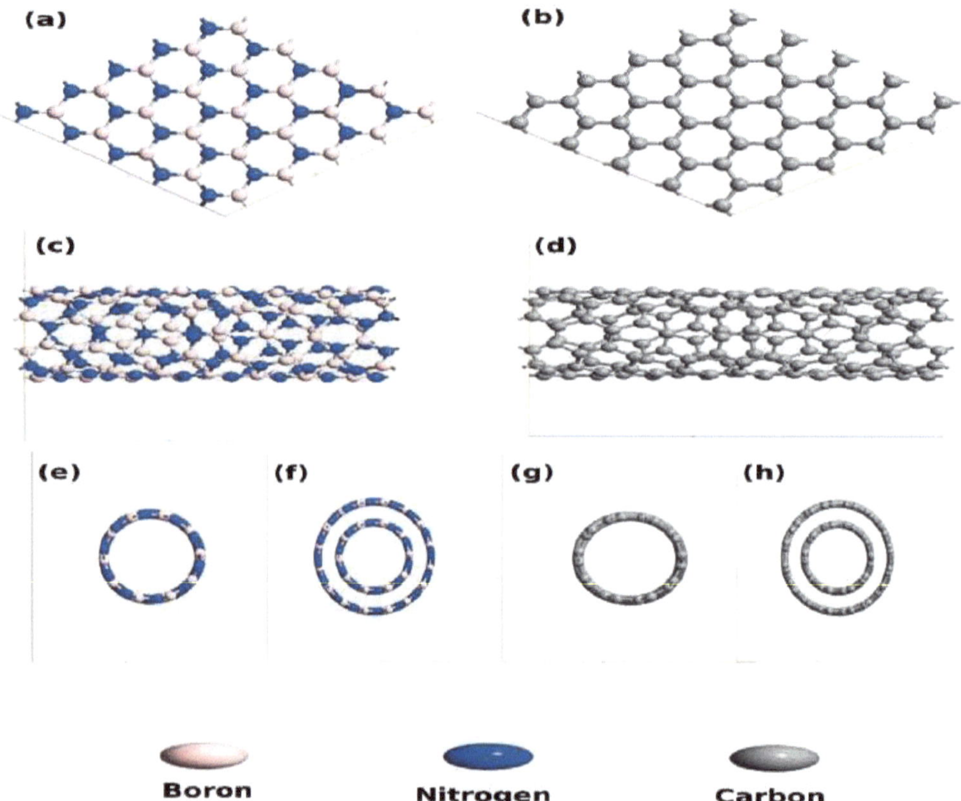

Fig. (1). Boron nitride and carbon nanostructures. **(a)**, **(c)**, **(e)**, and **(f)** are Boron nitride nanosheet, side view of single-walled boron nitride nanotube, front-view of single-walled boron nitride nanotube, and double-walled boron nitride nanotube. Similarly, **(b)**, **(d)**, **(g)**, and **(h)** are graphene, side view of single-walled carbon nanotube, front-view of single-walled carbon nanotube, and double-walled carbon nanotube.

CARBON NANOSTRUCTURES

Many research papers have reported large hydrogen storage capacity in materials like metal hydrides and complex metal hydrides [88 - 90]. But adsorption of hydrogen in such materials mainly happens through the chemisorption mechanism, which results in high absorption energy. Such high adsorption energy leads to poor reversibility avoiding the desorption process [83, 84]. Carbon nanomaterials like graphene and carbon nanotubes are other promising materials for solid-state hydrogen storage as they possess a large specific surface area, high polarity, and low mass density [93]. Graphene is a single layer hexagonal lattice of sp^2 hybridized carbon atoms. It is semi-metallic and is a zero bandgap semiconductor (see Fig. **2a**). Early research reports wide applications of graphene in the field of nanoscience. It has high thermal stability, flexibility, conductivity, and storage capacity [94]. Theoretical and experimental investigations report graphene as a good choice for gas adsorption and desorption due to its low binding energy value of approximately 0.2 eV, which can be easily reversed [95 - 97].

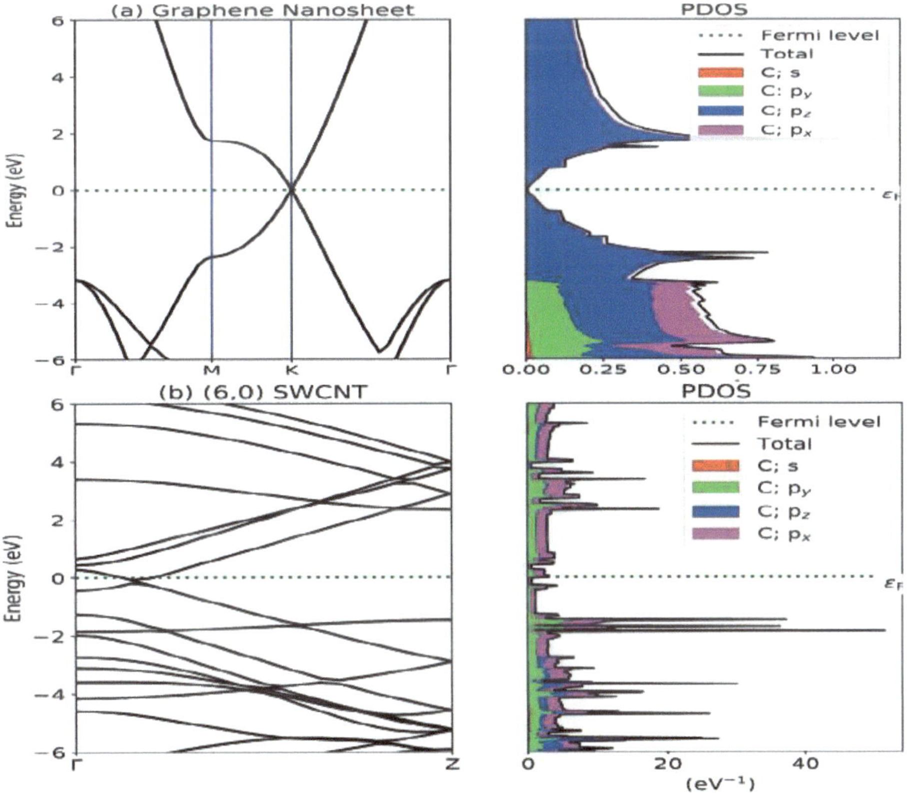

Fig. (2). (a) Electronic band structure and Projected Density of states for monolayer graphene (b) Electronic band structure and Projected density of states for (6,0) SWCNT.

Carbon nanotubes (CNTs) are another form of carbon-nanomaterials whose structure is the same as rolling graphene sheets in some specific direction. Still, in this case, the carbon atoms are sp^3 hybridized. Based on formation, CNTs are of two types: single-walled carbon nanotubes (SWCNTs) and multi-walled carbon nanotubes (MWCNTs). The electronic profile of the single-walled carbon nanotubes(SWCNTs) depends on the chirality (n, m) and diameter of the tube [98]. Fig. (**2b**) shows the band structure and projected density of states (PDOS) of a typical SWCNT. The modification of the electronic and mechanical properties opens the door for SWCNT's broad applications in making neon devices like nano detectors [99, 100], nano transistors [101, 102]. Many researchers also reported CNTs as a promising candidate from gas sensing devices as it shows good selectivity and sensitivity on different gases [103, 104]. This section briefly discusses the previous work on hydrogen absorption of graphene sheets and CNTs.

GRAPHENE NANOSHEETS

Hydrogen adsorption in graphene takes place either by physisorption or chemisorption. The physical absorption of hydrogen in graphene shows weak adsorption capacity. Some researchers even suggest that it is unsafe due to the adsorbent's fragile stability, making it impossible for long-term storage [105, 106]. Theoretical investigations reported the maximum gravimetric density (GD) value for graphene due to physical and chemical adsorption as 3.3% and 8.3%, respectively [107]. These GD values for graphene change with temperature and pressure. Lai-Peng *et al.* [108] investigated the hydrogen adsorption behavior for graphene at the temperatures above critical values, taking pressure reference from zero to 100 KPa. The authors reported an increase in adsorption capacity with pressure but decreased when temperature increased. A similar result was also reported by Kim *et al.* [109]. A theoretical investigation on a bilayer graphene system based on the post-Hartree-Fork/empirical potentials reports enhancement of the storage capacity with increased inter-layer spacing [110]. The inter-layer separation of 6 Å to 8 Å increases the GD by 30% to 40% compared to monolayer graphene. Sarah *et al.* [111] theoretically investigated the influence of curvature on hydrogen storage for monolayer graphene. The investigation revealed on adsorption of hydrogen more preferably at the local curvature with maximum convex and reported the high tendency of the hydrogen atoms adsorbed at the minima local curvature to desorb at low temperature.

Storage capacity for pristine graphene reported so far from the theoretical and experimental analysis shows low values [108]. Having weak binding energies (0.01eV to 0.06 eV for physical adsorption and 0.67 eV to 0.77 eV for chemical adsorption) is the main reason for poor storage capacity values. Structural defects, alkali decoration, and doping enhanced the hydrogen storage capacity. The hydrogen molecules are adsorbed on pyridinic N-doped graphene by physisorption and chemisorption mechanism. Through the forms process, the hydrogen binding energy is 0.64 eV. Whereas when the hydrogen gets adsorbed through the chemisorption process, the binding energy is 2.03 eV almost three times higher binding energy than pristine graphene [112]. From theoretical analysis, Z. M. Ao *et al.* reported the storage capacity value of 5.13 wt% for aluminum-doped graphene structure at ambient temperature and pressure, close to the US DOE value [105]. When decorated with Li atom in an external electric field and field-free condition, graphene with Stone-Wales defect exhibits high storage capacity within the average physisorption energy of 0.1-0.6 eV/H_2 with a desorption temperature 289 K [113]. Table **1** summarizes the details of the above investigation.

Table 1. Hydrogen storage capacity, physisorptionbinding energy, chemisorption binding energy, desorption temperature for different graphene-structure are summarized.N: Nitrogen, Al: Aluminum, T_D: Desorption temperature.

Graphene System	Storage Capacity (wt%)	Physisorption binding Energy (eV/H_2)	Chemisorption binding energy (eV/H)	T_D (K)	Ref.
Pristine-graphene	-	0.01-0.06	0.67-0.77	-	[108]
Pyridinic N-doped graphene	-	0.64	2.03	-	[112]
Al-doped-graphene	5.13	0.260	-	-	[105]
Li-decorated (55-77)-graphene	-	0.1-0.6	-	289	[114]

CARBON NANOTUBES

Many researchers have performed theoretical and experimental analyses on the hydrogen adsorption of CNTs [25, 39, 115, 116]. Early results show inconsistent storage values. Some researchers even reported CNTs as not good for hydrogen storage [117, 118]. An unclear adsorption mechanism(whether physisoroption or chemisorption) is the main factor for such inconvenient results. But at present, it is widely accepted that the hydrogen absorption mechanism in CNT is the co-existence of irreversible chemisorption and reversible physisorption [115, 119]. Through elastic recoil detection analysis, Safa *et al.* reported the predominance of the physisorption mechanism(adsorption through weak van der Waals interaction) for hydrogen adsorption at cryogenic temperature [120]. If the temperature ranges from 30°C to 100°C, hydrogen desorption is more common, but chemisorption is predominant at the temperature range from 100°C to 300°C. Several investigators have also reported the effect on hydrogen storage capacity by internal factors such as specific surface area, tube diameter, tube-curvature, *etc.*, and external factors

like measurement methods, temperature, pressure, *etc* [117, 121, 122].
Arellano *et al.* from the theoretical investigation of the hydrogen storage capacity
on SWCNTs of several diameters, reported the increase in molecular hydrogen
(H_2) binding energy with tube diameter decreases [32]. Their result agreed well
with Mpourmpakis *et al.* [123] and Kentaro *et al.* [124]. Wang *et al.* theoretically
analyzed different structures of fully hydrogenated CNTs [125]. They reported the
inverse square relation of hydrogenation energy and the tube diameter. The author
also reported that the armchair structure has more H_2 binding energy than the zig-
zag for the same diameter nanotube. The chirality of CNTs affects the storage
capacity only in the case of chemical adsorption(chemisorption) [119, 123].
However, in the physisorption mechanism, the effect of the tube's chirality is
negligible [115].

A comparative study of isolated SWCNTs and SWCNT bundle on hydrogen
adsorption capacity was investigated by Ghosh *et al.* through molecular dynamics
simulation [126]. The authors reported that an isolated SWCNT adsorption
capacity is significantly higher than that of the SWCNT bundles at lower
temperatures. Due to the smaller inter-tube spacing distance than the adsorbed
hydrogen layer's thickness, hydrogen molecules gets adsorbed in multilayers at
low temperatures. But at high temperatures, SWCNTs bundles show higher
hydrogen storage capacity than the isolated SWCNTs. At high temperatures, the
nanotubes adsorb only a single layer of hydrogen around them, and hence
adsorption within the interstitial space of the bundle becomes possible. The
storage capacity increases with an increase in the interstitial spacing of the tubes.
The authors further investigated the storage capacity for square array and
triangular array tubes and reported that square array tubes have higher storage
capacity than triangular array tubes. Muniz *et al.* also reported similar results
[119]. Through the grand canonical Monte Carlo (GCMC) simulation, Muniz *et
al.* theoretically investigate the hydrogen storage capacity for a structurally
optimized array of SWCNTs [127]. The authors reported that, at a pressure above
1 MPa, the square lattice structure has a higher storage capacity than the
triangular lattice structure. At the pressure below 1MPa, the triangular lattice
structure shows a higher storage capacity value. For the nanotubes of the same
diameter, the storage capacity in SWCNTs is always higher than the MWCNTs.
In the case of the same wall-to-wall spacing in MWNTs, increasing the number of
walls decreases storage capacity. But storage capacity is increased if the wall-to -
wall space increases, keeping the number of layers constant [128]. Many
researchers have also reported the effects of temperature and pressure on the H_2
storage of CNTs [129 - 132]. At constant pressure, the storage capacity decreases
with an increase in temperature, and at a constant temperature, the storage
capacity increases with an increase in pressure. Zhao *et al.* experimentally
investigated the hydrogen storage capacity by setting the reference temperatures at

77 K, 203 K, and 303 K, keeping the pressure at a constant value of 10 MPa [133]. The authors reported the decrease in storage capacity as the temperature increases, with the maximum storage value of 1.73% at 77 K. Poirier *et al.* in their work, also made a similar observation [134]. Considering moderate pressure of 4, 8, 12, 16, and 20 bar at room temperature, Kaskun *et al.* [135] experimentally studied the effect of pressure on hydrogen storage in Ni-MWCNTs and reported that the storage capacity increase with an increase in pressure which supports the result reported by Darkrim *et al.* [136].

Theoretical and experimental investigations on the H_2 storage of pristine CNTs give small storage capacity values [130, 137]. So far, different researchers have proposed various methods to enhance the storage capacity [138 - 140]. Doping is the most common method for carbon materials. From the theoretical calculation, Jung Hyun Cho *et al.* reported the storage capacity of 2.5 for Si-doped SWCNTs, which is double that of undoped SWCNTs [141]. Wang *et al.* reported a very high H_2 storage capacity value of 28 wt% at room temperature of Al-doped SWCNTs, significantly higher than the US DOE value [142]. From the first principle study, Surya *et al.* reported the storage capacity of 13.2 wt% for ammonia doped (5,5) SWCNT with small adsorption energy, which suits the release of hydrogen molecules at ambient conditions [143]. Table **2** summarizes the details of the above investigation.

Table 2. Hydrogen storage capacity, average adsorption energy, desorption temperature for different carbon nanotube structures are summarized. SWCNT: single-walled carbon nanotubes.

System	Storage Capacity (wt%)	Average Adsorption Energy(eV/H_2)	Desorption Temperature (K)	Ref.
Pristine-SWCNT	1.4	-	-	[141]
Si-doped-SWCNT	2.5	-	-	[141]
Al-(7,7)SWCNT	28	0.131	171	[142]
(5,5) SWCNT-5(NH$_3$+5 H$_2$)	8.18	0.091	148	[143]
(5,5) SWCNT-10 (NH$_3$+5 H$_2$)	13.2	0.092	148.5	[143]

BORON NITRIDE NANOSTRUCTURES

Boron nitride(BN) falls in the category of group III-V compounds. It exists in 3 phases:

  i. Hexagonal boron nitride (h-BN)
 ii. Cubic-boron nitride (c-BN)
iii. Wurtzite boron nitride (w-BN)

Hexagonal boron nitride is the most stable phase. Boron nitride nanotubes (BNNTs), boron nitride nanoribbons(BNNRs), and boron nitride nanosheets (BNNSs) are some common boron nitride nanostructures existing in the hexagonal phase. The boron(B)-nitrogen(N) bond length lies between 1.44-1.50 Å, the distance between B(N)-B(N) is 2.52 Å, and the bond angle between B---B(N-B-N) is 120°. Atomic structures of the BNNS and BNNTs are presented in Fig. (**1**).

Fig. **2(a, b)** presents the electronic profile of the boron nitride nanosheet and single-walled boron nitride nanotubes. In literature, the energy band gap of boron-nitride nanostructures lies between 3.6-6.2 eV obtained from different theoretical and experimental studies [144 - 148]. Using GGA-PBE as exchange-correlation functional, the energy bandgap for pristine boron nitride nanosheet falls to be 4.67 eV. By employing hybrid functional(HSE06), a bandgap tunes to 5.67 eV [146]. In most cases, the valence and conduction bands lie at the K-symmetry point, showing a direct bandgap nature see Fig. (**2a**). The valence band is populated by the N-2p orbitals and the conduction band by B-2p orbitals, as seen from the PDOS plot adjacent to the band plot for the boron nitride nanosheet. Similarly, Fig. (**3b**) presents the electronic band profiles for boron nitride nanotubes. A single-walled boron nitride nanotube has a direct bandgap of 3.96 eV at the gamma point, as shown in Fig. (**3b**). In the case of pristine boron nitride nanostructures, the analysis of the electronic profile reveals that N and B-2p orbitals, respectively, populate the valence and conduction bands. Multi-walled boron nitride nanotubes with different chiralities follow a similar trend [149, 150].

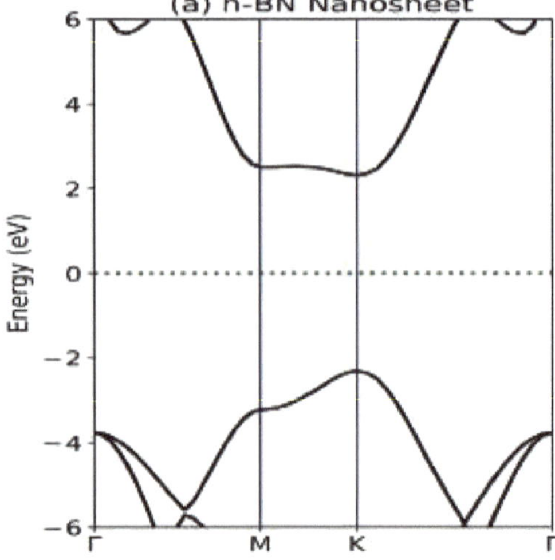
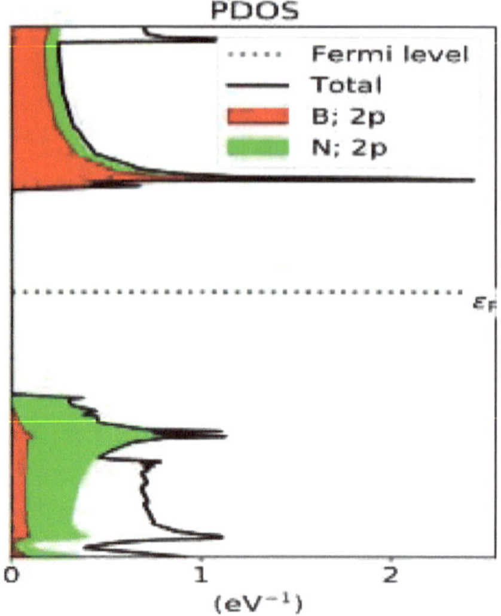

(Fig. 3) contd.....

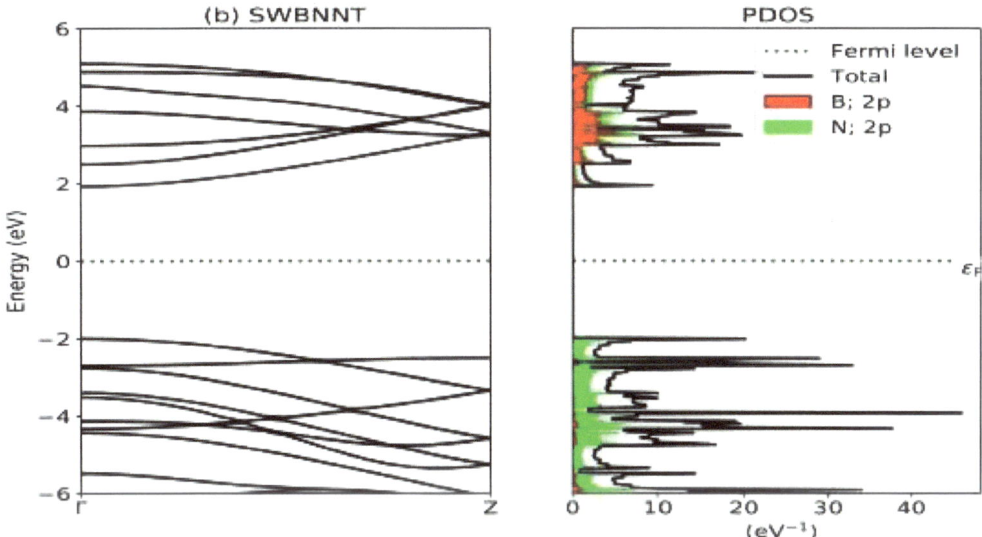

Fig. (3). Energy bandstructure and partial density of states for **(a)** pristine h-BN nanosheet and **(b)** single-walled boron nitride nanotube. In PDOS red and green color stacked represents boron(B) and nitrogen(N) 2p orbital contribution.

BORON NITRIDE(H-BN) NANOSHEETS

Boron nitride shares similarities in structural phases with carbon nanostructures. Like graphene nanosheets and carbon nanotubes (CNT), h-BN nanosheets and h-BN nanotubes(boron nitride nanotubes-BNNT) share similar structural properties. These nanomaterials have a high surface area to volume ratio and are lightweight, advantageous in many prospects for hydrogen storage. A common method of hydrogen molecule decoration on the pristine host materials is employed to examine their interactions. However, due to poor kinetics between pristine systems and H_2 molecules, the pristine systems cannot attain all the criteria put forward by the USDOE. Hydrogen storage on BNNSs was first experimentally reported by Lei *et al.* They reported a maximum storage capacity of 5.7 wt% under 5 MPa at room temperature. They also checked the desorption kinetics, where ~90% of the H_2 molecules releases at ambient conditions [151]. Chettri *et al.* reported the hydrogen storage capacity of BNNSs to be 6.7 wt% under the physisorption process with an average adsorption energy of 0.13 eV/H_2. The absence of hybridization between B, N-2p orbital of nanosheet and H-1s orbitals of H_2.> molecules hints at the adsorption process through weak van der Waals mechanism [146].

Similarly, anti-Kubas interaction during the hydrogen adsorption process was reported by J. Zhou *et al.*, highlighting the hydrogen storage prospect of h-BN nanosheets under the application of an external electric field normal to the plane of h-BN nanosheet. The H_2 molecule's storage could be tuned under applications of the externally applied field due to the polarisation of H_2 molecules when the molecules are oriented perpendicular to the plane of h-BN nanosheets [152]. They also reported the possibility of hydrogenation at different B/N and hollow sites as a mode to tune the energy bandgap of the *h*-BN nanosheet in their earlier theoretical work [153].

As most of the physical adsorption process is through weak van der Waals interaction and minimal charge transfer, the adsorption energy and the desorption temperature criteria are not met. Hence, some external influences like creating a vacancy, doping, decoration of atoms, applications of strain, external electric field enhance the hydrogen adsorption ability. This external influence increases the number of adsorbed hydrogen molecules and enhances the binding energy by increasing the charge transfers between the host materials and H_2 molecules. Tang *et al.* studied the effect of external electric field and biaxial strain on the hydrogen storage of Na-decorated boron nitride nanosheets. The external influence of field and biaxial strain increases the number of hydrogens adsorbed by the Na atoms. With a biaxial strain of 15% and external field strength of 5.14 nm^{-1}, Na-decorated BN-nanosheets meet the benchmark for hydrogen storage with an adsorption energy of 0.23 eV/ H_2 and hydrogen uptake capacity of 9 wt%. The external factor facilitates the flow of electrons from the hydrogen molecules to the Na-atom. Thus the results highlight that Na-decorated BNNSs are efficient in adsorbing many H_2 molecules [154]. A similar approach, where the external electric field is applied to tune the hydrogen adsorption kinetics of the boron nitride bilayers, was followed by Chettri et al., who reported the gravimetric density of 6.7 wt% on bilayer boron nitride with adsorption energy in the acceptable range. The external field helped to tune the desorption temperature for such a bilayer boron nitride system with the enhancement in adsorption energy [155].

Hydrogenated boron nitride nanosheet with lithium (Li) functionalization shows a promising response towards H_2 storage. On replacing H-atom with Li from the hydrogenated BN-sheet, the Li atom becomes cationic, and the Li-substituted region acts as a binding site for H_2 molecules. Thus, Li functionalization improves the hydrogen uptake ability of the hydrogenated BNNSs. The physisorption process dominates the hydrogen molecules adsorption [156]. Boron nitride atomic chains with different length decorated with Li atoms at both ends proves to be a promising candidate with an H_2 storage capacity of 29.2 wt% [157]. Another good approach is to tune the overall adsorption/desorption kinetics by functionalizing the boron nitride nanosheets by OLi_2 and CLi_3. Due to the strong polar nature of oxygen and lithium, carbon and lithium, cationic lithium helps absorb more H_2 molecules. As a result, the storage capacity increases to 6.80 wt% with desirable adsorption energy(0.15 eV per H_2) [158].

Hybrid structures have also garnered interest in checking their hydrogen storage abilities due to their unique physical and electronic properties. In hybrid systems, the charge distribution differs from its pristine constituent structures. As reported by Hussain *et al.*, lithium-decoration on hybrid domains of boron nitride and graphene show promising hydrogen storage capability with a gravimetric density of 8.7 wt%. The hybridization of Li-p and H-1s orbitals drives the hydrogen storage mechanism. Transfer of charge takes place from lithium to hydrogen molecules, thus tuning the electrostatic interactions. In this way, Li decoration tunes the hydrogen storage ability with an orbital interaction of the hybrid boron nitride and graphene domains [159].

Doping and decoration by foreign atoms have been considered another mode of enhancing and fulfilling the criteria for storing hydrogen molecules on the h-BN nanosheets. Ren *et al.* reported that Platinum and Palladium doped h-BN nanosheets could store hydrogen up to 4.5 wt% with an average adsorption energy of 1.010 and 0.705 eV per hydrogen molecule. The Kubas interaction and polarisation mechanism for H_2 adsorption on BN nanosheet are responsible for the higher adsorption energy [160]. Tokarev *et al.* presented a detailed study of pristine and oxygen-doped boron nitride for hydrogen storage. They reported that oxygen doping tunes the adsorption capability [161]. Recently, Rad *et al.* theoretically discussed the possibility of nickel decorated boron nitride nanoclusters as a hydrogen storage material. Decoration of nickel leads to the strong binding of the H_2 molecules in their vicinity, thus increasing the hydrogen storage capacity [162]. Kumar *et al.*, using the dispersion corrected DFT reported the enhanced H_2 storage ability of the carbon and oxygen co-doped BN-nanosheet [163]. Titanium decorated BN nanosheets have a storage capacity of 6 wt%. The adsorption energy lies between 0.3-0.7 eV/H_2 [164].

BORON NITRIDE NANOTUBES

BNNTs are superior to CNTs in thermal and mechanical stabilities, chemical inertness, and unique bonding nature, making them an exciting prospect for energy storage applications. The theoretical progress made to study the credibility of BNNTs to store H_2 molecules in recent years is quite outstanding. The hydrogen uptake capacity on (multi-walled and bamboo-like) BNNTs was first reported experimentally by Ma *et al.* The respective storage capacity of the BNNTs were 1.8 and 2.6 wt% under 10 MPa and ambient conditions [165]. Later, Jhi *et al.* performed a theoretical (DFT) study and reported that the modification of the sp^2 type bonding could lead to higher hydrogen storage with enhanced binding energy. Also, BNNTs have higher binding energy for H_2 molecules than CNTs [166]. Due to the bonding between B-N atoms, boron nitride nanotubes are preferably better hydrogen adsorption materials than carbon nanotubes (CNTs).

These bonds are covalent and partially ionic. In contrast, C-C bonds in the case of carbon nanostructures are covalent. Thus, ionic bonding plays a vital role in the adsorption of hydrogen molecules in BNNTs compared to CNTs. The DFT study performed by Mpourmpakis *et al.* validates the above result. They investigated the H_2 storage properties on (5,5) and (9,9) BNNTs and CNTs. Storage capacity and the adsorption energy were reportedly higher in BNNTs [167].

Ahadi *et al.* reported the hydrogen molecules' adsorption on the single, double and triple-walled BNNTs with different chirality and diameters using grand canonical Monte Carlo (GCMC) Simulation. They also focussed on the role of external pressure on the H_2 storage performance of all the BNNTs. Single and double-walled exhibit higher hydrogen molecule adsorption compared to triple-walled BNNTs. The percentage of hydrogen adsorption in the case of multi-walled BNNTs depends on the inner diameter. The larger the diameter, the smaller the curvature, and the higher the hydrogen molecule's binding. In single-walled BNNTs, adsorption is high when the nanotube diameter is small. The external pressure in units of MPa tunes the storage capacity of the BNNTs, similar to CNTs [168]. As reported by Wu *et al.*, for pristine BNNTs, the hydrogen molecules are adsorbed through surface interaction. The hydrogen molecules dissociate and get chemically adsorbed on the BNNTs when defects like Stone-Wales, substitutional, and vacancy are present [169 - 170]. The change in diameter leads to the shift in B-N bondlength.

Hence, a slight variation in the sp^2 type bonding, enhances boron and nitrogen atom reactivities, thus tuning the hydrogen adsorption ability on those B and N sites [171]. Han *et al.* also observed hydrogen molecule's interaction with single-walled BNNTs [172].

The mechanical properties of the BNNTs change under hydrogen molecule adsorption. Under hydrogen molecules' adsorption, the buckling strength reduces by an average of 14%, and with the temperature variation of 300-3000K, the buckling strength varies from 13-15% [178]. Single-walled BNNTs are decorated with rhodium, palladium, and nickel to increase the H_2 storage capacity. In this case, hybridization occurs between the d-orbital of the Rh, Pb, and Ni and the H-sigma orbital. Thus, the functionalization of BNNTs leads to an increase in average adsorption energy and the hydrogen uptake capacity [179]. Mananghaya *et al.* investigated BNNTs with vacancy defects and decorated by Transition Metals. Only titanium (Ti) and vanadium(V) functionalization were thermodynamically stable. A single Ti and V could bind seven H_2 with an uptake capacity of 7.2 wt% at ambient conditions [180]. However, chemisorbed hydrogen molecules are strongly bonded with the host materials and create hurdles in fast adsorption/desorption kinetics. With the help of metal-free reagents such as triflic acid, Bronsfed acid, the desorption of chemically adsorbed hydrogen molecules from boron nitride nanotubes was made possible at close to ambient temperature [181]. Recently, Banerjee *et al.* investigated the poly-lithiated(CLi$_2$) functionalized BNNTs for hydrogen storage application. Functionalization with different concentrations of CLi$_2$ enhances the E$_{ads}$ between 0.25-0.35 eV per H$_2$ with an uptake capacity of 4.41 wt% [182]. The hydrogen storage properties of some of the expanded, doped decorated boron nitride nanostructures are listed above in Table **3**.

Table 3. Hydrogen storage capacity, average adsorption energy, desorption temperature for different boron nitride nanostructures are summarized. BNNS: boron nitride nanosheet, BNNTs: boron nitride nanotubes, SWBNNT: single-walled boron nitride nanotubes.

Boron Nitride System	Storage Capacity (wt%)	Average Adsorption Energy(eV/H$_2$)	Desorption Temperature (K)	Ref.
Expanded hexagonal boron nitride (eh-BN)	2.46	-	243	[173]
Co-doped h-BNNS	12.94	0.18	243	[174]
Pt-doped h-BNNS	4.93	0.23	271	[174]
Porous BNNS	7.50	0.16		
Ce-doped BNNTs	5.60	0.22	268	[175]
C-doped (BN)$_{12}$ cages	7.43	-	-	[176]
Ti decorated (8,0) SWBNNT	3.9-5.7	-	-	[177]

CONCLUDING REMARKS

This review article summarizes the progress of utilizing the carbon and boron nitride nanostructures for hydrogen molecule storage by employing a first principle study. As revealed from the theoretical studies, hydrogen molecules make weak bonds with pristine carbon and boron nitride nanomaterials. Alternative approaches like creating a vacancy, doping, and decoration of atoms by applying strain, an external electric field on the carbon and boron nitride nanomaterials, are promising to enhance hydrogen adsorption ability. Theoretical results suggest the adsorption of hydrogen molecules on the host materials through physisorption and chemisorption processes. In doping and decoration transition metals on carbon and boron nitride nanostructures, the d-orbital electrons of TMs hybridize with the sigma-orbital of hydrogen molecules. Such kind of interaction is known as Kubas-interaction. The host materials are decorated with lightweight alkali and alkaline earth metals atoms to overcome the low adsorption energy crisis. Another approach is where poly-lithiated molecules decoration on pristine carbon and boron nitride nanostructures tunes the adsorption/desorption kinetics of the host materials. The alkali and alkaline earth metals, when decorated, transfer charge to the host materials and becomecationic, leading to a higher rate of H_2 adsorption. Carbon and oxygen doped on boron nitride nanostructures also show a good response towards hydrogen storage. Most of the theoretical calculation on the carbon and boron nitride nanostructures shows a potential to be utilized for onboard hydrogen storage for mobile application. However, for hydrogen storage, only a few experimental studies are conducted. So, it is the right moment to take these theoretical works as guidance to experimentally synthesize the hydrogen storage material for mobile applications under ambient conditions.

CONSENT FOR PUBLICATION

Not applicable.

CONFLICT OF INTEREST

The author declares no conflict of interest, financial or otherwise.

ACKNOWLEDGEMENTS

D. P. Rai acknowledges a core research grant from the Department of Science and Technology SERB (CRG DST-SERB, New Delhi, India) *via* sanction no. CRG/2018/00009(VER-1).

REFERENCES

[1]　H., Ritchie; M., Roser Energy *Our World in Data,* **2020**.

[2]　Depcik, C.; Cassady, T.; Collicott, B.; Burugupally, S.P.; Li, X.; Alam, S.S.; Arandia, J.R.; Hobeck, J. Comparison of lithium ion Batteries, hydrogen fueled combustion Engines, and a hydrogen fuel cell in powering a small Unmanned Aerial Vehicle. *Energy Convers. Manage.,* **2020**, *207*, 112514.
[http://dx.doi.org/10.1016/j.enconman.2020.112514]

[3]　Liu, X.; Reddi, K.; Elgowainy, A.; Lohse-Busch, H.; Wang, M.; Rustagi, N. Comparison of well-to-wheels energy use and emissions of a hydrogen fuel cell electric vehicle relative to a conventional gasoline-powered internal combustion engine vehicle. *Int. J. Hydrogen Energy,* **2020**, *45*(1), 972-983.
[http://dx.doi.org/10.1016/j.ijhydene.2019.10.192]

[4]　Naqvi, S.R.; Hussain, T.; Luo, W.; Ahuja, R. Metallized siligraphene nanosheets (SiC_7) as high capacity hydrogen storage materials. *Nano Res.,* **2018**, *11*(7), 3802-3813.
[http://dx.doi.org/10.1007/s12274-017-1954-z]

[5]　Schlapbach, L.; Züttel, A. Hydrogen-storage materials for mobile applications.*Materials for Sustainable Energy: A Collection of Peer-Reviewed Research and Review Articles from Nature Publishing Group*; World Scientific Publishing Co., **2010**, pp. 265-270.
[http://dx.doi.org/10.1142/9789814317665_0038]

[6]　Obodo, K.O.; Ouma, C.N.M.; Bessarabov, D. Low-temperature water electrolysis, **2021**.
[http://dx.doi.org/10.1016/B978-0-12-822813-5.00003-5]

[7]　Obodo, K.O.; Ouma, C.N.M.; Modisha, P.M.; Bessarabov, D. Density functional theory calculation of Ti_3C_2 MXene monolayer as catalytic support for platinum towards the dehydrogenation of

methylcyclohexane. *Appl. Surf. Sci.,* **2020**, *529*, 147186.
[http://dx.doi.org/10.1016/j.apsusc.2020.147186]

[8]   Ouma, C.N.M.; Obodo, K.O.; Modisha, P.M.; Bessarabov, D. *"Si, P, S and Se surface additives as catalytic activity boosters for dehydrogenation of methylcyclohexane to toluene - a liquid organic hydrogen carrier system,"*; Materials Chemistry and Physics Under-revi, **2021**, pp. 1-19.

[9]   Züttel, A. *Materials for hydrogen storage,* **2003**.
[http://dx.doi.org/10.1016/S1369-7021(03)00922-2]

[10]   Magneville, B.; Gentilleau, B.; Villalonga, S.; Nony, F.; Galiano, H. Modeling, parameters identification and experimental validation of composite materials behavior law used in 700 bar type IV hydrogen high pressure storage vessel. *Int. J. Hydrogen Energy,* **2015**, *40*(38), 13193-13205.
[http://dx.doi.org/10.1016/j.ijhydene.2015.06.121]

[11]   Zheng, J.; Liu, X.; Xu, P.; Liu, P.; Zhao, Y.; Yang, J. Development of high pressure gaseous hydrogen storage technologies. *Int. J. Hydrogen Energy,* **2012**, *37*(1), 1048-1057.
[http://dx.doi.org/10.1016/j.ijhydene.2011.02.125]

[12]   Lutz, M.; Bhouri, M.; Linder, M.; Bürger, I. Adiabatic magnesium hydride system for hydrogen storage based on thermochemical heat storage: Numerical analysis of the dehydrogenation. *Appl. Energy,* **2019**, *236*, 1034-1048.
[http://dx.doi.org/10.1016/j.apenergy.2018.12.038]

[13]   Yanxing, Z.; Maoqiong, G.; Yuan, Z.; Xueqiang, D.; Jun, S. Thermodynamics analysis of hydrogen storage based on compressed gaseous hydrogen, liquid hydrogen and cryo-compressed hydrogen. *Int. J. Hydrogen Energy,* **2019**, *44*(31), 16833-16840.
[http://dx.doi.org/10.1016/j.ijhydene.2019.04.207]

[14]   Rai, D.P.; Vu, T.V.; Laref, A.; Joshi, H.; Patra, P.K. Promising optoelectronic response of 2D monolayer MoS2: A first principles study. *Chem. Phys.,* **2020**, *538*, 110824.
[http://dx.doi.org/10.1016/j.chemphys.2020.110824]

[15]   Rai, D.P.; Vu, T.V.; Laref, A.; Hossain, M.A.; Haque, E.; Ahmad, S.; Khenata, R.; Thapa, R.K. Electronic properties and low lattice thermal conductivity (κ_l) of mono-layer (ML) MoS$_2$: FP-LAPW incorporated with spin–orbit coupling (SOC). *RSC Advances,* **2020**, *10*(32), 18830-18840.
[http://dx.doi.org/10.1039/D0RA02585B]

[16]   Hussain, T.; Chakraborty, S.; Ahuja, R. Metal-functionalized silicene for efficient hydrogen storage. *ChemPhysChem,* **2013**, *14*(15), 3463-3466.
[http://dx.doi.org/10.1002/cphc.201300548] [PMID: 24009141]

[17]   Hussain, T.; Mortazavi, B.; Bae, H.; Rabczuk, T.; Lee, H.; Karton, A. Enhancement in hydrogen storage capacities of light metal functionalized Boron–Graphdiyne nanosheets. *Carbon,* **2019**, *147*, 199-205.
[http://dx.doi.org/10.1016/j.carbon.2019.02.085]

[18]   Hussain, T.; Chakraborty, S.; De Sarkar, A.; Johansson, B.; Ahuja, R. Enhancement of energy storage capacity of Mg functionalized silicene and silicane under external strain. *Appl. Phys. Lett.,* **2014**, *105*(12), 123903.
[http://dx.doi.org/10.1063/1.4896503]

[19]   Hussain, T.; Sarkar, A.D.; Maark, T.A.; Sun, W.; Ahuja, R. Strain and doping effects on the energetics of hydrogen desorption from the MgH$_2$ (001) surface. *Europhys. Lett.,* **2013**, *101*(2), 27006. [Europhysics Letters].
[http://dx.doi.org/10.1209/0295-5075/101/27006]

[20]   Rahimi, R.; Solimannejad, M. BC3 graphene-like monolayer as a drug delivery system for nitrosourea anticancer drug: A first-principles perception. *Appl. Surf. Sci.,* **2020**, *525*, 146577.
[http://dx.doi.org/10.1016/j.apsusc.2020.146577]

[21]   Varunaa, R.; Ravindran, P. Potential hydrogen storage materials from metal decorated 2D-C$_2$N: an *ab*

initio study. *Phys. Chem. Chem. Phys.,* **2019**, *21*(45), 25311-25322.
[http://dx.doi.org/10.1039/C9CP05105H] [PMID: 31701096]

[22]　Qin, G.; Cui, Q.; Yun, B.; Sun, L.; Du, A.; Sun, Q. High capacity and reversible hydrogen storage on two dimensional C 2 N monolayer membrane. *Int. J. Hydrogen Energy,* **2018**, *43*(21), 9895-9901.
[http://dx.doi.org/10.1016/j.ijhydene.2018.04.065]

[23]　Jalili, S.; Molani, F.; Schofield, J. Ti-coated BC $_2$ N nanotubes as hydrogen storage materials. *Can. J. Chem.,* **2013**, *91*(7), 598-604.
[http://dx.doi.org/10.1139/cjc-2012-0386]

[24]　Lim, K. L.; Kazemian, H.; Yaakob, Z.; Daud, W. R. W. *Solid-state materials and methods for hydrogen storage: A critical review,* **2010**.
[http://dx.doi.org/10.1002/ceat.200900376]

[25]　Niemann, M.U.; Srinivasan, S.S.; Phani, A.R.; Kumar, A.; Goswami, D.Y.; Stefanakos, E.K. Nanomaterials for hydrogen storage applications: A review. *J. Nanomater.,* **2008**, *2008*(1), 1-9.
[http://dx.doi.org/10.1155/2008/950967]

[26]　Kumar, A.; Vyas, N.; Ojha, A.K. Hydrogen storage in magnesium decorated boron clusters (Mg$_2$Bn, n = 4–14): A density functional theory study. *Int. J. Hydrogen Energy,* **2020**, *45*(23), 12961-12971.
[http://dx.doi.org/10.1016/j.ijhydene.2020.03.018]

[27]　Samadizadeh, M.; Peyghan, A.A.; Rastegar, S.F. DFT studies of Hydrogen adsorption and dissociation on MgO nanotubes. *Main Group Chem.,* **2016**, *15*(2), 107-116.
[http://dx.doi.org/10.3233/MGC-150189]

[28]　Tabtimsai, C.; Ruangpornvisuti, V.; Tontapha, S.; Wanno, B. A DFT investigation on group 8B transition metal-doped silicon carbide nanotubes for hydrogen storage application. *Appl. Surf. Sci.,* **2018**, *439*, 494-505.
[http://dx.doi.org/10.1016/j.apsusc.2017.12.255]

[29]　Goudarziafshar, H.; Abdolmaleki, M.; Moosavi-zare, A.R.; Soleymanabadi, H. Hydrogen storage by Ni-doped silicon carbide nanocage: A theoretical study. *Physica E,* **2018**, *101*, 78-84.
[http://dx.doi.org/10.1016/j.physe.2018.03.001]

[30]　Wang, Y.; Ji, Y.; Li, M.; Yuan, P.; Sun, Q.; Jia, Y. Li and Ca Co-decorated carbon nitride nanostructures as high-capacity hydrogen storage media. *J. Appl. Phys.,* **2011**, *110*(9), 094311.
[http://dx.doi.org/10.1063/1.3656454]

[31]　Liu, Z.; Liu, S.; Er, S. Hydrogen storage properties of Li-decorated B2S monolayers: A DFT study. *Int. J. Hydrogen Energy,* **2019**, *44*(31), 16803-16810.
[http://dx.doi.org/10.1016/j.ijhydene.2019.04.234]

[32]　Arellano, J.S.; Molina, L.M.; Rubio, A.; López, M.J.; Alonso, J.A. Interaction of molecular and atomic hydrogen with (5,5) and (6,6) single-wall carbon nanotubes. *J. Chem. Phys.,* **2002**, *117*(5), 2281-2288.
[http://dx.doi.org/10.1063/1.1488595]

[33]　Wang, L.; Chen, X.; Du, H.; Yuan, Y.; Qu, H.; Zou, M. First-principles investigation on hydrogen storage performance of Li, Na and K decorated borophene. *Appl. Surf. Sci.,* **2018**, *427*, 1030-1037.
[http://dx.doi.org/10.1016/j.apsusc.2017.08.126]

[34]　Song, N.; Wang, Y.; Gao, H.; Jiang, W.; Zhang, J.; Xu, B.; Sun, Q.; Jia, Y. Electric field improved hydrogen storage of Ca-decorated monolayer MoS2. *Phys. Lett. A,* **2015**, *379*(9), 815-819.
[http://dx.doi.org/10.1016/j.physleta.2014.12.045]

[35]　Iijima, S.; Ajayan, P.M.; Ichihashi, T. Growth model for carbon nanotubes. *Phys. Rev. Lett.,* **1992**, *69*(21), 3100-3103.
[http://dx.doi.org/10.1103/PhysRevLett.69.3100] [PMID: 10046725]

[36]　Novoselov, K.S.; Geim, A.K.; Morozov, S.V.; Jiang, D.; Zhang, Y.; Dubonos, S.V.; Grigorieva, I.V.; Firsov, A.A. Electric field effect in atomically thin carbon films. *Science,* **2004**, *306*(5696), 666-669.
[http://dx.doi.org/10.1126/science.1102896] [PMID: 15499015]

[37]   Bolotin, K.I.; Sikes, K.J.; Jiang, Z.; Klima, M.; Fudenberg, G.; Hone, J.; Kim, P.; Stormer, H.L. Ultrahigh electron mobility in suspended graphene. *Solid State Commun.,* **2008**, *146*(9-10), 351-355.
[http://dx.doi.org/10.1016/j.ssc.2008.02.024]

[38]   Avouris, P.; Xia, F. *Graphene applications in electronics and photonics,* **2012**.
[http://dx.doi.org/10.1557/mrs.2012.206]

[39]   Cheng, J.; Zhang, L.; Ding, R.; Ding, Z.; Wang, X.; Wang, Z.; Fang, X. Influence of chemical potential on the computer simulation of hydrogen storage in single-walled carbon nanotube array. *Comput. Mater. Sci.,* **2008**, *44*(2), 601-604.
[http://dx.doi.org/10.1016/j.commatsci.2008.04.022]

[40]   Mahmood, M.; Casciano, D.A.; Mocan, T.; Iancu, C.; Xu, Y.; Mocan, L.; Iancu, D.T.; Dervishi, E.; Li, Z.; Abdalmuhsen, M.; Biris, A.R.; Ali, N.; Howard, P.; Biris, A.S. Cytotoxicity and biological effects of functional nanomaterials delivered to various cell lines. *J. Appl. Toxicol.,* **2010**, *30*(1), 74-83.
[http://dx.doi.org/10.1002/jat.1475] [PMID: 19760634]

[41]   Khazaei, A.; Rad, M.N.S.; Borazjani, M.K. Organic functionalization of single-walled carbon nanotubes (SWCNTs) with some chemotherapeutic agents as a potential method for drug delivery. *Int. J. Nanomedicine,* **2010**, *5*(1), 639-645.
[http://dx.doi.org/10.2147/IJN.S11146] [PMID: 20856839]

[42]   Vardharajula, S. *Functionalized carbon nanotubes: Biomedical applications,* **2012**.
[http://dx.doi.org/10.2147/IJN.S35832]

[43]   Kuila, T.; Bose, S.; Mishra, A. K.; Khanra, P.; Kim, N. H.; Lee, J. H. *Chemical functionalization of graphene and its applications,* **2012**.
[http://dx.doi.org/10.1016/j.pmatsci.2012.03.002]

[44]   Liu, Z.; Robinson, J.T.; Tabakman, S.M.; Yang, K.; Dai, H. Carbon materials for drug delivery & cancer therapy. *Mater. Today,* **2011**, *14*(7-8), 316-323.
[http://dx.doi.org/10.1016/S1369-7021(11)70161-4]

[45]   Balandin, A.A. *Thermal properties of graphene and nanostructured carbon materials*; , **2011**.
[http://dx.doi.org/10.1038/nmat3064]

[46]   Nika, D.; Balandin, A. Phonon Transport in Graphene 1203.4282, 1-41.**2012**

[47]   Shahil, K.M.F.; Balandin, A.A. Thermal properties of graphene and multilayer graphene: Applications in thermal interface materials. *Solid State Commun.,* **2012**, *152*(15), 1331-1340.
[http://dx.doi.org/10.1016/j.ssc.2012.04.034]

[48]   Chopra, N.G.; Luyken, R.J.; Cherrey, K.; Crespi, V.H.; Cohen, M.L.; Louie, S.G.; Zettl, A. Boron nitride nanotubes. *Science,* **1995**, *269*(5226), 966-967.
[http://dx.doi.org/10.1126/science.269.5226.966] [PMID: 17807732]

[49]   Rubio, A.; Corkill, J.L.; Cohen, M.L. Theory of graphitic boron nitride nanotubes. *Phys. Rev. B Condens. Matter,* **1994**, *49*(7), 5081-5084.
[http://dx.doi.org/10.1103/PhysRevB.49.5081] [PMID: 10011453]

[50]   Blase, X.; Rubio, A.; Louie, S.G.; Cohen, M.L. Stability and band gap constancy of boron nitride nanotubes. *Europhys. Lett.,* **1994**, *28*(5), 335-340.
[http://dx.doi.org/10.1209/0295-5075/28/5/007]

[51]   Moon, H.H.W.H. "Theoretical study of defects of BN nanotubes: a molecular-mechanics study," *Phys. E Low Dimens. Syst. Nanostruct.,* **2005**, *28*, 419-422.

[52]   Choyal, V.; Choyal, V.K.; Kundalwal, S.I. Effect of atom vacancies on elastic and electronic properties of transversely isotropic boron nitride nanotubes: A comprehensive computational study. *Comput. Mater. Sci.,* **2019**, *156*, 332-345.
[http://dx.doi.org/10.1016/j.commatsci.2018.10.013]

[53]   Choyal, V.; Kundalwal, S.I. Effect of Stone–Wales defects on the mechanical behavior of boron

nitride nanotubes. *Acta Mech.,* **2020**, *231*(10), 4003-4018.
[http://dx.doi.org/10.1007/s00707-020-02748-x]

[54]   Novoselov, K.S.; Jiang, D.; Schedin, F.; Booth, T.J.; Khotkevich, V.V.; Morozov, S.V.; Geim, A.K. Two-dimensional atomic crystals. *Proc. Natl. Acad. Sci. USA,* **2005**, *102*(30), 10451-10453.
[http://dx.doi.org/10.1073/pnas.0502848102] [PMID: 16027370]

[55]   Vatanparast, M.; Shariatinia, Z. Hexagonal boron nitride nanosheet as novel drug delivery system for anticancer drugs: Insights from DFT calculations and molecular dynamics simulations. *J. Mol. Graph. Model.,* **2019**, *89*, 50-59.
[http://dx.doi.org/10.1016/j.jmgm.2019.02.012] [PMID: 30870649]

[56]   Permyakova, E.S.; Sukhorukova, I.V.; Antipina, L.Y.; Konopatsky, A.S.; Kovalskii, A.M.; Matveev, A.T.; Lebedev, O.I.; Golberg, D.V.; Manakhov, A.M.; Shtansky, D.V. Synthesis and Characterization of Folate Conjugated Boron Nitride Nanocarriers for Targeted Drug Delivery. *J. Phys. Chem. C,* **2017**, *121*(50), 28096-28105.
[http://dx.doi.org/10.1021/acs.jpcc.7b10841]

[57]   Weng, Q.; Wang, B.; Wang, X.; Hanagata, N.; Li, X.; Liu, D.; Wang, X.; Jiang, X.; Bando, Y.; Golberg, D. Highly water-soluble, porous, and biocompatible boron nitrides for anticancer drug delivery. *ACS Nano,* **2014**, *8*(6), 6123-6130.
[http://dx.doi.org/10.1021/nn5014808] [PMID: 24797563]

[58]   Sargazi-Avval, H.; Yoosefian, M.; Ghaffari-Moghaddam, M.; Khajeh, M.; Bohlooli, M. Potential applications of armchair, zigzag, and chiral boron nitride nanotubes as a drug delivery system: Letrozole anticancer drug encapsulation. *Appl. Phys., A Mater. Sci. Process.,* **2021**, *127*(4), 257.
[http://dx.doi.org/10.1007/s00339-021-04402-2]

[59]   Xie, Y.; Huo, Y.P.; Zhang, J.M. First-principles study of CO and NO adsorption on transition metals doped (8,0) boron nitride nanotube. *Appl. Surf. Sci.,* **2012**, *258*(17), 6391-6397.
[http://dx.doi.org/10.1016/j.apsusc.2012.03.048]

[60]   Amiri Fadradi, M.; Movlarooy, T. Ab initio study of adsorption of CO on BNNTs: For gas nanosensor applications. *Mater. Chem. Phys.,* **2018**, *215*, 360-367.
[http://dx.doi.org/10.1016/j.matchemphys.2018.04.102]

[61]   Phalinyot, S.; Tabtimsai, C.; Wanno, B. Nitrogen monoxide storage and sensing applications of transition metal–doped boron nitride nanotubes: a DFT investigation. *Struct. Chem.,* **2019**, *30*(6), 2135-2149.
[http://dx.doi.org/10.1007/s11224-019-01339-4]

[62]   Rakhshi, M.; Mohsennia, M.; Rasa, H.; Sameti, M.R. First-principle study of ammonia molecules adsorption on boron nitride nanotubes in presence and absence of static electric field and ion field. *Vacuum,* **2018**, *155*, 456-464.
[http://dx.doi.org/10.1016/j.vacuum.2018.06.047]

[63]   Cho, H.; Kim, J.H.; Hwang, J.H.; Kim, C.S.; Jang, S.G.; Park, C.; Lee, H.; Kim, M.J. Single- and double-walled boron nitride nanotubes: Controlled synthesis and application for water purification. *Sci. Rep.,* **2020**, *10*(1), 7416.
[http://dx.doi.org/10.1038/s41598-020-64096-z] [PMID: 32366898]

[64]   e, S.; Geng, R.; Zhu, Z.; Xie, L.; Lu, W.; Li, C.; Yao, Y. Large-scale fabrication of boron nitride nanotubes and their application in thermoplastic polyurethane based composite for improved thermal conductivity. *Ceram. Int.,* **2018**, *44*(18), 22794-22799.
[http://dx.doi.org/10.1016/j.ceramint.2018.09.069]

[65]   Li, C.; Long, X.; e, S.; Zhang, Q.; Li, T.; Wu, J.; Yao, Y. Magnesium-induced preparation of boron nitride nanotubes and their application in thermal interface materials. *Nanoscale,* **2019**, *11*(24), 11457-11463.
[http://dx.doi.org/10.1039/C9NR03915E] [PMID: 31188376]

[66]   e, S.; Geng, R.; Zhu, Z.; Xie, L.; Lu, W.; Li, C.; Yao, Y. Large-scale fabrication of boron nitride

nanotubes and their application in thermoplastic polyurethane based composite for improved thermal conductivity. *Ceram. Int.,* **2018**, *44*(18), 22794-22799.
[http://dx.doi.org/10.1016/j.ceramint.2018.09.069]

[67] Hao, S.; Zhou, G.; Duan, W.; Wu, J.; Gu, B.L. Tremendous spin-splitting effects in open boron nitride nanotubes: application to nanoscale spintronic devices. *J. Am. Chem. Soc.,* **2006**, *128*(26), 8453-8458.
[http://dx.doi.org/10.1021/ja057420e] [PMID: 16802810]

[68] Moradian, R.; Azadi, S. Magnetism in defected single-walled boron nitride nanotubes. *Europhys. Lett.,* **2008**, *83*(1), 17007.
[http://dx.doi.org/10.1209/0295-5075/83/17007]

[69] Xie, Y.; Zhang, J.M. First-principles study on substituted doping of BN nanotubes by transition metals V, Cr and Mn. *Comput. Theor. Chem.,* **2011**, *976*(1-3), 215-220.
[http://dx.doi.org/10.1016/j.comptc.2011.08.031]

[70] Wei, X.; Wang, M.S.; Bando, Y.; Golberg, D. Electron-beam-induced substitutional carbon doping of boron nitride nanosheets, nanoribbons, and nanotubes. *ACS Nano,* **2011**, *5*(4), 2916-2922.
[http://dx.doi.org/10.1021/nn103548r] [PMID: 21425863]

[71] Malek, A.; Movlarooy, T.; Pilehrood, S.H. Ground-State Magnetic Phase in Transition-Metal-Doped Boron Nitride Nanosheet With (5,0) Chirality. *IEEE Magn. Lett.,* **2019**, *10*, 1-4.
[http://dx.doi.org/10.1109/LMAG.2019.2951085]

[72] Chettri, B. *Induced magnetic states upon electron–hole injection at B and N sites of hexagonal boron nitride bilayer: A density functional theory study,* **2021**.
[http://dx.doi.org/10.1002/qua.26680]

[73] Roy, D.; Hossain, M.K.; Hasan, S.M.; Khanom, S.; Hossain, M.A.; Ahmed, F. Half-metallic ferromagnetism induced by TM-TM atom pair co-doping in 2D hexagonal boron nitride monolayer: A first principle study. *Mater. Sci. Eng. B,* **2021**, *271*, 115247.
[http://dx.doi.org/10.1016/j.mseb.2021.115247]

[74] Kohn, W.; Sham, L.J. Self-consistent equations including exchange and correlation effects. *Phys. Rev.,* **1965**, *140*(4A), A1133-A1138.
[http://dx.doi.org/10.1103/PhysRev.140.A1133]

[75] Smidstrup, S.; Markussen, T.; Vancraeyveld, P.; Wellendorff, J.; Schneider, J.; Gunst, T.; Verstichel, B.; Stradi, D.; Khomyakov, P.A.; Vej-Hansen, U.G.; Lee, M.E.; Chill, S.T.; Rasmussen, F.; Penazzi, G.; Corsetti, F.; Ojanperä, A.; Jensen, K.; Palsgaard, M.L.N.; Martinez, U.; Blom, A.; Brandbyge, M.; Stokbro, K. QuantumATK: an integrated platform of electronic and atomic-scale modelling tools. *J. Phys. Condens. Matter,* **2020**, *32*(1), 015901.
[http://dx.doi.org/10.1088/1361-648X/ab4007] [PMID: 31470430]

[76] Segall, M.D.; Lindan, P.J.D.; Probert, M.J.; Pickard, C.J.; Hasnip, P.J.; Clark, S.J.; Payne, M.C. First-principles simulation: ideas, illustrations and the CASTEP code. *J. Phys. Condens. Matter,* **2002**, *14*(11), 2717-2744.
[http://dx.doi.org/10.1088/0953-8984/14/11/301]

[77] Guest, M.F.; Bush, I.J.; Van Dam, H.J.J.; Sherwood, P.; Thomas, J.M.H.; Van Lenthe, J.H.; Havenith, R.W.A.; Kendrick, J. The GAMESS-UK electronic structure package: algorithms, developments and applications. *Mol. Phys.,* **2005**, *103*(6-8), 719-747.
[http://dx.doi.org/10.1080/00268970512331340592]

[78] Delley, B. Time dependent density functional theory with DMol³. *J. Phys. Condens. Matter,* **2010**, *22*(38), 384208.
[http://dx.doi.org/10.1088/0953-8984/22/38/384208] [PMID: 21386542]

[79] Hafner, J. *Materials simulations using VASP-a quantum perspective to materials science,* **2007**.
[http://dx.doi.org/10.1016/j.cpc.2007.02.045]

[80] Perdew, J.P.; Burke, K.; Ernzerhof, M. Generalized gradient approximation made simple. *Phys. Rev.*

Lett., **1996**, *77*(18), 3865-3868.
[http://dx.doi.org/10.1103/PhysRevLett.77.3865] [PMID: 10062328]

[81]　Jackson, K.; Pederson, M.R. Accurate forces in a local-orbital approach to the local-density approximation. *Phys. Rev. B Condens. Matter,* **1990**, *42*(6), 3276-3281.
[http://dx.doi.org/10.1103/PhysRevB.42.3276] [PMID: 9995841]

[82]　Grimme, S. Semiempirical GGA-type density functional constructed with a long-range dispersion correction. *J. Comput. Chem.,* **2006**, *27*(15), 1787-1799.
[http://dx.doi.org/10.1002/jcc.20495] [PMID: 16955487]

[83]　Lee, K.; Murray, É.D.; Kong, L.; Lundqvist, B.I.; Langreth, D.C. Higher-accuracy van der Waals density functional. *Phys. Rev. B Condens. Matter Mater. Phys.,* **2010**, *82*(8), 081101.
[http://dx.doi.org/10.1103/PhysRevB.82.081101]

[84]　Liu, Y.; Liu, W.; Wang, R.; Hao, L.; Jiao, W. Hydrogen storage using Na-decorated graphyne and its boron nitride analog. *Int. J. Hydrogen Energy,* **2014**, *39*(24), 12757-12764.
[http://dx.doi.org/10.1016/j.ijhydene.2014.06.107]

[85]　Hussain, T.; Islam, M.S.; Rao, G.S.; Panigrahi, P.; Gupta, D.; Ahuja, R. Hydrogen storage properties of light metal adatoms (Li, Na) decorated fluorographene monolayer. *Nanotechnology,* **2015**, *26*(27), 275401.
[http://dx.doi.org/10.1088/0957-4484/26/27/275401] [PMID: 26066734]

[86]　Weng, Q.; Zeng, L.; Chen, Z.; Han, Y.; Jiang, K.; Bando, Y.; Golberg, D. Hydrogen Storage in Carbon and Oxygen Co☐Doped Porous Boron Nitrides. *Adv. Funct. Mater.,* **2021**, *31*(4), 2007381.
[http://dx.doi.org/10.1002/adfm.202007381]

[87]　Panigrahi, P.; Naqvi, S.R.; Hankel, M.; Ahuja, R.; Hussain, T. Enriching the hydrogen storage capacity of carbon nanotube doped with polylithiated molecules. *Appl. Surf. Sci.,* **2018**, *444*, 467-473.
[http://dx.doi.org/10.1016/j.apsusc.2018.02.040]

[88]　Schneemann, A.; White, J.L.; Kang, S.; Jeong, S.; Wan, L.F.; Cho, E.S.; Heo, T.W.; Prendergast, D.; Urban, J.J.; Wood, B.C.; Allendorf, M.D.; Stavila, V. Nanostructured Metal Hydrides for Hydrogen Storage. *Chem. Rev.,* **2018**, *118*(22), 10775-10839.
[http://dx.doi.org/10.1021/acs.chemrev.8b00313] [PMID: 30277071]

[89]　Bellosta von Colbe, J.; Ares, J-R.; Barale, J.; Baricco, M.; Buckley, C.; Capurso, G.; Gallandat, N.; Grant, D.M.; Guzik, M.N.; Jacob, I.; Jensen, E.H.; Jensen, T.; Jepsen, J.; Klassen, T.; Lototskyy, M.V.; Manickam, K.; Montone, A.; Puszkiel, J.; Sartori, S.; Sheppard, D.A.; Stuart, A.; Walker, G.; Webb, C.J.; Yang, H.; Yartys, V.; Züttel, A.; Dornheim, M. Application of hydrides in hydrogen storage and compression: Achievements, outlook and perspectives. *Int. J. Hydrogen Energy,* **2019**, *44*(15), 7780-7808.
[http://dx.doi.org/10.1016/j.ijhydene.2019.01.104]

[90]　El Kharbachi, A.; Dematteis, E.M.; Shinzato, K.; Stevenson, S.C.; Bannenberg, L.J.; Heere, M.; Zlotea, C.; Szilágyi, P.Á.; Bonnet, J-P.; Grochala, W.; Gregory, D.H.; Ichikawa, T.; Baricco, M.; Hauback, B.C. Metal Hydrides and Related Materials. Energy Carriers for Novel Hydrogen and Electrochemical Storage. *J. Phys. Chem. C,* **2020**, *124*(14), 7599-7607.
[http://dx.doi.org/10.1021/acs.jpcc.0c01806]

[91]　Sakintuna, B.; Lamari-Darkrim, F.; Hirscher, M. *Metal hydride materials for solid hydrogen storage: A review,* **2007**.
[http://dx.doi.org/10.1016/j.ijhydene.2006.11.022]

[92]　Michel, K.J.; Ozoliņš, V. Recent advances in the theory of hydrogen storage in complex metal hydrides. *MRS Bull.,* **2013**, *38*(6), 462-472.
[http://dx.doi.org/10.1557/mrs.2013.130]

[93]　Panella, B.; Hirscher, M.; Roth, S. Hydrogen adsorption in different carbon nanostructures. *Carbon,* **2005**, *43*(10), 2209-2214.
[http://dx.doi.org/10.1016/j.carbon.2005.03.037]

[94] Obeng, Y.; Srinivasan, P. Graphene: Is it the future for semiconductors? An overview of the material, devices, and applications. *Electrochem. Soc. Interface,* **2011**, *20*(1), 47-52.
[http://dx.doi.org/10.1149/2.F05111if]

[95] Thierfelder, C.; Witte, M.; Blankenburg, S.; Rauls, E.; Schmidt, W.G. Methane adsorption on graphene from first principles including dispersion interaction. *Surf. Sci.,* **2011**, *605*(7-8), 746-749.
[http://dx.doi.org/10.1016/j.susc.2011.01.012]

[96] Ohba, T.; Takase, A.; Ohyama, Y.; Kanoh, H. Grand canonical Monte Carlo simulations of nitrogen adsorption on graphene materials with varying layer number. *Carbon,* **2013**, *61*, 40-46.
[http://dx.doi.org/10.1016/j.carbon.2013.04.061]

[97] Bagsican, F.R.; Winchester, A.; Ghosh, S.; Zhang, X.; Ma, L.; Wang, M.; Murakami, H.; Talapatra, S.; Vajtai, R.; Ajayan, P.M.; Kono, J.; Tonouchi, M.; Kawayama, I. Adsorption energy of oxygen molecules on graphene and two-dimensional tungsten disulfide. *Sci. Rep.,* **2017**, *7*(1), 1774.
[http://dx.doi.org/10.1038/s41598-017-01883-1] [PMID: 28496178]

[98] Rai, D.P.; Singh, Y.T.; Chettri, B.; Houmad, M.; Patra, P.K. A theoretical investigation of electronic and optical properties of (6,1) single-wall carbon nanotube (SWCNT). *Carbon Letters,* **2021**, *31*(3), 441-448.
[http://dx.doi.org/10.1007/s42823-020-00172-8]

[99] He, X.; Fujimura, N.; Lloyd, J.M.; Erickson, K.J.; Talin, A.A.; Zhang, Q.; Gao, W.; Jiang, Q.; Kawano, Y.; Hauge, R.H.; Léonard, F.; Kono, J. Carbon nanotube terahertz detector. *Nano Lett.,* **2014**, *14*(7), 3953-3958.
[http://dx.doi.org/10.1021/nl5012678] [PMID: 24875576]

[100] Xiao, L.; Zhang, Y.; Wang, Y.; Liu, K.; Wang, Z.; Li, T.; Jiang, Z.; Shi, J.; Liu, L.; Li, Q.; Zhao, Y.; Feng, Z.; Fan, S.; Jiang, K. A polarized infrared thermal detector made from super-aligned multiwalled carbon nanotube films. *Nanotechnology,* **2011**, *22*(2), 025502.
[http://dx.doi.org/10.1088/0957-4484/22/2/025502] [PMID: 21135478]

[101] Tahaei, S.H.; Ghoreishi, S.S.; Yousefi, R.; Aderang, H. A computational study of a carbon nanotube junctionless tunneling field-effect transistor (CNT-JLTFET) based on the charge plasma concept. *Superlattices Microstruct.,* **2019**, *125*, 168-176.
[http://dx.doi.org/10.1016/j.spmi.2018.11.004]

[102] Pourian, P.; Yousefi, R.; Ghoreishi, S.S. Effect of uniaxial strain on electrical properties of CNT-based junctionless field-effect transistor: Numerical study. *Superlattices Microstruct.,* **2016**, *93*, 92-100.
[http://dx.doi.org/10.1016/j.spmi.2016.03.014]

[103] Wang, Y.; Yeow, J.T.W. A review of carbon nanotubes-based gas sensors. *J. Sens.,* **2009**, *2009*, 1-24.
[http://dx.doi.org/10.1155/2009/493904]

[104] Zhang, W-D.; Zhang, W.H. Carbon nanotubes as active components for gas sensors. *J. Sens.,* **2009**, *2009*, 1-16.
[http://dx.doi.org/10.1155/2009/160698]

[105] Ao, Z.M.; Jiang, Q.; Zhang, R.Q.; Tan, T.T.; Li, S. Al doped graphene: A promising material for hydrogen storage at room temperature. *J. Appl. Phys.,* **2009**, *105*(7), 074307.
[http://dx.doi.org/10.1063/1.3103327]

[106] Liu, W.; Zhao, Y.H.; Nguyen, J.; Li, Y.; Jiang, Q.; Lavernia, E.J. Electric field induced reversible switch in hydrogen storage based on single-layer and bilayer graphenes. *Carbon,* **2009**, *47*(15), 3452-3460.
[http://dx.doi.org/10.1016/j.carbon.2009.08.012]

[107] Tozzini, V.; Pellegrini, V. Prospects for hydrogen storage in graphene. *Phys. Chem. Chem. Phys.,* **2013**, *15*(1), 80-89.
[http://dx.doi.org/10.1039/C2CP42538F] [PMID: 23165421]

[108] Ma, L.P.; Wu, Z.S.; Li, J.; Wu, E.D.; Ren, W.C.; Cheng, H.M. Hydrogen adsorption behavior of

graphene above critical temperature. *Int. J. Hydrogen Energy,* **2009**, *34*(5), 2329-2332.
[http://dx.doi.org/10.1016/j.ijhydene.2008.12.079]

[109]   Kim, B.H.; Hong, S.J.; Baek, S.J.; Jeong, H.Y.; Park, N.; Lee, M.; Lee, S.W.; Park, M.; Chu, S.W.;
Shin, H.S.; Lim, J.; Lee, J.C.; Jun, Y.; Park, Y.W. N-type graphene induced by dissociative H_2
adsorption at room temperature. *Sci. Rep.,* **2012**, *2*(1), 690.
[http://dx.doi.org/10.1038/srep00690] [PMID: 23012645]

[110]   Patchkovskii, S.; Tse, J.S.; Yurchenko, S.N.; Zhechkov, L.; Heine, T.; Seifert, G. Graphene
nanostructures as tunable storage media for molecular hydrogen. *Proc. Natl. Acad. Sci. USA,* **2005**,
102(30), 10439-10444.
[http://dx.doi.org/10.1073/pnas.0501030102] [PMID: 16020537]

[111]   Goler, S.; Coletti, C.; Tozzini, V.; Piazza, V.; Mashoff, T.; Beltram, F.; Pellegrini, V.; Heun, S.
Influence of graphene curvature on hydrogen adsorption: Toward hydrogen storage devices. *J. Phys.
Chem. C,* **2013**, *117*(22), 11506-11513.
[http://dx.doi.org/10.1021/jp4017536]

[112]   Wu, S.; Fan, K.; Wu, M.; Yin, G. Effect of nitrogen doping and external electric field on the
adsorption of hydrogen on graphene. *Eur. Phys. J. Appl. Phys.,* **2016**, *75*(1), 10402.
[http://dx.doi.org/10.1051/epjap/2016160059]

[113]   Kim, D.; Lee, S.; Hwang, Y.; Yun, K.H.; Chung, Y.C. Hydrogen storage in Li dispersed graphene with
Stone–Wales defects: A first-principles study. *Int. J. Hydrogen Energy,* **2014**, *39*(25), 13189-13194.
[http://dx.doi.org/10.1016/j.ijhydene.2014.06.163]

[114]   Kim, D.; Lee, S.; Hwang, Y.; Yun, K.H.; Chung, Y.C. Hydrogen storage in Li dispersed graphene with
Stone–Wales defects: A first-principles study. *Int. J. Hydrogen Energy,* **2014**, *39*(25), 13189-13194.
[http://dx.doi.org/10.1016/j.ijhydene.2014.06.163]

[115]   Barghi, S.H.; Tsotsis, T.T.; Sahimi, M. Chemisorption, physisorption and hysteresis during hydrogen
storage in carbon nanotubes. *Int. J. Hydrogen Energy,* **2014**, *39*(3), 1390-1397.
[http://dx.doi.org/10.1016/j.ijhydene.2013.10.163]

[116]   Reyhani, A.; Mortazavi, S.Z.; Nozad Golikand, A.; Moshfegh, A.Z.; Mirershadi, S. The effect of
various acids treatment on the purification and electrochemical hydrogen storage of multi-walled
carbon nanotubes. *J. Power Sources,* **2008**, *183*(2), 539-543.
[http://dx.doi.org/10.1016/j.jpowsour.2008.05.039]

[117]   Chen, Y.; Liu, Q.; Yan, Y.; Cheng, X.; Liu, Y. Influence of sample cell physisorption on
measurements of hydrogen storage of carbon materials using a Sieverts apparatus. *Carbon,* **2010**,
48(3), 714-720.
[http://dx.doi.org/10.1016/j.carbon.2009.10.016]

[118]   Liu, C.; Chen, Y.; Wu, C.Z.; Xu, S.T.; Cheng, H.M. Hydrogen storage in carbon nanotubes revisited.
Carbon, **2010**, *48*(2), 452-455.
[http://dx.doi.org/10.1016/j.carbon.2009.09.060]

[119]   Muniz, A.R.; Meyyappan, M.; Maroudas, D. On the hydrogen storage capacity of carbon nanotube
bundles. *Appl. Phys. Lett.,* **2009**, *95*(16), 163111.
[http://dx.doi.org/10.1063/1.3253711]

[120]   Safa, S.; Larijani, M.M.; Fathollahi, V.; Kakuee, O.R. Investigating hydrogen storage behavior of
carbon nanotubes at ambient temperature and above by ion beam analysis. *Nano,* **2010**, *5*(6), 341-347.
[http://dx.doi.org/10.1142/S1793292010002256]

[121]   Sudibandriyo, M.; PDK Wulan, P.; Prasodjo, P. Adsorption capacity and its dynamic behavior of the
hydrogen storage on carbon nanotubes. *International Journal of Technology,* **2015**, *6*(7), 1128-1136.
[http://dx.doi.org/10.14716/ijtech.v6i7.1747]

[122]   Bianco, S.; Giorcelli, M.; Musso, S.; Castellino, M.; Agresti, F.; Khandelwal, A.; Lo Russo, S.;
Kumar, M.; Ando, Y.; Tagliaferro, A. Hydrogen adsorption in several types of carbon nanotubes. *J.*

Nanosci. Nanotechnol., **2010**, *10*(6), 3860-3866.
[http://dx.doi.org/10.1166/jnn.2010.1972] [PMID: 20355380]

[123] Mpourmpakis, G.; Froudakis, G.E.; Lithoxoos, G.P.; Samios, J. Effect of curvature and chirality for hydrogen storage in single-walled carbon nanotubes: A Combined *ab initio* and Monte Carlo investigation. *J. Chem. Phys.,* **2007**, *126*(14), 144704.
[http://dx.doi.org/10.1063/1.2717170] [PMID: 17444729]

[124] Doi, K.; Nakano, H.; Ohta, H.; Tachibana, A. First-Principle Molecular-Dynamics Study of Hydrogen and Aluminium Nanowires in Carbon Nanotubes. *Mater. Sci. Forum,* **2007**, *539-543*, 1409-1414.
[http://dx.doi.org/10.4028/www.scientific.net/MSF.539-543.1409]

[125] Wang, Y.J.; Wang, L.Y.; Wang, S.B.; Wu, L.; Jiao, Q.Z. Hydrogen storage of carbon nanotubes: Theoretical studies. *Adv. Mat. Res.,* **2011**, *179-180*, 722-727.
[http://dx.doi.org/10.4028/www.scientific.net/AMR.179-180.722]

[126] Ghosh, S.; Padmanabhan, V. Hydrogen storage capacity of bundles of single-walled carbon nanotubes with defects. *Int. J. Energy Res.,* **2017**, *41*(8), 1108-1117.
[http://dx.doi.org/10.1002/er.3691]

[127] Minami, D.; Ohkubo, T.; Kuroda, Y.; Sakai, K.; Sakai, H.; Abe, M. Structural optimization of arranged carbon nanotubes for hydrogen storage by grand canonical Monte Carlo simulation. *Int. J. Hydrogen Energy,* **2010**, *35*(22), 12398-12404.
[http://dx.doi.org/10.1016/j.ijhydene.2010.08.082]

[128] Chen, Y.; Liu, B.; Wu, J.; Huang, Y.; Jiang, H.; Hwang, K. Mechanics of hydrogen storage in carbon nanotubes. *J. Mech. Phys. Solids,* **2008**, *56*(11), 3224-3241.
[http://dx.doi.org/10.1016/j.jmps.2008.07.007]

[129] Tang, C.; Man, C.; Chen, Y.; Yang, F.; Luo, L.; Liu, Z-F.; Mei, J.; Lau, W-M.; Wong, K-W. Realizing the Storage of Pressurized Hydrogen in Carbon Nanotubes Sealed with Aqueous Valves. *Energy Technol. (Weinheim),* **2013**, *1*(5-6), 309-312.
[http://dx.doi.org/10.1002/ente.201300039]

[130] Liu, Z.; Xue, Q.; Ling, C.; Yan, Z.; Zheng, J. Hydrogen storage and release by bending carbon nanotubes. *Comput. Mater. Sci.,* **2013**, *68*, 121-126.
[http://dx.doi.org/10.1016/j.commatsci.2012.09.025]

[131] Zubizarreta, L.; Gomez, E.I.; Arenillas, A.; Ania, C.O.; Parra, J.B.; Pis, J.J. H_2 storage in carbon materials. *Adsorption,* **2008**, *14*(4-5), 557-566.
[http://dx.doi.org/10.1007/s10450-008-9116-y]

[132] Ariharan, A.; Viswanathan, B.; Nandhakumar, V. Hydrogen sorption in phosphorous substituted carbon material. *Indian Journal of Chemistry - Section A Inorganic, Physical, Theoretical and Analytical Chemistry,* **2015**, *54A*(12), 1423-1433.

[133] Zhao, T.; Ji, X.; Jin, W.; Yang, W.; Li, T. Hydrogen storage capacity of single-walled carbon nanotube prepared by a modified arc discharge. *Fuller. Nanotub. Carbon Nanostruct.,* **2017**, *25*(6), 355-358.
[http://dx.doi.org/10.1080/1536383X.2017.1305358]

[134] Poirier, E.; Chahine, R.; Bénard, P.; Cossement, D.; Lafi, L.; Mélançon, E.; Bose, T.K.; Désilets, S. Storage of hydrogen on single-walled carbon nanotubes and other carbon structures. *Appl. Phys., A Mater. Sci. Process.,* **2004**, *78*(7), 961-967.
[http://dx.doi.org/10.1007/s00339-003-2415-y]

[135] Kaskun, S.; Kayfeci, M. The synthesized nickel-doped multi-walled carbon nanotubes for hydrogen storage under moderate pressures. *Int. J. Hydrogen Energy,* **2018**, *43*(23), 10773-10778.
[http://dx.doi.org/10.1016/j.ijhydene.2018.01.084]

[136] Darkrim, F.L.; Malbrunot, P.; Tartaglia, G.P. Review of hydrogen storage by adsorption in carbon nanotubes. *Int. J. Hydrogen Energy,* **2002**, *27*(2), 193-202.
[http://dx.doi.org/10.1016/S0360-3199(01)00103-3]

[137] Park, C.; Keane, M.A. Controlled growth of highly ordered carbon nanofibers from Y zeolite supported nickel catalysts. *Langmuir,* **2001**, *17*(26), 8386-8396.
[http://dx.doi.org/10.1021/la0110390]

[138] Liu, C.; Fan, Y.Y.; Liu, M.; Cong, H.T.; Cheng, H.M.; Dresselhaus, M.S. Hydrogen storage in single-walled carbon nanotubes at room temperature. *Science,* **1999**, *286*(5442), 1127-1129.
[http://dx.doi.org/10.1126/science.286.5442.1127] [PMID: 10550044]

[139] Lee, J.H.; Rhee, K.Y.; Park, S.J. Effects of cryomilling on the structures and hydrogen storage characteristics of multi-walled carbon nanotubes. *Int. J. Hydrogen Energy,* **2010**, *35*(15), 7850-7857.
[http://dx.doi.org/10.1016/j.ijhydene.2010.05.083]

[140] Stadie, N.P.; Purewal, J.J.; Ahn, C.C.; Fultz, B. Measurements of hydrogen spillover in platinum doped superactivated carbon. *Langmuir,* **2010**, *26*(19), 15481-15485.
[http://dx.doi.org/10.1021/la9046758] [PMID: 20187626]

[141] Cho, J.H.; Yang, S.J.; Lee, K.; Park, C.R. Si-doping effect on the enhanced hydrogen storage of single walled carbon nanotubes and graphene. *Int. J. Hydrogen Energy,* **2011**, *36*(19), 12286-12295.
[http://dx.doi.org/10.1016/j.ijhydene.2011.06.110]

[142] X. CHENG, H. ZHANG, and Y. TANG, "VERY HIGH HYDROGEN STORAGE CAPACITY OF AL-ADSORBED SINGLE-WALLED CARBON NANOTUBE (SWCNT): MULTI-LAYERED STRUCTURE OF HYDROGEN MOLECULES. *Int. J. Mod. Phys. B,* **2013**, *27*(14), 1350061.
[http://dx.doi.org/10.1142/S0217979213500616]

[143] Surya, V.J.; Iyakutti, K.; Rajarajeswari, M.; Kawazoe, Y. First-Principles Study on Hydrogen Storage in Single Walled Carbon Nanotube Functionalized with Ammonia. *J. Comput. Theor. Nanosci.,* **2010**, *7*(3), 552-557.
[http://dx.doi.org/10.1166/jctn.2010.1393]

[144] Topsakal, M.; Aktürk, E.; Ciraci, S. First-principles study of two- and one-dimensional honeycomb structures of boron nitride. *Phys. Rev. B Condens. Matter Mater. Phys.,* **2009**, *79*(11), 115442.
[http://dx.doi.org/10.1103/PhysRevB.79.115442]

[145] Siahlo, A.I.; Poklonski, N.A.; Lebedev, A.V.; Lebedeva, I.V.; Popov, A.M.; Vyrko, S.A.; Knizhnik, A.A.; Lozovik, Y.E. Structure and energetics of carbon, hexagonal boron nitride, and carbon/hexagonal boron nitride single-layer and bilayer nanoscrolls. *Phys. Rev. Mater.,* **2018**, *2*(3), 036001.
[http://dx.doi.org/10.1103/PhysRevMaterials.2.036001]

[146] Chettri, B.; Patra, P.K.; Hieu, N.N.; Rai, D.P. Hexagonal boron nitride (h -BN) nanosheet as a potential hydrogen adsorption material: A density functional theory (DFT) study. *Surf. Interfaces,* **2021**, *24*, 101043.
[http://dx.doi.org/10.1016/j.surfin.2021.101043]

[147] Weng, Q.; Kvashnin, D.G.; Wang, X.; Cretu, O.; Yang, Y.; Zhou, M.; Zhang, C.; Tang, D.M.; Sorokin, P.B.; Bando, Y.; Golberg, D. Tuning of the Optical, Electronic, and Magnetic Properties of Boron Nitride Nanosheets with Oxygen Doping and Functionalization. *Adv. Mater.,* **2017**, *29*(28), 1700695.
[http://dx.doi.org/10.1002/adma.201700695] [PMID: 28523720]

[148] Baumeier, B.; Krüger, P.; Pollmann, J. Structural, elastic, and electronic properties of SiC, BN, and BeO nanotubes. *Phys. Rev. B Condens. Matter Mater. Phys.,* **2007**, *76*(8), 085407.
[http://dx.doi.org/10.1103/PhysRevB.76.085407]

[149] Serhan, M.; Abusini, M.; Almahmoud, E.; Omari, R.; Al-Khaza'leh, K.; Abu-Farsakh, H.; Ghozlan, A.; Talla, J. The electronic properties of different chiralities of defected boron nitride nanotubes: Theoretical study. *Computational Condensed Matter,* **2020**, *22*, e00439.
[http://dx.doi.org/10.1016/j.cocom.2019.e00439]

[150] Jhi, S.H.; Roundy, D.J.; Louie, S.G.; Cohen, M.L. Formation and electronic properties of double-

walled boron nitride nanotubes. *Solid State Commun.,* **2005,** *134*(6), 397-402.
[http://dx.doi.org/10.1016/j.ssc.2005.02.007]

[151] Lei, W.; Zhang, H.; Wu, Y.; Zhang, B.; Liu, D.; Qin, S.; Liu, Z.; Liu, L.; Ma, Y.; Chen, Y. Oxygen-doped boron nitride nanosheets with excellent performance in hydrogen storage. *Nano Energy,* **2014,** *6,* 219-224.
[http://dx.doi.org/10.1016/j.nanoen.2014.04.004]

[152] Zhou, J.; Wang, Q.; Sun, Q.; Jena, P.; Chen, X.S. Electric field enhanced hydrogen storage on polarizable materials substrates. *Proc. Natl. Acad. Sci. USA,* **2010,** *107*(7), 2801-2806.
[http://dx.doi.org/10.1073/pnas.0905571107] [PMID: 20133647]

[153] Chettri, B.; Patra, P.K.; Srivastava, S.; Lalhriatzuala, L.; Zadeng, L.; Rai, D.P. Electronic Properties of Hydrogenated Hexagonal Boron Nitride (h-BN): DFT Study. *Senhri Journal of Multidisciplinary Studies,* **2019,** *4*(2), 72-79.
[http://dx.doi.org/10.36110/sjms.2019.04.02.008]

[154] Tang, C.; Zhang, X.; Zhou, X. Most effective way to improve the hydrogen storage abilities of Na-decorated BN sheets: applying external biaxial strain and an electric field. *Phys. Chem. Chem. Phys.,* **2017,** *19*(7), 5570-5578.
[http://dx.doi.org/10.1039/C6CP07433B] [PMID: 28165071]

[155] Chettri, B.; Patra, P. K.; Rai, D. P. *Enhanced H$_2$ storage capacity of the bilayer hexagonal Boron Nitride(h-BN) incorporating Van der Waals interaction under applied external electric field,* **2021.**

[156] Banerjee, P.; Pathak, B.; Ahuja, R.; Das, G.P. First principles design of Li functionalized hydrogenated h-BN nanosheet for hydrogen storage. *Int. J. Hydrogen Energy,* **2016,** *41*(32), 14437-14446.
[http://dx.doi.org/10.1016/j.ijhydene.2016.02.113]

[157] Wang, Y.; Wang, F.; Xu, B.; Zhang, J.; Sun, Q.; Jia, Y. Theoretical prediction of hydrogen storage on Li-decorated boron nitride atomic chains. *J. Appl. Phys.,* **2013,** *113*(6), 064309.
[http://dx.doi.org/10.1063/1.4790868]

[158] Naqvi, S.R.; Rao, G.S.; Luo, W.; Ahuja, R.; Hussain, T. Hexagonal Boron Nitride (h-BN) Sheets Decorated with OLi, ONa, and Li$_2$F Molecules for Enhanced Energy Storage. *ChemPhysChem,* **2017,** *18*(5), 513-518.
[http://dx.doi.org/10.1002/cphc.201601063] [PMID: 28098421]

[159] Hu, Z.Y.; Shao, X.; Wang, D.; Liu, L.M.; Johnson, J.K. A first-principles study of lithium-decorated hybrid boron nitride and graphene domains for hydrogen storage. *J. Chem. Phys.,* **2014,** *141*(8), 084711.
[http://dx.doi.org/10.1063/1.4893177] [PMID: 25173034]

[160] Ren, J.; Zhang, N.; Zhang, H.; Peng, X. First-principles study of hydrogen storage on Pt (Pd)-doped boron nitride sheet. *Struct. Chem.,* **2015,** *26*(3), 731-738.
[http://dx.doi.org/10.1007/s11224-014-0531-2]

[161] Tokarev, A.; Kjeang, E.; Cannon, M.; Bessarabov, D. Theoretical limit of reversible hydrogen storage capacity for pristine and oxygen-doped boron nitride. *Int. J. Hydrogen Energy,* **2016,** *41*(38), 16984-16991.
[http://dx.doi.org/10.1016/j.ijhydene.2016.07.010]

[162] Rad, A.S.; Ayub, K. Enhancement in hydrogen molecule adsorption on B12N12 nano-cluster by decoration of nickel. *Int. J. Hydrogen Energy,* **2016,** *41*(47), 22182-22191.
[http://dx.doi.org/10.1016/j.ijhydene.2016.08.158]

[163] Kumar, E.M.; Sinthika, S.; Thapa, R. First principles guide to tune h-BN nanostructures as superior light-element-based hydrogen storage materials: role of the bond exchange spillover mechanism. *J. Mater. Chem. A Mater. Energy Sustain.,* **2015,** *3*(1), 304-313.
[http://dx.doi.org/10.1039/C4TA04706K]

[164] Shevlin, S.A.; Guo, Z.X. Transition-metal-doping-enhanced hydrogen storage in boron nitride systems. *Appl. Phys. Lett.,* **2006**, *89*(15), 153104.
[http://dx.doi.org/10.1063/1.2360232]

[165] Ma, R.; Bando, Y.; Zhu, H.; Sato, T.; Xu, C.; Wu, D. Hydrogen uptake in boron nitride nanotubes at room temperature. *J. Am. Chem. Soc.,* **2002**, *124*(26), 7672-7673.
[http://dx.doi.org/10.1021/ja026030e] [PMID: 12083917]

[166] Jhi, S.H.; Kwon, Y.K. Hydrogen adsorption on boron nitride nanotubes: A path to room-temperature hydrogen storage. *Phys. Rev. B Condens. Matter Mater. Phys.,* **2004**, *69*(24), 245407.
[http://dx.doi.org/10.1103/PhysRevB.69.245407]

[167] Mpourmpakis, G.; Froudakis, G. E. *Why boron nitride nanotubes are preferable to carbon nanotubes for hydrogen storage?. An ab initio theoretical study,* **2007**.
[http://dx.doi.org/10.1016/j.cattod.2006.09.023]

[168] Ahadi, Z.; Shadman, M.; Yeganegi, S.; Asgari, F. *Hydrogen adsorption capacities of multi-walled boron nitride nanotubes and nanotube arrays: A grand canonical Monte Carlo study,* **2012**.
[http://dx.doi.org/10.1007/s00894-011-1316-9]

[169] Song, E.H.; Yoo, S.H.; Kim, J.J.; Lai, S.W.; Jiang, Q.; Cho, S.O. External electric field induced hydrogen storage/release on calcium-decorated single-layer and bilayer silicene. *Phys. Chem. Chem. Phys.,* **2014**, *16*(43), 23985-23992.
[http://dx.doi.org/10.1039/C4CP02638A] [PMID: 25285782]

[170] Xiu-Ying, L.; Chao-Yang, W.; Yong-Jian, T.; Wei-Guo, S.; Wei-Dong, W.; Jia-Jing, X. Theoretical studies on hydrogen adsorption of single-walled boron-nitride and carbon nanotubes using grand canonical Monte Carlo method. *Physica B,* **2009**, *404*(14-15), 1892-1896.
[http://dx.doi.org/10.1016/j.physb.2008.11.073]

[171] Wu, X.; Yang, J.; Hou, J.G.; Zhu, Q. Defects-enhanced dissociation of H_2 on boron nitride nanotubes. *J. Chem. Phys.,* **2006**, *124*(5), 054706.
[http://dx.doi.org/10.1063/1.2162897] [PMID: 16468900]

[172] Han, S.S.; Kang, J.K.; Lee, H.M.; van Duin, A.C.T.; Goddard, W.A., III Theoretical study on interaction of hydrogen with single-walled boron nitride nanotubes. II. Collision, storage, and adsorption. *J. Chem. Phys.,* **2005**, *123*(11), 114704.
[http://dx.doi.org/10.1063/1.1999629] [PMID: 16392580]

[173] Fu, P.; Wang, J.; Jia, R.; Bibi, S.; Eglitis, R.I.; Zhang, H.X. Theoretical study on hydrogen storage capacity of expanded h-BN systems. *Comput. Mater. Sci.,* **2017**, *139*, 335-340.
[http://dx.doi.org/10.1016/j.commatsci.2017.08.015]

[174] Aal, S.A.; Alfuhaidi, A.K. Enhancement of hydrogen storage capacities of Co and Pt functionalized h-BN nanosheet: Theoretical study. *Vacuum,* **2021**, *183*, 109838.
[http://dx.doi.org/10.1016/j.vacuum.2020.109838]

[175] Zhang, Z.W.; Zheng, W.T.; Jiang, Q. Hydrogen adsorption on Ce/BNNT systems: A DFT study. *Int. J. Hydrogen Energy,* **2012**, *37*(6), 5090-5099.
[http://dx.doi.org/10.1016/j.ijhydene.2011.12.036]

[176] Wu, H.Y.; Fan, X.F.; Kuo, J.L.; Deng, W.Q. Carbon doped boron nitride cages as competitive candidates for hydrogen storage materials. *Chem. Commun. (Camb.),* **2010**, *46*(6), 883-885.
[http://dx.doi.org/10.1039/B911503J] [PMID: 20107638]

[177] Durgun, E.; Jang, Y.R.; Ciraci, S. Hydrogen storage capacity of Ti-doped boron-nitride and B / Be - substituted carbon nanotubes. *Phys. Rev. B Condens. Matter Mater. Phys.,* **2007**, *76*(7), 073413.
[http://dx.doi.org/10.1103/PhysRevB.76.073413]

[178] Ebrahimi-Nejad, S.; Shokuhfar, A. Compressive buckling of open-ended boron nitride nanotubes in hydrogen storage applications. *Physica E,* **2013**, *50*, 29-36.
[http://dx.doi.org/10.1016/j.physe.2013.02.021]

[179] Zhang, L.P.; Wu, P.; Sullivan, M.B. Hydrogen adsorption on Rh, Ni, and Pd functionalized single-walled boron nitride nanotubes. *J. Phys. Chem. C,* **2011**, *115*(10), 4289-4296.
[http://dx.doi.org/10.1021/jp1078554]

[180] Mananghaya, M.; Yu, D.; Santos, G.N. Hydrogen adsorption on boron nitride nanotubes functionalized with transition metals. *Int. J. Hydrogen Energy,* **2016**, *41*(31), 13531-13539.
[http://dx.doi.org/10.1016/j.ijhydene.2016.05.225]

[181] Roy, L.; Bhunya, S.; Paul, A. A Metal-Free Strategy to Release Chemisorbed H_2 from Hydrogenated Boron Nitride Nanotubes. *Angew. Chem.,* **2014**, *126*(46), 12638-12643.
[http://dx.doi.org/10.1002/ange.201403610]

[182] Panigrahi, P.; Kumar, A.; Bae, H.; Lee, H.; Ahuja, R.; Hussain, T. Capacity enhancement of polylithiated functionalized boron nitride nanotubes: an efficient hydrogen storage medium. *Phys. Chem. Chem. Phys.,* **2020**, *22*(27), 15675-15682.
[http://dx.doi.org/10.1039/D0CP01237H] [PMID: 32618312]

CHAPTER 2

Suitable Nanomaterials for Retinal Implant Technique and Future Trends

Mohan L.Verma[1] and Ashish Tiwari[1,*]

[1] *Computational Nanomaterial Lab, Department of Applied Physics and Department of Electronics and Telecommunication, Shri Shankaracharya Technical Campus, Bhilai, Chhattisgarh, India*

Abstract: Artificial vision for blind patients suffering from retinal diseases has shown promising results in the last two decades, especially after the advancement in CMOS technology. In the modern era, two types of retinal implant techniques are very popular, one is the epiretinal implant and the other is the subretinal implant technique. Even though the method of data processing is different in the above-mentioned techniques, utilization of appropriate nanomaterial for the durability of the implant has always been a major concern. Materials such as titanium nitride (TiN), iridium oxide (IrOx), platinum grey, and carbon nanotube (CNT) were employed in recent years in many retinal prosthetic projects. Manufacturing of stimulating electrodes and coating of electronic devices to avoid infiltrations are the two important applications where nanomaterials are utilized in the retinal implant system. This chapter discusses the important and desired physical properties of nanomaterials viz. conductivity, tensile strength, absorption of photons, and adsorption of water molecules for the subretinal implant technique. Since the implant is located inside the retina, the isolated and corrosive environment is the main challenge. This study is based on the first-principles of density functional theory (DFT). Considering the recent advancements, the materials are comparatively analyzed, and new nanomaterial is also suggested.

Keywords: Absorption, Adsorption, Artificial vision, Conductivity, CMOS, Corrosive, Density Functional Theory, Electrode, Nanomaterial, Retinal implant, Tensile strength.

INTRODUCTION

The visual prosthesis is the art of substituting the damaged retinal cells with some electronic aid causing the electrical elicitation resulting in vision.

* **Corresponding author Ashish Tiwari**: Computational Nanomaterial Lab, Department of Applied Physics and Department of Electronics and Telecommunication, Shri Shankaracharya Technical Campus, Bhilai, Chhattisgarh, India; Tel: +919303452648; E-mail: tiwari.ashish99@gmail.com

Dibya Prakash Rai (Ed.)

The damage in retinal cells is mainly caused by the diseases like retinitis pigmentosa and age-related macular degeneration [1]. The researchers across the globe initially targeted the identification of the phosphene area including its behaviour, mapping of retinal tissues, and methods of processing the image data. Different combinations of electrode were utilized in the past to generate visual sensations. Microphotodiodes, charged-coupled-devices (CCD), and now CMOS image sensors are frequently utilized for retinal implant applications. Before going into the details further, it is important to understand the functioning of an eye [2].

The neuroepithelium sensor, which is present in the retina, processes the tremendous complex information received through the excellent neuroprocessor called the retina that results in vision. In the retinal implant technique the damaged retinal cells (rods and cones) are electrically stimulated through elicitation of nerves resulting in artificial vision. The 130 million photoreceptor cells with 1.2 million retinal ganglion cells (RGC) convert the chromatic and achromatic color images into a chemical and electrical signal. This conversion results in spatial and temporal resolution of the images. After this, the electrical form of the information is carried towards the cortex of the brain through geniculation muscles and humans see the image [1].

The categories of the visual prosthesis are:

Visual Cortical Implant

This is one of the conventional approaches of retinal implant and the largest number of blind patients were treated using this approach. The diseased visual pathways, glaucoma, and diabetic retinopathy responsible for damaging the inner structure of the retina are bypassed in this approach (Fig. **1**). It bypasses all diseased visual pathways, glaucoma, and diabetic retinopathy which can damage the inner retina. Despite these advantages, the visual cortical implant has a very high safety threshold than the other approaches of the visual prosthesis. Due to this high threshold, sometimes the patients face an unhealthy environment leading to death. Moreover, in this approach, the signal is processed outside the retina through signal processing devices; hence its efficacy and placement of electrodes inside the visual cortex are the other significant challenges of this approach [1].

Optic Nerve Implant

The Optic nerve implants, on the other hand, have the advantages of higher density resolution and compact size. It is safer than the visual cortical implant and the invasiveness of this approach is lesser too. As shown in Fig. (**2**), the optic nerve is responsible to carry the information to the brain; hence, in this approach the rehabilitation becomes viable. However, the adaptability of this approach is discouraged due to the blur vision perception, size, and position of the implant to capture images. In comparison to the retinal implant technique, this approach is more challenging and dangerous because of its position. The elicitation of visual spikes resulting in vision rebuilding is still a matter of discussion [1].

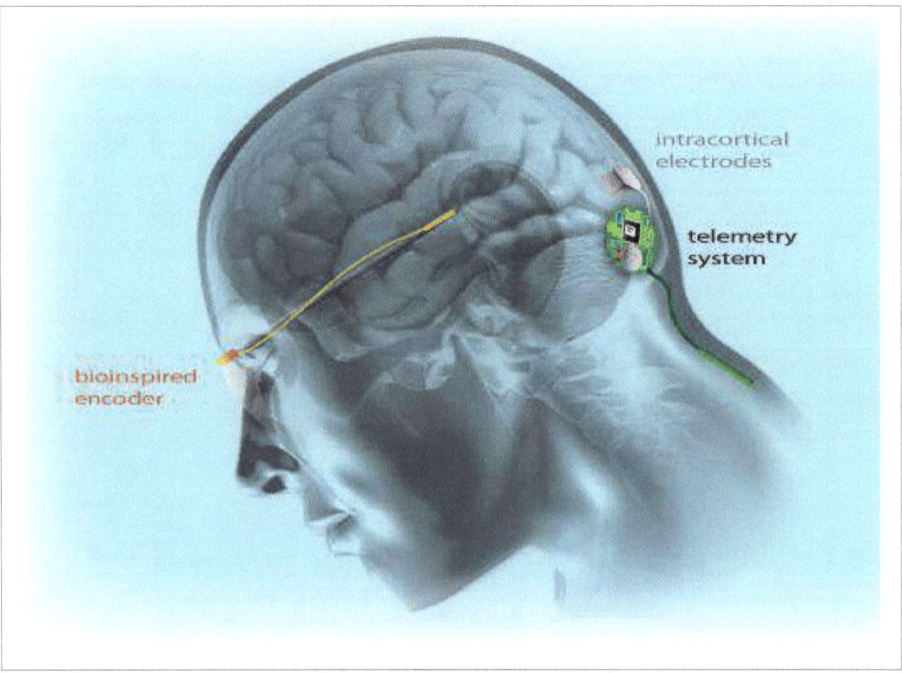

Fig. (1). Visual cortex implants position with intracortical electrodes.

Retinal Implant (Epiretinal and Subretinal Implant)

The distinguished approach of epiretinal and subretinal implant is depicted in Fig. (**3**). In epiretinal implants, the device is placed over the retinal tissues and image data is processed via external devices whereas in subretinal implants, the damaged retinal tissues are itself replaced with the implanting device such as CMOS image sensor. Both the techniques have their own advantages and disadvantages, such as in epiretinal implant it is easier to produce the high resolutions and surgical requirements are minimal. The major challenge is to provoke the inner retinal through the outer signal excitation [3]. The subretinal implant has the advantages of less mechanical fixation and decreased stimulation current. Isolated environment generating the heating effect and placement of implanting device inside the retinal are the main challenges in subretinal implant [4]. An example of the subretinal implant is shown in Fig. (**4**). This chapter is based on exploring the technique of subretinal implants.

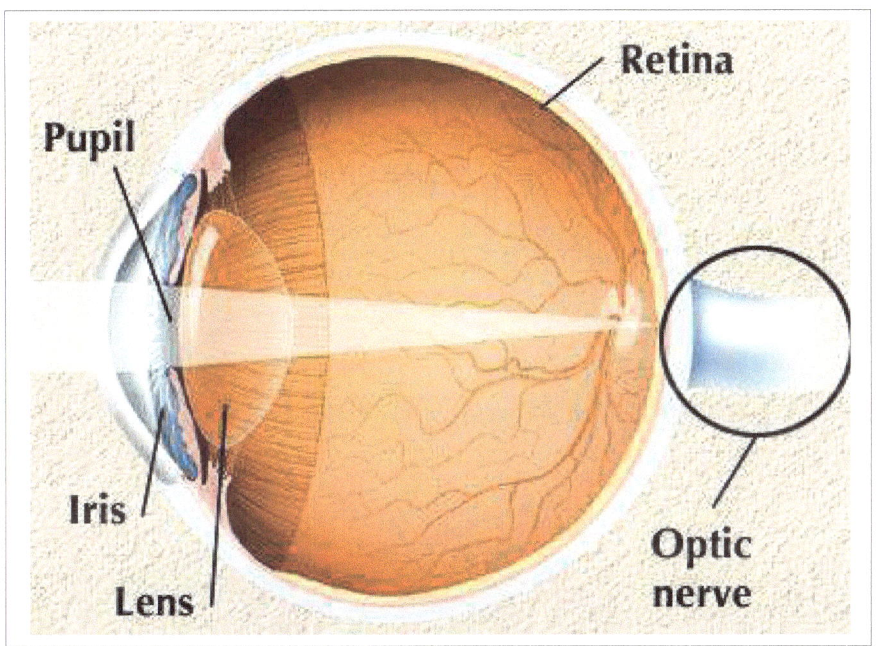

Fig. (2). Position of optic nerve implant.

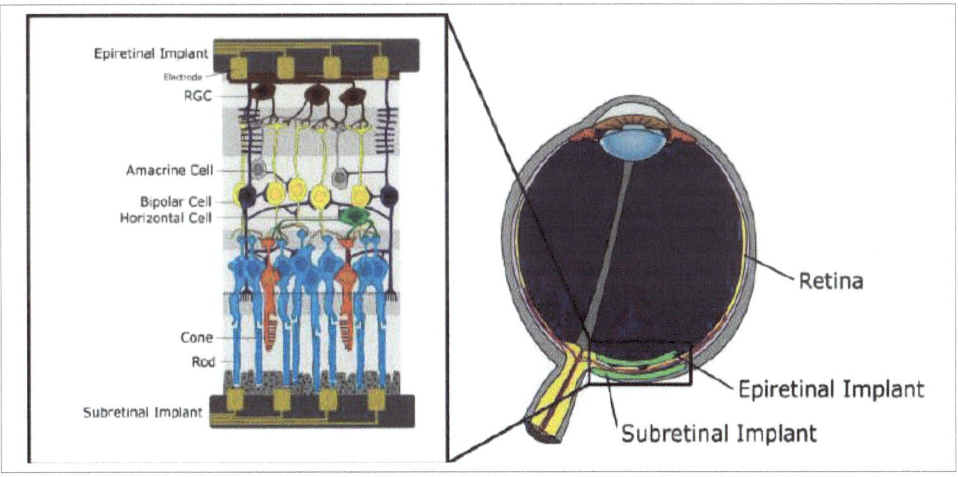

Fig. (3). Epiretinal vs. Subreitnal implant.

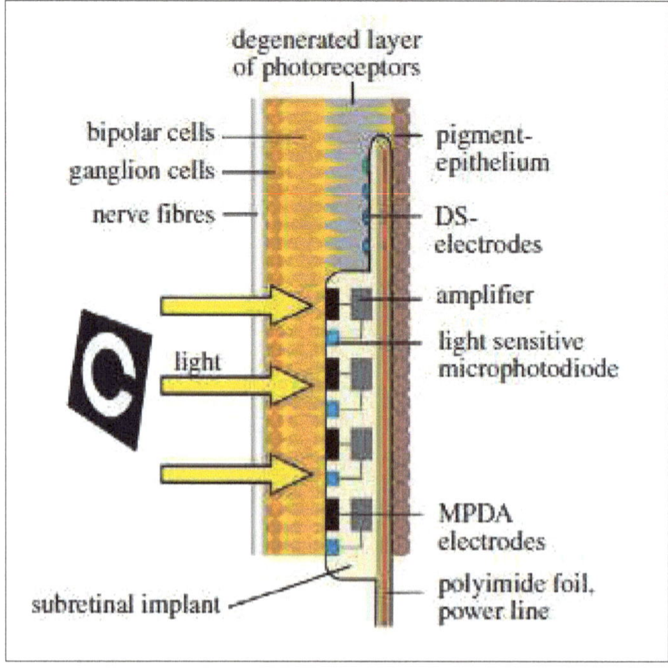

Fig. (4). Position of electrode, amplifier and other electronic circuits in subretinal implant.

The electrode design techniques of the subretinal implant were initially based on the employment of microphotodiodes in close proximity [5]. However, a larger area with low-resolution output was the problem in this approach. Later, with the advent of the charged coupled device (CCD), the problem of low-resolution images was solved. CCD provided a high dynamic range, high fill factor, improved image quality, and lower noise outputs. However, high power consumption, off-chip functionality, higher cost, and lower speed of imaging were the limitations of the CCDs [6].

After the year 2000, especially, the CMOS-based devices superseded the CCDs. Lower cost, lower power consumption, on-chip functionality, and higher speed with compatible imaging properties gave an edge to CMOS image sensor (CIS). Though CIS is more prone to noise than the CCD, however, with the improvement in data converter circuits, this problem was limited to a considerable range. The principle and working of CIS are shown in Fig. (**5**). The photodiode is responsible for the collection and conversion of the photon into electrons. The reset transistor controls the operation of photodiode. In the first phase, the charges are collected known as the reset phase followed by the integration phase where the electrons are converted into voltage [7].

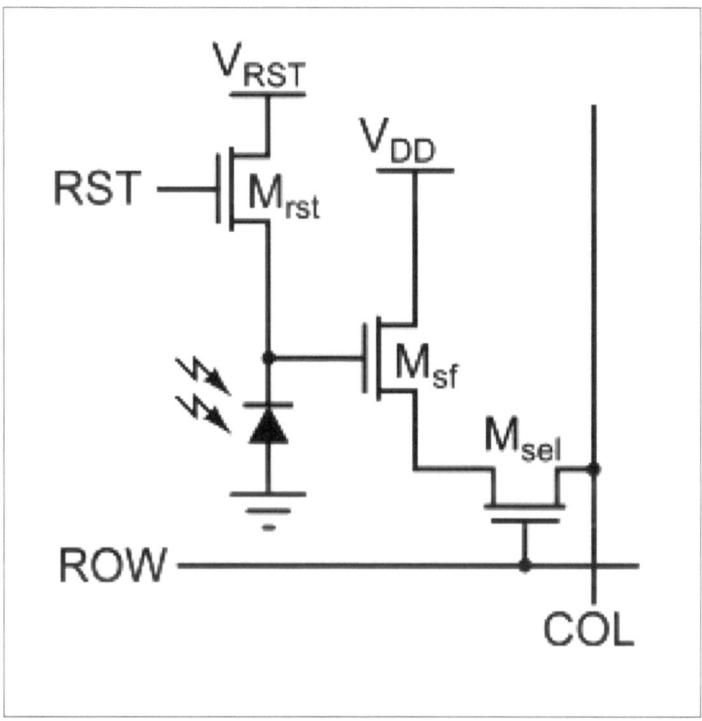

Fig. (5). Basic schematic of three transistor CMOS image sensor.

The basic components of a commercial subretinal implant are shown in Fig. (**6**) [8]. Following are the major points that need to be considered while designing the subretinal implant:

1. If the stimulus electrode (Fig. **6**) is charged and discharged continuously by monophased voltage pulses then a visual perception is created with the help of delayed spikes in the RGC.
2. The stimulus electrode is driven by the CIS. The basic details and schematic of the CIS have already been explained earlier in this text.
3. It is required that the overall efficiency (that includes imaging performance, noise performance, and biocompatibility) of the CIS must be high so that the implant can sustain itself for a prolonged period.

The placement of subretinal electrodes inside the retinal tissues generates some fundamental challenges. The first challenge is the generation of heat due to an isolated environment [9 - 12]. Second is the improper deposition of a number of photons on the surface of an electrode, and third is the infiltration of electronic circuits [13 - 16]. Hence effective surface area which is related to the charge density eventually, conductivity, thermal properties, optical properties, tensile strength, and bio-compatibility are the desired parameters for a subretinal implant [17, 18]. Considering the challenges and prerequisites of subretinal implant, titanium nitride (TiN), platinum grey, iridium oxide (IrOx), and carbon nanotube (CNT) have been frequently utilized in the past as electrode materials. The recent advancement in material design and application, ZrN and HfN have also shown promising results especially in terms of corrosive and mechanical properties [19 - 22]. This chapter provides a brief idea and comparative study about the required properties of nanomaterials used in subretinal implant applications.

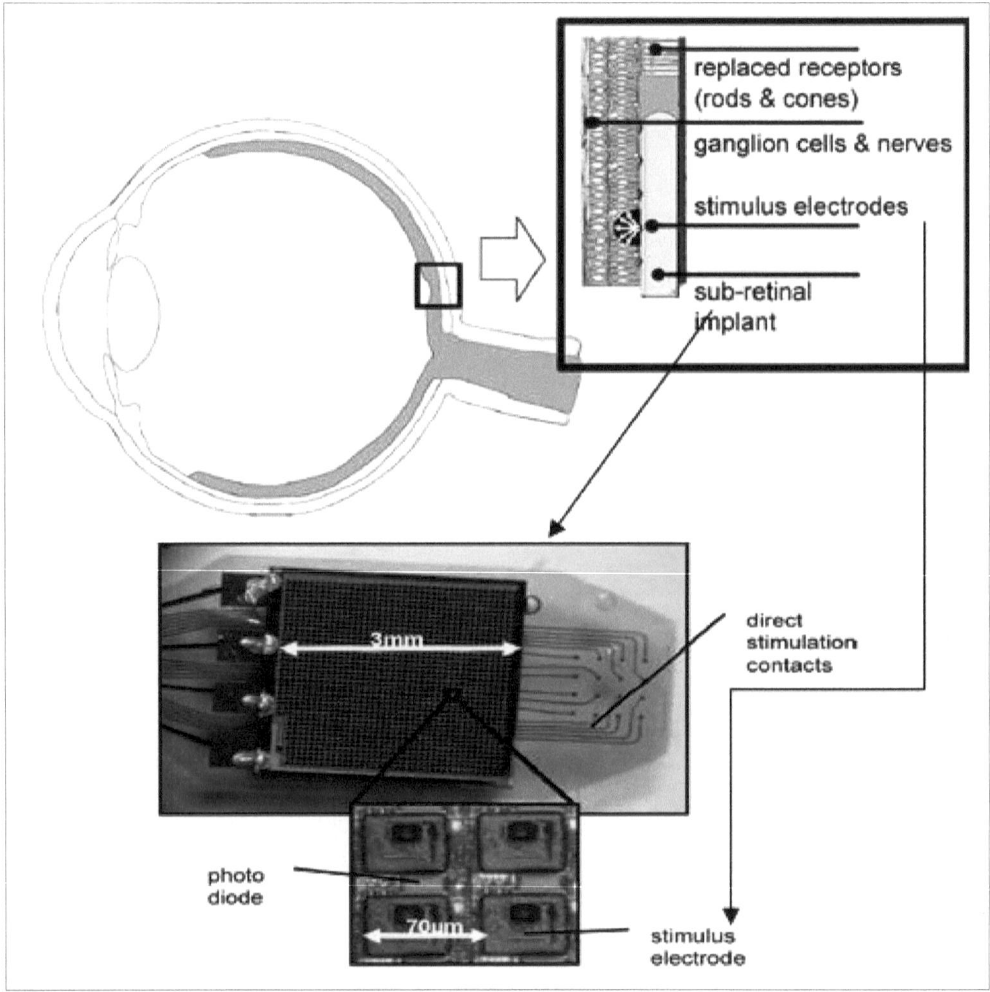

Fig. (6). The position of the subretinal chip and photographs of the mounted retinal implant chip with magnified highlighted area.

DESIRED PROPERTIES OF AN ELECTRODE MATERIAL

Structural Properties

A stable crystal structure is desired for the efficient operation of the implant. The implementation is done on the "Spanish Initiative for Electronic Simulations with Thousands of Atoms" (SIESTA). Fig. (7) shown below is the optimized structure of TiN, ZrN,and HfN materials [23]. Since these materials belong to the same space group Fm-3m, the common procedure is usually adopted for their optimization. The bond length, lattice constant, and ground state energy of ZrN is the highest among these materials. This indicates that the structural stability of ZrN is superior p [24 - 26]. The factor of incompressibility is the measure of sustaining the pressure upon volume on each side of the material is also better in ZrN [27].

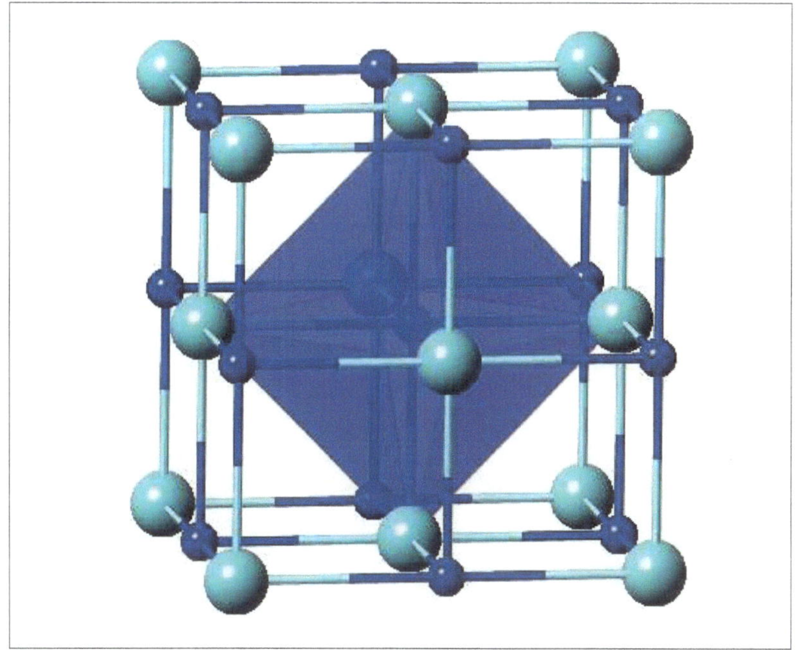

Fig. (7). Optimized crystal structures of TiN, ZrN, and HfN.

Threshold Charge Density

Charge density is an important electrode property in the design of the subretinal implant. It is observed that the electrode with a small size requires a higher charge density to match the power requirements in comparison with the larger electrodes.

TiN and IrO_2 are the popular choices for subretinal electrode design due to their higher charge density values. The reported values of charge density for TiN is in between 0.2 to 0.6 mC/cm² whereas it is 0.26 to 0.32 mC/cm² for IrO_2 [28].

The charge distribution plot of TiNis shown in Fig. (**8**). The charges are dispersed across the N molecule strongly (shown in the pink color) and the nature of the bonding is ionic [29, 30].

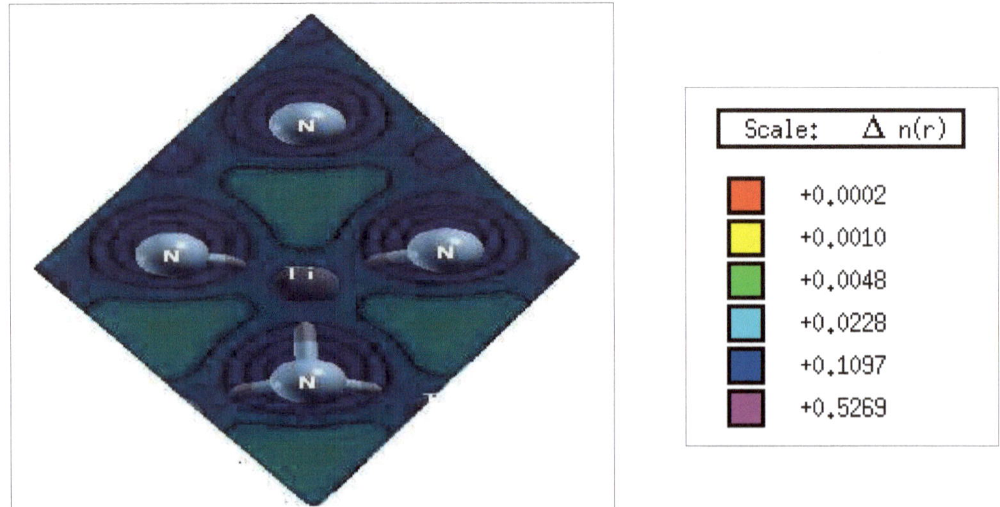

Fig. (8). Charge distribution plot of TiN with color codes.

Thermal Properties

Cohesive energy and Enthalpy of formation give the basic idea about the thermal stability of any material [31, 32]. The negative values of these properties indicate that a strong bond exists between the atoms of the material. The higher the negative values the better the bonding strength. The negative value of Enthalpy also shows that the nature of the compound is exothermic. In one of the recent study, we found that HfN has better thermal stability in comparison with ZrN and TiN [27]. Following formulas are used for the calculation of cohesive energy and formation of enthalpy:

$$E_{coh} = \frac{1}{n}(E_{MN-bulk} - E_{M} - E_{N}) \qquad (1)$$

$$\Delta H = \Delta U = \frac{1}{n}(E_{MN-bulk} - E_{M-bulk} - E_{N-bulk}) \qquad (2)$$

where 'n' is the number of atoms, E_{bulk} is the total energy of system, E_M is the energy of a single transition metal atom and E_N is the energy of atom nitrogen.

The values of these thermal properties are shown in Table **1** [27]. As already stated that the subretinal implant faces the challenge of heat dissipation due to the isolated environment, hence thermal stability becomes very important for the operation of subretinal implant over a prolonged period. It can be concluded from the table that the thermal stability of HfN is greater than the other two.

Table 1. Thermal properties of TiN, ZrN and HfN.

Material	Cohesive Energy(eV/atom)Ecoh	Formation Enthalpy(eV/atom)ΔH
TiN	-18.38	-5.03
ZrN	-19.04	-5.02
HfN	-19.11	-5.16

Electronic Properties

The electronic properties of any material give a clear idea about the physical and chemical behavior. Fig. (**9**) shows the band plots and density of states (DOS) curves for the TiN and ZrN [33, 34]. The d orbitals of Ti and Zr contribute the most while the p orbitals for N have the primary contribution [27]. The distinct peaks are the illustration of the fact that the behaviour of TiN and ZrN is metalloid with a strong covalent bond among the elements.

Optical Properties

As already stated, optical properties are very important for any material to be used for retinal implant applications. Especially, in the case of a subretinal implant the stimulation electrode is placed inside the retina.

The problem of improper light on the surface of the material is always there. Ideally, the absorption of the light must be high and the reflectance must be low for any material used in the retinal implant application. Following equations are frequently used in the calculation of optical properties [35, 36]:

$$\varepsilon(\omega) = \varepsilon_1(\omega) + i\varepsilon_2(\omega) \tag{3}$$

where ε_1 (real) and ε_2 (imaginary) are:

$$\varepsilon_1 = 1 + \frac{8\pi^2 e^2}{m^2} \Sigma_{Vb.Cb} \int_{BZ} d^3k \, \frac{2}{2\pi} \frac{|eM_{CbVb}(K)|^2}{[E_{cb}(k)-E_{vb}(k)]} \frac{h^3}{[E_{cb}(k)-E_{vb}(k)]-h^2\omega^2} \tag{4}$$

$$\varepsilon_2 = \frac{4\pi^2}{m^2\omega^2} \Sigma_{Vb.Cb} \int_{BZ} d^3k \, \frac{2}{2\pi} |eM_{CbVb}(K)|^2 \delta[E_{cb}(k) - E_{vb}(k) - h\omega] \tag{5}$$

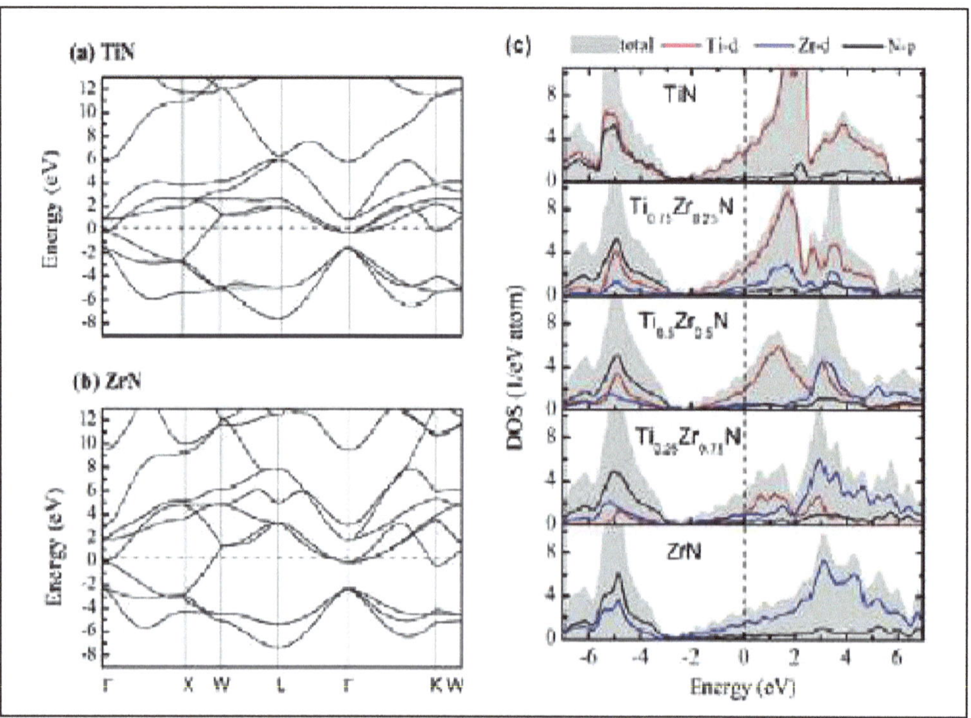

Fig. (9). Electronic properties of TiN and ZrN, (a) Band plot of TiN, (b) Band plot of ZrN, (c) Density of state curves.

The optical properties shown in Fig. (**10**) depict that the behaviour of TiN and ZrN is metallic since the negative curves were observed [37 - 40]. The peaks of TiN for genarlizedgradiant approximations (GGA) are higher than the local density approximations (LDA) [41]. However, ZrN shows a better behaviour than the TiN with a higher positive dielectric function for the same energy values.

The reflectivity curves are shown in Fig. (**10c**) indicated the different behaviour of TiN and ZrN with a change in the temperature value [42]. Ideally for a subretinal implant the value of the reflectance must be on the lower side to ensure the maximum availability of the optical signal on the surface of the material. If the reflectivity of any material is high then the optical signal will get a reflection from the surface of the material. The figure suggests that at the lower values of the temperature the reflectivity of the TiN is more negative than the ZrN.

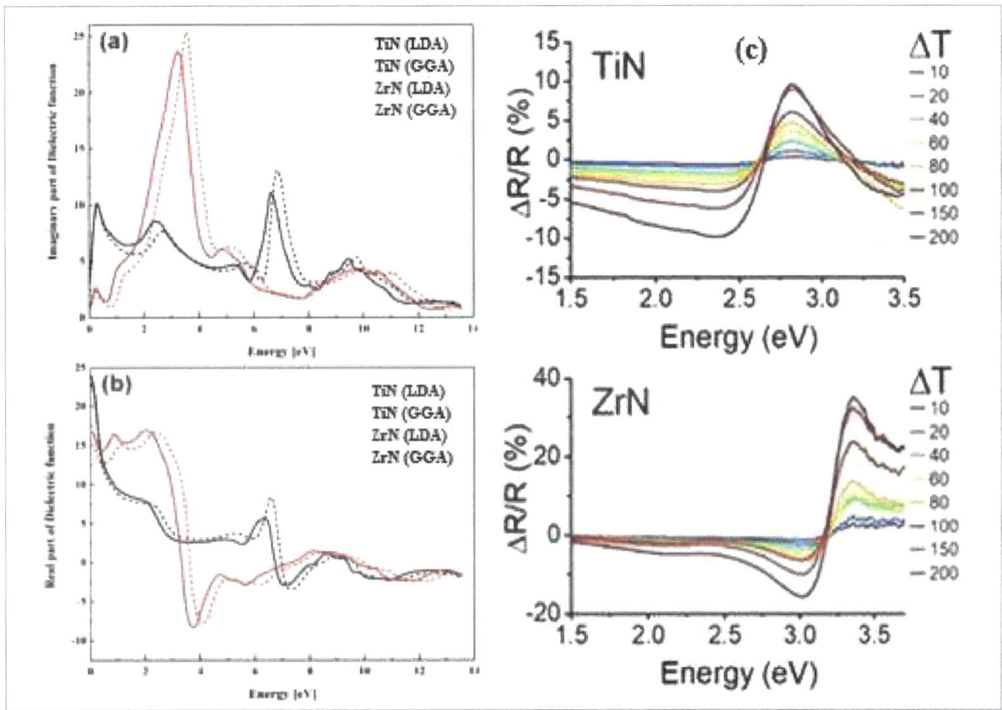

Fig. (10). Optical properties of TiN and ZrN, (a) Imaginary part of dielectric properties, (b) Real part of dielectric properties, (c) Reflectance of TiN and ZrN under different temperature.

However, with an increase in temperature the reflectivity increases, and at around 3 V it attains the positive value, still TiN shows better reflectance performance than ZrN.

Adsorption Properties

As stated earlier, infiltration of electronic equipment in the corrosive environment of inter retinal cells is one of the major obstacles in the performance of image sensors.

It is important for any material to show the least reactivity with the water molecule and to ensure that adsorption properties are studied. In one of the researches, TiN has been studied under the different surface positions of (1 0 0), (1 1 0) and (1 1 1) [43]. Figs. (**11 - 13**) show the adsorption of a single water molecule on the surface of TiN.

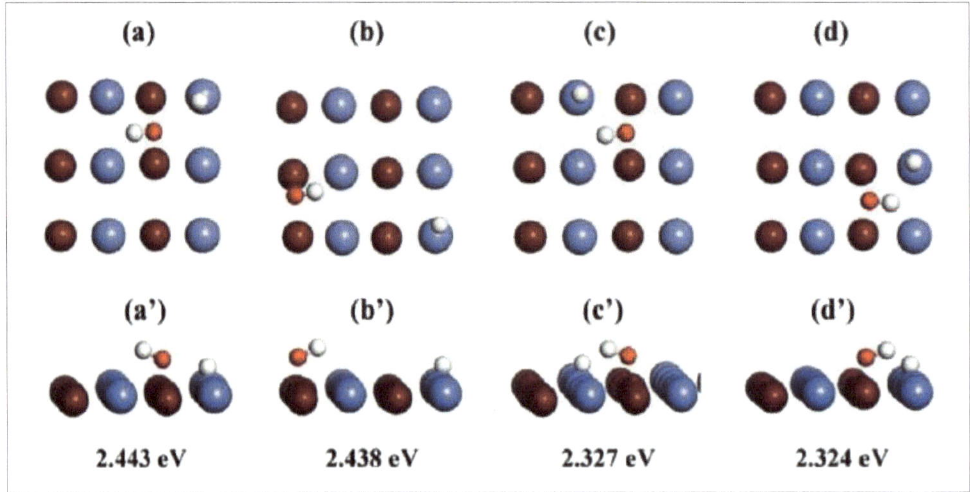

Fig. (11). Adsorption energy of water molecule on the surface of TiN (1 0 0).

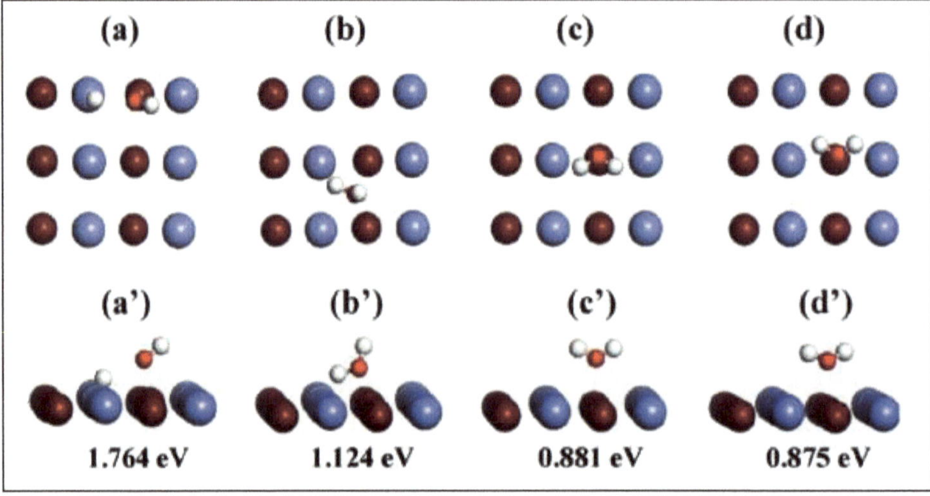

Fig. (12). Adsorption energy of water molecule on the surface of TiN (1 1 0).

Notably, the negative adsorption energy of the water molecule shows its instability on the surface of the TiN. This property is desired for any material to be embraced as a coating material or electrode material in bioelectronic applications such as subretinal implant in this case. The most negative adsorption value is observed for the configuration of (100). In one of our studies related to the cubical metal nitrides, HfN showed better anticorrosive attributes [27]. The adsorption property characterizes the material to be hydrophobic/phillic depending upon the application.

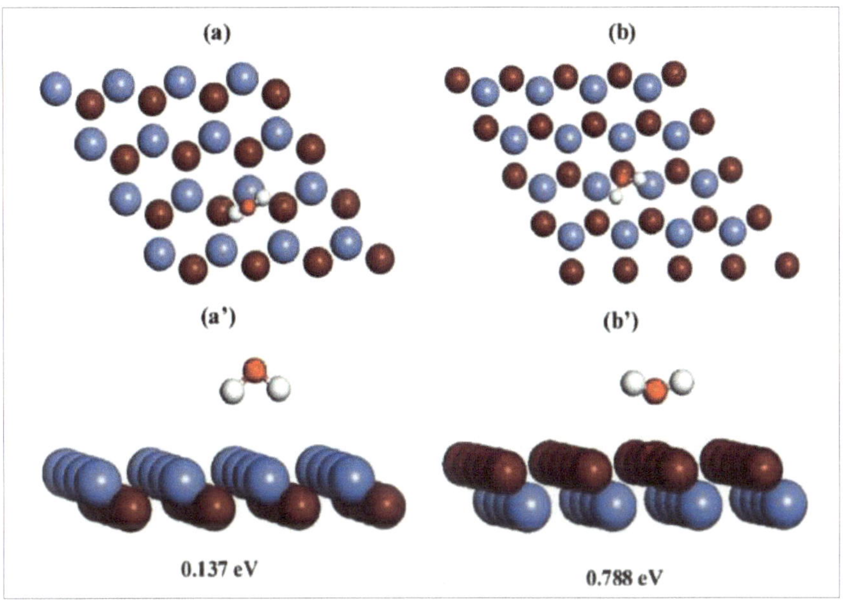

Fig. (13). Adsorption energy of water molecule on the surface of TiN (1 1 1).

OUR RECENT CONTRIBUTION IN ELECTRODE MATERIAL MODELING

In the past few years, we have been investigating the attributes of electrode materials so that appropriate material for the retinal implants can be identified. In one of our recent works [27], the supercell structures of cubical transition metal nitrides(TiN, ZrN, and HfN) as shown in Fig. (14) were studied in detail. Through the analysis of cohesive energy and formation enthalpy, the thermal stability and chemical reactivity of HfNwas found to be superior to TiN and ZrN.

The analysis of important optical properties of absorption coefficient, optical conductivity, and reflectivity showed that the surface of HfN material is less reflective as well as more conducive. The analysis of Young's modulus, shear modulus, and Poisson's ratio ensured the mechanical stability of the HfN material. Even the adsorption of the water molecule was the least stable at the surface of HfN.

The major drawback found in the above study was the overlapped bandgap in all the structures of TiN, ZrN, and HfN. Hence in another work [44], we investigated the transformation of the TiN monolayer into the HfN monolayer, as shown in Fig. (15). With all other characteristics, an increased electronic conductivity with tunable bandgap was reported in this work. The obtained tunable bandgap is certainly useful in the subretinal implant technique as the biological constraints of controlled voltage swing, charge density, and flow of electrons can be easily manageable and achieved.

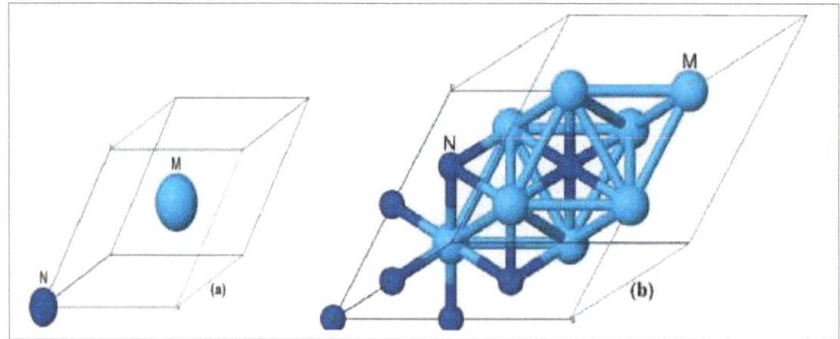

Fig. (14). Molecular structure of **(a)** Primitive cell, **(b)** Supercell (2×2×2) of cubical transistion metal system.

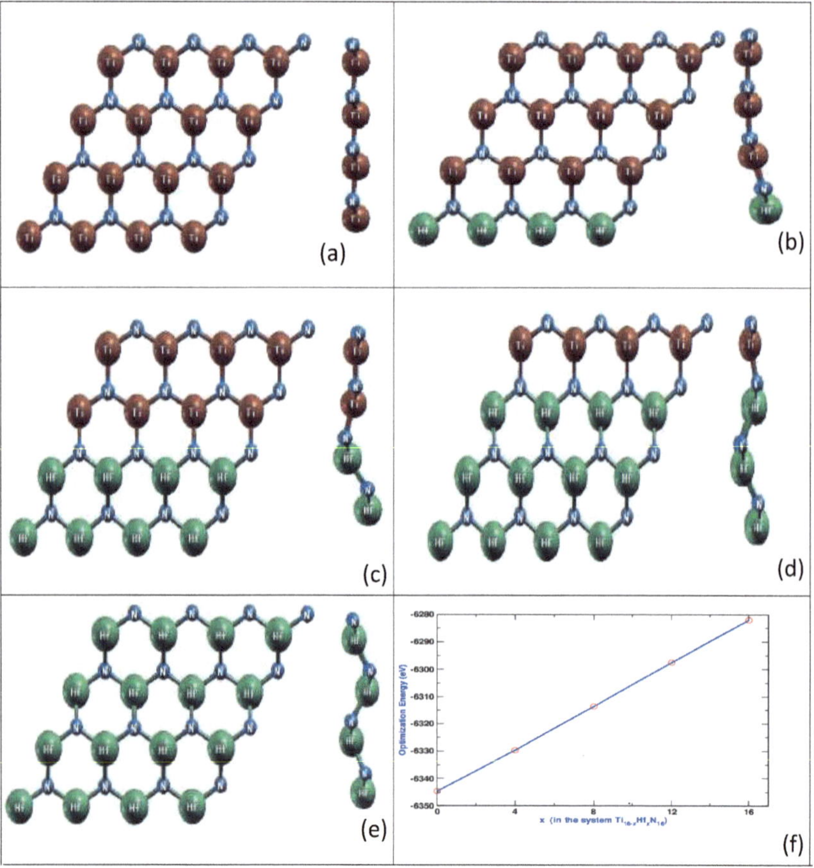

Fig. (15). Structure of mono layer systems **(a)** $Ti_{16}Hf_0N_{16}$**(b)** $Ti_{12}Hf_4N_{16}$**(c)** $Ti_8Hf_8N_{16}$**(d)** $Ti_4Hf_{12}N_{16}$**(e)** $Ti_0Hf_{16}N_{16}$ and **(f)** optimization energy vs composition plot of the system $Ti_{16-x}Hf_xN_{16}$ ($x\{0,4,16\}$). The side view is also shown in the same figure.

CONCLUSION

It can be concluded from this study that the material for subretinal implant application must be able to sustain under corrosive as well as isolated environment. It should have high absorption coefficient and lower reflectivity constant. The metallic behavior works in the favor of such materials. Higher charge density is required so that effective surface area requirement could be met. TiN and CNT are popular materials for retinal implant applications, however, with excellent physical, chemical, optical, corrosive, and tensile strength properties, ZrN and HfN are emerging as popular alternative.

CONSENT FOR PUBLICATION

Not applicable.

CONFLICT OF INTEREST

The author declares no conflict of interest, financial or otherwise.

ACKNOWLEDGEMENTS

We gratefully acknowledge the kind support of the management of Shri Shankaracharya Technical Campus-SSGI. Helpful discussions with Prof. Ravindra Pandey (Michigan Technological University, USA) are acknowledged.

REFERENCES

[1]   Tiwari, A.; Talwekar, R. H. **2019**.
        [http://dx.doi.org/10.1080/03772063.2017.1417750]

[2]   Zrenner, E. Will Retinal Implants Restore Vision?
        [http://dx.doi.org/10.1126/science.1067996]

[3]   Chang, Y.-C.; Weiland, J. D.; Humayun, M. S. **2018**.
        [http://dx.doi.org/10.1007/978-3-319-67260-1_2]

[4]   Niu, J.; Liu, Y.; Ren, Q.; Zhou, Y.; Zhou, Y.; Niu, S. Vision Implants: An Electrical Device Will Bring Light to the Blind. *Sci. China, Ser. F. Inf. Sci.,* **2008**, *51*(1), 101-110.
        [http://dx.doi.org/10.1007/s11432-007-0072-z]

[5]   Sachs, H.G.; Schanze, T.; Wilms, M.; Rentzos, A.; Brunner, U.; Gekeler, F.; Hesse, L. Subretinal implantation and testing of polyimide film electrodes in cats. *Graefes Arch. Clin. Exp. Ophthalmol.,* **2005**, *243*(5), 464-468.
        [http://dx.doi.org/10.1007/s00417-004-1049-x] [PMID: 15578200]

[6]   Bigas, M.; Cabruja, E.; Forest, J.; Salvi, J. Review of CMOS Image Sensors. *Microelectronics,* **2006**, *37*(5), 433-451.
        [http://dx.doi.org/10.1016/j.mejo.2005.07.002]

[7]   Tiwari, A.; Talwekar, R.H. Analysis and Mathematical Modelling of Charge Injection Effect for Efficient Performance of CMOS Imagers and CDS Circuit. *IET Circuits Dev. Syst.,* **2020**, *14*(7), 1038-1048.
        [http://dx.doi.org/10.1049/iet-cds.2020.0096]

[8]   Graf, H.G.; Harendt, C.; Engelhardt, T.; Scherjon, C.; Warkentin, K.; Richter, H.; Burghartz, J.N. High Dynamic Range CMOS Imager Technologies for Biomedical Applications. *IEEE J. Solid-State Circuits,* **2009**, *44*(1), 281-289.
        [http://dx.doi.org/10.1109/JSSC.2008.2007437]

[9]   Chow, A. Y.; Chow, V. Y. *Subretinal Electrical Stimulation of the Rabbit Retina,* **1997**.
        [http://dx.doi.org/10.1016/S0304-3940(97)00185-7]

[10]   Chow, A.Y.; Chow, V.Y.; Packo, K.H.; Pollack, J.S.; Peyman, G.A.; Schuchard, R. The artificial silicon retina microchip for the treatment of vision loss from retinitis pigmentosa. *Arch. Ophthalmol.,* **2004**, *122*(4), 460-469.
        [http://dx.doi.org/10.1001/archopht.122.4.460] [PMID: 15078662]

[11]   Peyman, G.; Chow, A.Y.; Liang, C.; Chow, V.Y.; Perlman, J.I.; Peachey, N.S. Subretinal semiconductor microphotodiode array. *Ophthalmic Surg. Lasers,* **1998**, *29*(3), 234-241.
        [http://dx.doi.org/10.3928/1542-8877-19980301-10] [PMID: 9547778]

[12]   Ortmanns, M. *Florian Gekeler, Karin Kobuch, Hartmut Normann Schwahn, Alfred Stett, Kei Shinoda, E. Z. "Subretinal Electrical Stimulation of the Rabbit Retina with Acutely Implanted Electrode Arrays*; Springer-Verlag, **2004**.

[13]   Chader, G.J.; Weiland, J.; Humayun, M.S. Artificial vision: needs, functioning, and testing of a retinal electronic prosthesis. *Prog. Brain Res.,* **2009**, *175*, 317-332.
        [http://dx.doi.org/10.1016/S0079-6123(09)17522-2] [PMID: 19660665]

[14]   Zrenner, E. *The Subretinal Implant: Can Microphotodiode Arrays Replace Degenerated Retinal Photoreceptors to Restore Vision?*; , **2002**.

[15]   Hoffmann, F. *Entwicklung Eines Dynamischen Modells Der Stoftproduktion Yon Zuckerruben Mit Hilfe Der Multiplen Regressionsanalyse.,* **1980**.

[16]   Werginz, P.; Benav, H.; Zrenner, E.; Rattay, F. *Modeling the Response of ON and OFF Retinal Bipolar Cells during Electric Stimulation.,* **2015**.
        [http://dx.doi.org/10.1016/j.visres.2014.12.002]

[17]   Birkholz, M.; Ehwald, K.E.; Wolansky, D.; Costina, I.; Baristiran-Kaynak, C.; Fröhlich, M.; Beyer, H.; Kapp, A.; Lisdat, F. Corrosion-Resistant Metal Layers from a CMOS Process for Bioelectronic Applications. *Surf. Coat. Tech.,* **2010**, *204*(12–13), 2055-2059.
        [http://dx.doi.org/10.1016/j.surfcoat.2009.09.075]

[18]   Hämmerle, H.; Kobuch, K.; Kohler, K.; Nisch, W.; Sachs, H.; Stelzle, M. Biostability of micro-photodiode arrays for subretinal implantation. *Biomaterials,* **2002**, *23*(3), 797-804.
        [http://dx.doi.org/10.1016/S0142-9612(01)00185-5] [PMID: 11771699]

[19]   Menghani, J.; Pai, K.B.; Totlani, M.K. Corrosion and Wear Behaviour of ZrN Thin Films. *Tribol. -. Mater. Surfaces Interfaces,* **2011**, *5*(3), 122-128.
        [http://dx.doi.org/10.1179/1751584X11Y.0000000013]

[20]   Yu, S.; Zeng, Q.; Oganov, A.R.; Frapper, G.; Huang, B.; Niu, H.; Zhang, L. First-Principles Study of Zr-N Crystalline Phases: Phase Stability, Electronic and Mechanical Properties. *RSC Advances,* **2017**, *7*(8), 4697-4703.
        [http://dx.doi.org/10.1039/C6RA27233A]

[21]   Guo, F.; Wang, J.; Du, Y.; Wang, J.; Shang, S.L.; Li, S.; Chen, L. First-Principles Study of Adsorption and Diffusion of Oxygen on Surfaces of TiN, ZrN and HfN. *Appl. Surf. Sci.,* **2018**, *452*, 457-462.
        [http://dx.doi.org/10.1016/j.apsusc.2018.05.052]

[22]   Patsalas, P. Zirconium Nitride: A Viable Candidate for Photonics and Plasmonics? In: *Thin Solid Films*; Elsevier B.V, **2019**.
        [http://dx.doi.org/10.1016/j.tsf.2019.137438]

[23]   Dr. Mohan L Verma, Lecture -7 : Optimization of Materials using SIESTA.

https://www.youtube.com/watch?v=1kH3seyDRrk

[24]   Solar, M. The SIESTA Method for Ab Initio Order-N Materials. *J. Phys. Condens. Matter,* **2002**, *14*(11), 2745-2779.
[http://dx.doi.org/10.1088/0953-8984/14/11/302]

[25]   Zhang, R.Q.; Zhang, Q.Z.; Zhao, M.W. A Scheme for the Economical Use of Numerical Basis Sets in Calculations with SIESTA. *Theor. Chem. Acc.,* **2004**, *112*(3), 158-162.
[http://dx.doi.org/10.1007/s00214-004-0598-8]

[26]   Ordejón, P; Artacho, E.; Cachau, R. Linear Scaling DFT Calculations with Numerical Atomic Orbitals. *MRS Online Proceedings Library (2001),* **2004**, *677, 96.*
[http://dx.doi.org/10.1557/PROC-677-AA9.6]

[27]   Tiwari, A.; Talwekar, R.H.; Verma, M.L. Talwekar, R.H.; Verma, M. L. First-Principles Study on the Structural, Electronic, Optical, Mechanical, and Adsorption Properties of Cubical Transition Metal Nitrides MN (M=Ti, Zr and Hf). *J. Electron. Mater.,* **2021**, *50*(6), 3312-3325.
[http://dx.doi.org/10.1007/s11664-021-08814-x]

[28]   Eleftheriou, C.G.; Zimmermann, J.B.; Kjeldsen, H.D.; David-Pur, M.; Hanein, Y.; Sernagor, E. Carbon nanotube electrodes for retinal implants: A study of structural and functional integration over time. *Biomaterials,* **2017**, *112*, 108-121.
[http://dx.doi.org/10.1016/j.biomaterials.2016.10.018] [PMID: 27760395]

[29]   Amorim, R.G.; Zhong, X.; Mukhopadhyay, S.; Pandey, R.; Rocha, A.R.; Karna, S.P. Strain- and electric field-induced band gap modulation in nitride nanomembranes. *J. Phys. Condens. Matter,* **2013**, *25*(19), 195801.
[http://dx.doi.org/10.1088/0953-8984/25/19/195801] [PMID: 23604312]

[30]   Azam, S.; Bila, J.; Kamarudin, H.; Reshak, A.H.; Reshak, A.H. Electronic Structure, Electronic Charge Density and Optical Properties of 3-Methyl-1,4-Dioxo-1, 4-Dohydronaphthalen-2-Yl-Sulfanyl (C13H10O4S) Detection of Cardiac Arrhythmia Patterns by Adaptive Analysis View Project Electronic Structure, Electronic Charge Density and Optical Properties of 3-Methyl-1,4-Dioxo-1, 4-Dohydronaphthalen-2-Yl-Sulfanyl (C 13 H 10 O 4 S). , **2014**; 2, .

[31]   Galperin, Y. *Introduction to modern solid state physics,* http://wwwgradinetti.org/teaching/chem 121/assests/

[32]   Shigemi, A.; Wada, T. Enthalpy of Formation of Various Phases and Formation Energy of Point Defects in Perovskite-Type NaNbO3 by First-Principles Calculation., **2004**.
[http://dx.doi.org/10.1143/JJAP.43.6793]

[33]   Kumar, M.; Ishii, S.; Umezawa, N.; Nagao, T. NAGAO, T. Band Engineering of Ternary Metal Nitride System Ti1-x ZrxN for Plasmonic Applications. *Opt. Mater. Express,* **2015**, *6*(1), 29.
[http://dx.doi.org/10.1364/OME.6.000029]

[34]   Hinuma, Y.; Pizzi, G.; Kumagai, Y.; Oba, F.; Tanaka, I. Band Structure Diagram Paths Based on Crystallography. *Comput. Mater. Sci.,* **2017**, *128*, 140-184.
[http://dx.doi.org/10.1016/j.commatsci.2016.10.015]

[35]   Zhao, L.; Zhang, X.; Fan, C.; Liang, Z.; Han, P. First-Principles Study on the Structural, Electronic and Optical Properties of BiOX (X=Cl, Br, I) Crystals. *Physica B,* **2012**, *407*(17), 3364-3370.
[http://dx.doi.org/10.1016/j.physb.2012.04.039]

[36]   Pourghazi, A.; Dadsetani, M. Electronic and Optical Properties of BaTe, BaSe and BaS from First Principles. *Physica B,* **2005**, *370*(1–4), 35-45.
[http://dx.doi.org/10.1016/j.physb.2005.08.032]

[37]   Martin, C.; Miller, K.H.; Makino, H.; Craciun, D.; Simeone, D.; Craciun, V. Optical properties of Ar ions irradiated nanocrystalline ZrC and ZrN thin films. *J. Nucl. Mater.,* **2017**, *488*, 16-21.
[http://dx.doi.org/10.1016/j.jnucmat.2017.02.041] [PMID: 32020950]

[38]   Patsalas, P.; Kalfagiannis, N.; Kassavetis, S. Optical Properties and Plasmonic Performance of

Titanium Nitride. *Materials (Basel),* **2015**, *8*(6), 3128-3154.
[http://dx.doi.org/10.3390/ma8063128]

[39]   Karlsson, B.; Shimshock, R.P.; Seraphin, B.O.; Haygarth, J.C.; Karlsson, B. Optical Properties of Cvd-Coated Tin, Zrn and Hfn. *Phys. Scr.,* **1982**, *25*(6), 775-779.
[http://dx.doi.org/10.1088/0031-8949/25/6A/030]

[40]   Gueddaoui, H.; Maabed, S.; Schmerber, G.; Guemmaz, M.; Parlebas, J.C. Structural and Optical Properties of Vanadium and Hafnium Nitride Nanoscale Films: Effect of Stoichiometry. *Eur. Phys. J. B,* **2007**, *60*(3), 305-312.
[http://dx.doi.org/10.1140/epjb/e2006-00350-9]

[41]   Farzan, M.; Elahi, S.; Salehi, H.; Abolhasani, M.R. A Comparison of The~Structural, Electronic, Optical and Elastic Properties of Wurtzite, Zinc-Blende and Rock Salt TlN: A DFT Study. *Acta Phys. Pol. A,* **2016**, *130*(3), 758-768.
[http://dx.doi.org/10.12693/APhysPolA.130.758]

[42]   Diroll, B.; Saha, S.; Shalaev, V.; Boltasseva, A.; Schaller, R. *Broadband Ultrafast Dynamics of Refractory Metals: TiN and ZrN,* **2020**.
[http://dx.doi.org/10.1002/adom.202000652]

[43]   Sanyal, S.; Waghmare, U.V.; Ruud, J.A. Adsorption of Water on TiN (1 0 0), (1 1 0) and (1 1 1) Surfaces: A First-Principles Study. *Appl. Surf. Sci.,* **2011**, *257*(15), 6462-6467.
[http://dx.doi.org/10.1016/j.apsusc.2011.02.042]

[44]   Verma, M.L.; Tiwari, A. Theoretical Investigation on Structural Transformation of TiN to HfN Monolayer: A   First   Principles   Study.   *Chem.   Phys.   Lett.,*   **2021**,   *781*, 138992.https://doi.org/https://doi.org/10.1016/j.cplett.2021.138992
[http://dx.doi.org/10.1016/j.cplett.2021.138992]

CHAPTER 3

Recent Advancements in the Design of Electrode Materials for Rechargeable Batteries

Shivani Gupta[1], Abhishek Kumar Gupta[1,*], Sarvesh Kumar Gupta[1] and **Mohan L Verma[2,*]**

[1] *Department of Physics and Material Science, Madan Mohan Malaviya University of Technology, Gorakhpur (U. P.),273010, India*

[2] *Department of Applied Physics, SSGI, Shri Shankaracharya Technical Campus, Junwani, Bhilai, Chhattisgarh490020, India*

Abstract: As the world progresses towards sustainable energy and concomitant decarbonization of its electrical supply, modern civilization is approaching the fourth industrial revolution with a boom of digital devices and innovative technologies. As a result, the demand for high-performance batteries has skyrocketed, and many research initiatives for the design and development of high-performance rechargeable batteries are being taken. With the incremental standardization of battery designs, enhancement in their performance mainly relies on technical advancement in key electrode materials (cathode and anode materials). This chapters reviews the state-of-the-art materials used in fabricating electrodes, including a description of their structures and storage mechanisms and modification of commonly used materials for electrode working either in the solid-state or in the solution for aqueous or non-aqueous electrolytes. Based on the appropriate metrics such as operating voltage, specific energy, capacity, cyclic stability and life cycle, the performance of different electrodes has also been assessed. Along with the recent advancement, pertaining limitations are briefly covered and analyzed with some viable solutions in the pursuit of cathode and anode materials with fast kinetics, high voltage, and long cycle life.

Keywords: Electrode materials, Life cycle, Operating voltage, Rechargeable batteries, Specific energy, Sustainable energy.

INTRODUCTION

Lithium-ion batteries (LIBs) is one of the most promising secondary sources for portable electronic devices such as laptops, mobile, camera, *etc* [1]. LIBs got

* **Corresponding authors Mohan Lal Verma and Abhishek Kumar Gupta**: Department of Applied Physics, FET-SSGI, Shri Shankaracharya Technical Campus, Junwani, Bhilai, Chhattisgarh 490020, India and Department of Physics and Material Science, Madan Mohan Malaviya University of Technology, Gorakhpur (U. P.),273010, India; Tel: +919303452648; E-mails: akgupta.mmmut@gmail.com and drmohanlv@gmail.com

Dibya Prakash Rai (Ed.)

commercialized in 1992 as the first rechargeable battery to be used in future electronic devices [2]. Its components included carbon as the negative electrode, a non-aqueous electrolyte, and lithium cobaltate ($LiCoO_2$) as the positive electrode [3]. Recently rechargeable batteries have gained tremendous attention for application in hybrid electric vehicles [4]. With the increase in demand, the need for advanced battery technology is accelerating at a rapid pace. Batteries that could give high power density, high energy density and are safe have become a requirement in the present scenario [5]. The better performance of the battery is limited by expensive positive electrodes; which is one of the key element of the battery [6, 7]. Since 1980, Goodenough has devoted enormous efforts to formulate oxides based on transition metal elements with a particular emphasis on those materials whose structure supports higher mobility of Li^+ ions [6]. The journey of discovery of electrode materials started with the layered structure of $LiCoO_2$ in 1980 marked by two important milestones: one in 1986 with the discovery of $LiMn_2O_4$ spinel structure and the other in 1997 for the $LiMPO_4$ (M = Fe, Mn, etc.) olivine family [2]. For systems with high energy density, mainly layered structure cathodes are employed, whereas spinel and olivine families are preferred ones for Li-ion batteries with high-power density [7]. High specific capacity, safety, good chemical stability, environment friendliness, and cost-effectiveness are some of the specific properties of the lithium compounds required for suitable insertion of lithium-ions [8]. This chapter primarily focuses on the pros and cons of cathode and anode materials used in LIBs such as $LiCoO_2$ (LCO), layered $LiNiO_2$ (LNO), layered NCA, spinel $LiMn_2O_4$ (LMO), olivine $LiFePO_4$ (LFP), hard carbon, soft carbon, carbon nanotubes (CNTs) and graphene. By keeping some very important physical parameters in mind, comparative study of these materials as the active electrochemical element is discussed in this chapter. From the perspective of transportation of Li-ion in these lattices, it is our goal to address different types of electrode materials, their structural and synthesis challenges, efficiency, electrochemical stability window and safety issues [9].

MECHANISMS OF ENERGY STORAGE IN LI-ION BATTERIES

There are three main components in a battery: one positive electrode (cathode), one negative electrode (anode) and an ion-conducting medium in between the electrodes called the electrolyte. When a battery charges cations like Na^+, Li^+, H^+ from one electrode migrate *via* the electrolyte and are inserted or absorbed on another electrode causing the electrochemical system to polarise [1]. The discharge is a reversible depolarization process in which ions desorb or get extracted from one electrode and are transferred to the other. Electricity is produced as electrons are generated due to the movement of ions passing *via* an external circuit. In LIBs, lithium ions leave the cathode material and go to the anode through the electrolyte and separator. At an anode, lithium ions penetrate the anode material structure by different mechanisms resulting in the storage of energy between the electrodes [1]. Lithium ions from the anode flow through the electrolyte to the cathode during LIB discharging, generating energy to power the batteries. While charging and discharging, electrons generated are forced to flow through the external circuit in the direction opposite to that of Li-ion. Ions flowing oppositely through the electrolyte and electrons flowing through the external circuit are related processes, which implies that if one stops, the other automatically will get abandoned. To put it another way, if ions can't go through the electrolyte because the battery is completely discharged, electrons can't move through the outside circuit either, and the battery can't power the device. This is the complete energy storage mechanism in lithium-ion batteries (Fig. **1**) [10].

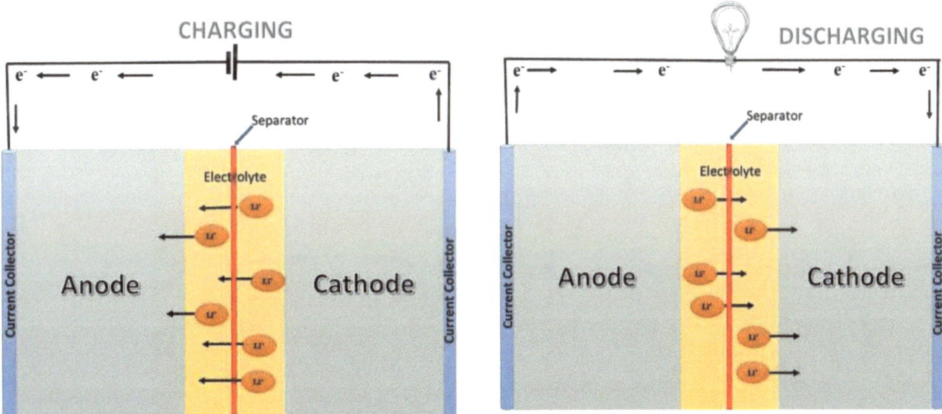

Fig. (1). Mechanism of charging and discharging of Li-ion battery.

The performance of LIB is highly governed by anode and cathode material. Any rechargeable cell relies on reversible chemical processes at the electrodes. A membrane that is permeable to electrolytes physically separates the anode and cathode. The role of electrolyte is also important as it mainly decides the ionic conductivity and electrochemical window of a battery. Electrolytes could either be liquid or solid. Typically solid electrolytes are coupled with liquid or gaseous electrodes. Solid electrodes and solid electrolyte is hardly used together because of the challenging solid-solid interface. Electrolytes must not allow electrons to pass through them such that electrons are forced to move *via* an external circuit to power an electronic device [1].

CATHODE MATERIALS

The positive electrode, often known as the cathode, is one of the most important components that govern the operation of batteries. Cathode materials must have a stable crystalline structure for a broad range of compositions. Some undesirable phase changes and significant compositional changes may occur due to oxidation reactions during the charging of batteries [1]. It is a very challenging task to select structurally stable and appropriate cathode materials [11]. Whilst discharging as the transportation of lithium ions back to the cathode material happens, transition metal ions are reduced by the electrons from the anode. The rates of these two methods along with the accessibility of lithium ions at the electrode-electrolyte interface regulate the overall discharge current. The performance of the cathode to the large extent depend on its morphology and intrinsic electrochemical properties because the exchange of lithium ions occurs only at the interface of electrode and electrolyte [7]. The three classes of cathode material used in LIBs are 2-D layered structure, 3-D framework and 1-D olivine family, as shown in Fig. (**2**). They are classified into these three groups based on the diffusion pathway of ions into the electrode materials [10].

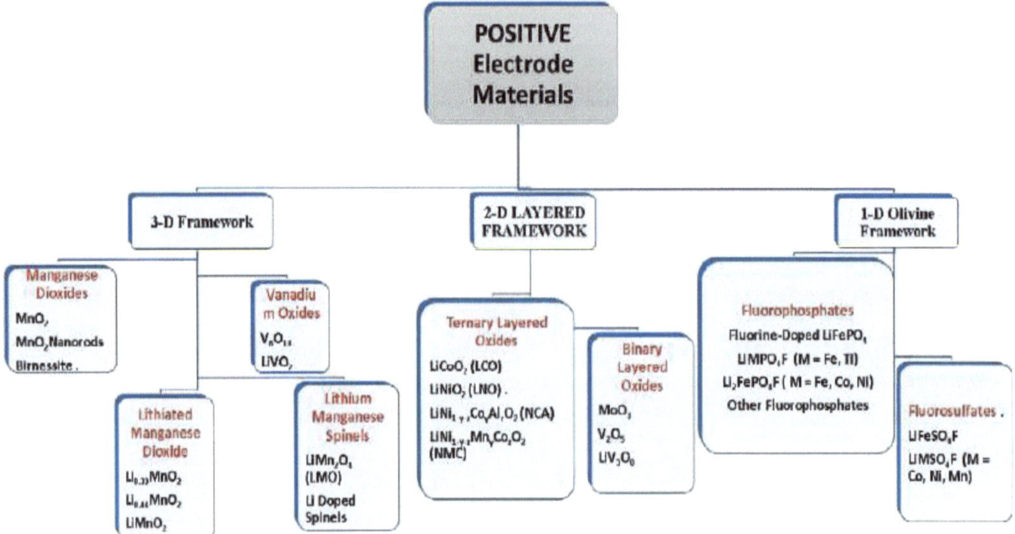

Fig. (2). Classification of positive electrode materials for LIBs.

Layered LiCoO$_2$ (LCO)

"The structure of $LiCoO_2$ oxides is similar to that of layered α-$NaFeO_2$ (space group R3m, No. 166) in which oxygen ions are packed in a cubical arrangement such that Co and Li-ions occupy the octahedral sites of alternating layers with ABCABC kind of stacking sequence" [13]. Goodenough *et al.* discovered $LiCoO_2$ (LCO) to be used as cathode material is now regarded as the prototypal cathode oxide that has rapid charge-discharge occurring in it within the potential range of 3.6-4.2 V [12]. Because of the strong bonding of cobalt in CoO_2 layers and the weaker bonding of lithium, smooth intercalation and deintercalation of Li-ion into the framework of lithium cobalt oxide is achievable, resulting in a few hundred charging and discharging cycles [10]. Direct cobalt-cobalt interaction in LCO provides high values of electronic conductivity and no cobalt ion migration in lithium sites with a high diffusion coefficient. The main problem with lithium cobalt oxide is the fact that only half of lithium ions can be extracted and inserted reversibly.

C. Julien *et al.* [10] showed that the charge-discharge cycle of $Li//LiCoO_2$ cell is for different phases of $LixCoO_2$ with x varying from 0 to 1. The capacity of the cell working between 3.6 to 4.2 V is somewhat constant whereas the capacity of the cell working between 3.6-4.5 V reduced dramatically with the number of cycles. $Li//LiCoO_2$ cell's practical capacity is thus limited to 140mAh/g [13]. The problem with LCO is its degradation due to the dissolution of metal ions in the electrolytic solution that leads to the release of oxygen.

G. G. Amatucci in his work [15] showed the plot of loss of cobalt as a function of cycling voltage. As the voltage increases loss of Co ions also increases and at 4.5 V, a sharp increase in dissolution of cobalt was observed. Hence, it was concluded that the decrease in capacity with cycling beyond the typical reversible voltage of 4.2V is indeed related to the dissolution of cobalt in the electrolyte [14]. One of the most effective techniques to prevent the breakdown of the cathode is its surface modification by coating it with metal oxides such as ZrO_2, Al_2O_3, TiO_2, *etc* [15]. Another possible solution is using a solid electrolyte rather than using a traditional liquid electrolyte [12]. In the case of all-solid-state batteries, solid electrolyte material can provide lithium-ion transport but cobalt ions are much bigger and cannot pass through this electrolyte so such a problem can be solved in all-solid-state batteries [15].

Layered $LiNiO_2$ (LNO)

Co is a toxic element in $LiCoO_2$ that can be replaced by non-toxic Nickel and forms a layered structure of $LiNiO_2$ (LNO) [3]. The working voltage for $LiNiO_2$ is greater than 3.5V and its charge-discharge curve displays a couple of plateaus. LNO based batteries have been observed to have 150mAh/g of rechargeable capacity within the range of 2.5-4.2V [3]. So, from the point of view of capacity increasing, LNO seems to be more promising combined with the advantages of nickel being cheaper than cobalt. However there are a few drawbacks of LNO, (i) it is difficult to synthesise $LiNiO_2$ such that the valence state of all the nickel ions is Ni^{3+} and nickel ions are distributed in a very ordered manner with no mixing of Ni^{3+} and Li^+ ions in the lithium plane,(ii) there occurs phase transition in LNO while charging and discharging which is irreversible, (iii) at high temperature there is exothermic oxygen release, (iv) in the charged state there are certain safety issues [12]. Hence for commercial lithium-ion batteries, LNO does not seem to be a promising electrode [9]. However, these drawbacks of LCO and LNO can be overcome by using mixed $LiNi_{1-y}Co_yO_2$ phases rather than a pure phase. An example of mixed $LiNi_{1-y}Co_yO_2$ is $LiNi_{0.85}Co_{0.15}O_2$ solid solution that exhibited outstanding cyclability and almost 180mAh/g reversible capacity with appealing electrochemical properties [3].

Layered NCA

Partial substitution of nickel ions to cobalt is not only the solution, some more alternatives are available like; doping with aluminium, chromium and some other material also stabilizes the system [17 - 19]. The most famous representative of such material is nickel-rich materials stabilized with aluminium popularly known as NCA [12]. NCA comprise 80% of nickel, 15% of cobalt, and 5% of aluminium. Although aluminium reduces capacity slightly, it stabilises the system over a greater number of cycles. It also improves thermal stability, which improves the battery's overall electrochemical performance. [16]. Structurally and thermally stable mixed materials' unique composition $LiNi_{0.80}Co_{0.15}Al_{0.15}O_2$ is being commercially used in an 85-kWh battery pack that drives Tesla's EV Roadster at high speed [17]. NCA-type electrodes are utilized in SAFT commercial batteries globally for application in a variety of fields ranging from military to space [13].

Spinel $LiMn_2O_4$ (LMO)

Lithium magnesium oxide with spinel structure was founded by J.B Goodenough three years after the discovery of LCO [18]. LMO has a cubic spinel structure as shown in (Fig. **3**). The spinel structure is primarily characterized by elementary groups as follows: (1) MnO_6 octahedra, in which magnese is surrounded by 6 atoms of oxygen. (2) LiO_4 tetrahedra, in which one lithium-ion is surrounded by 4 oxygen atoms. These two elementary cells combine to give a cubic spinel structure and provide 3D ionic and electronic transport. So, it can be considered an alternative to LCO as it increases safety, reduces cost, and reduces toxicity by replacing toxic cobalt elements [19]. Moreover, LCO is a layered structure that provides 2D ionic and electronic transport but in the case of the spinel structure, ions and electrons can move in all three directions *i.e.*, 3D transport.

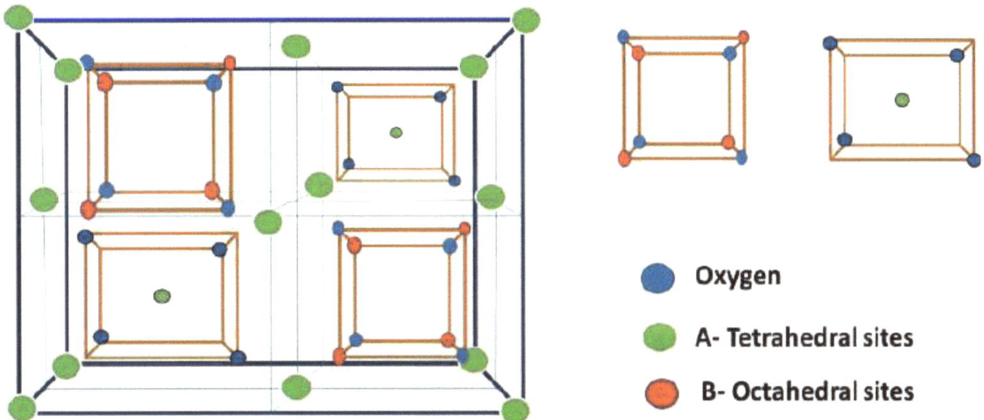

Fig. (3). Schematic representation of cubic spinel AB_2O_4 structure [13].

However, there are some major challenges with LMO: (1) preparation of fine quality samples is difficult, (2) even though in the 4V domain JT distortion of LMO spinel is not expected, due to local discharge conditions it has been observed that there is some form of tetragonal $Li_2Mn_2O_4$ layer at the surface of LMO particles, and (3) upon insertion of more than one Li per Mn, the transformation of $LiMn_2O_4$ in the tetragonal phase is observed. The theoretical capacity of LMO is limited to 140mAh/g as its tetragonal and cubic phases coexist [13]. The capacity fade process in LMO is concerned with substantial volume expansion linked with structural changes which are problematic. Several studies have been carried out to solve the capacity fading issue of LMO by substituting Li for Mn. Enhancement in the stability of $LiMn_2O_4$ spinel has been achieved with the reduction of Mn^{3+} concentration after the substitution with monovalent ions. "In the system Li-Mn-O, $Li_{1+x}Mn_2-xO_4$ spinel's exist as a continuous series of solid solution in the range $0.00 < x < 0.33$" [13]. Generally, electrochemical performance is found to be improved with Li-rich spinel [20]. As a result, defect spinel's $Li_{1-\delta}Mn_{2-\delta}O_4$ have gained enormous attention as a positive electrode material. Defect spinels can reversibly intercalate and deintercalated lithium smoothly. The high capacity (>230 mAh /g) of the defective $Li_{1-\delta}Mn_{2-\delta}O_4$ manganospinels is the result of vacancies at both tetrahedral and octahedral manganese sites [23, 24].

Olivine LiFePO$_4$ (LFP)

$LiFePO_4$ (LFP) materials are a member of the olivine family and have an orthorhombic crystal structure (Pnma space group, No. 62) [21]. "It consists of a distorted hexagonal close-packed (hcp) oxygen framework containing Li and Fe located in half the octahedral sites and P ions in one-eighth of the tetrahedral sites " [9]. The phospho-olivine $LiFePO_4$ (LFP) has demonstrated a theoretical capacity of 170mAh/g and hence is considered a potential cathode material to be used in rechargeable batteries. Moreover, being inexpensive and non-toxic it has an edge over cobalt-oxide-based materials for application at a large scale like in electric vehicles. Although LFP has low electronic conductivity of about 10^{-9} S/cm and the diffusion coefficients of lithium ions in LFP are also very low but if LFP particles are reduced to nanosize it would shorten the lithium-ion diffusion path in the electrode [13]. Also, the electrical conduction of LFP could be remarkably enhanced with the synthesis of carbon-coated LFP providing high rate capability. Hence it can be concluded the future of lithium iron phosphate technology is rather bright.

ANODE MATERIALS

In the case of cathodes, layered structures - lithium cobalt oxide or NCA or ternary NMC materials compete with spinel lithium manganese oxide and olivine lithium iron phosphate but in the case of negative electrode materials, carbon is dominating. Carbon provides a high specific capacity (more than 300 mAh per gram) and very low voltage about 0.1 volts, very close to lithium metal [22]. While carbon is the first negative electrode material used for commercial mass production of lithium-ion batteries, it's still one of the most widely used [1]. The reversible intercalation and deintercalation of lithium ions into graphite in the organic electrolyte was discovered by Besenhard and Eichinger in 1976 proposing this material as an anode [13]. The intercalation of lithium in graphite (Fig. **4**) can be given by the equation below:

Two categories of carbon have gained research interest for anode material: hard carbon which has the disordered orientation of crystallites and soft carbon (graphitized carbon) which has the orientation of crystallites in the same direction [13].

Hard Carbon

In comparison to soft carbon ; hard carbon provide higher capacity and reversibility. For example, the capacity of hard graphite is up to 600mAh/g [23]. Although with such a disordered arrangement hard carbon can accumulate more lithium and provide higher specific energy density, it suffers from poor rate capability because of many pores and voids associated with its random alignment [23].

$$xLi^+ + C_6(\text{in graphite}) + x\ e^- \leftrightarrow Li_xC_6$$

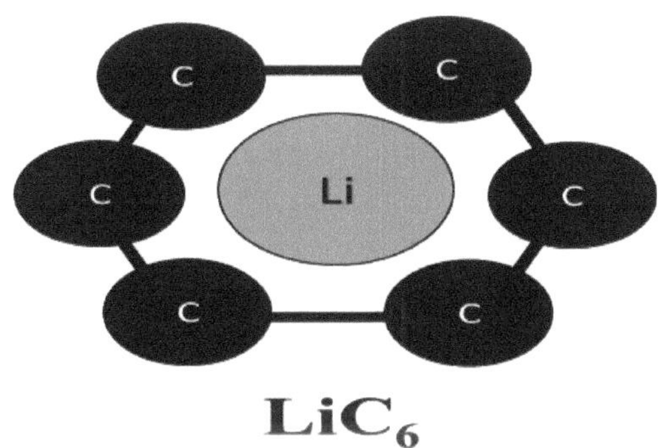

Fig. (4). Chemical formula for lithium accumulation by carbon. Six atoms of carbon can accumulate one ion of lithium.

Soft Carbon

Soft carbon anodes have more than 90% of coulombic efficiency and long cycle life. Corresponding to LiC_6 formation the specific capacity of soft carbon is obtained up to 400mAh/g. Hence. enormous efforts are being devoted to improve the capacity of soft carbon materials [24]. There are different types of soft carbon materials commercialised for the anode [25]. The popular ones are Mesophase-pitch-based carbon fibre (MCF), Massive Artificial Graphite (MAG), Partially Graphitized Porous Carbon (PGPC) and Mesocarbon Microbead (MCMB) [26].

Carbon Nanotubes

The future of different types of carbon materials is carbon nanomaterials [27]. The most famous are carbon nanotubes (CNTs) which show superior electronic conductivity and are thermally stable. Based on the thickness and number of co-axial layers CNTs can be broadly classified into two categories: single-walled nanotubes (SWCNTs) and multi-walled nanotubes (MWCNTs). SWCNTs prepared by laser vaporization yielded a very high capacity of about 1000 mAh/g, very close to the theoretically predicted value of 1110mAh/g [13]. MWCNTs gave very little capacity as compared to SWCNTs. The capacity of MWCNT is almost 250 mAh/g which increases up to 400 mAh/g after purification. However, the reversible capacity of chemically drilled MWCNT (DMWCNT) prepared by a solid-state process exceeded the value of 600mAh/g that remained constant up to

20 cycles [28]. Although CNTs show amazing properties, they are not yet extensively used in LIBs, as the preparation of CNTs without any defects and voids is difficult and costly.

Graphene

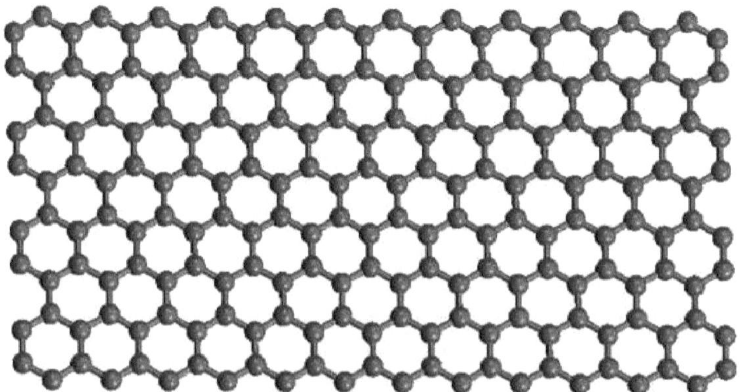

Fig. (5). A visual depiction of the structure of a layered, microscopic segment of graphene.

Graphene is one atomic sheet of graphite (Fig. **5**). Because of its mechanical robustness, excellent electronic conductivity, and wide surface area, it is an intriguing material for anodes [1]. When numerous graphene sheets are considered, the theoretical capacity exceeds that of graphite, despite the fact that the ability of a single layer of graphene to absorb lithium is smaller. Experimentally it has been observed graphene has more initial capacity than graphite [10]. The only drawback of graphene is its ageing upon cycling. For example, the specific discharge capacity delivered by oxidized graphene nanoribbons at its first cycle was 820 mAh/g which decreased to about 550 mAh/g at fifteen cycles [29].

Wang *et al* [30] have found a remarkable solution through the preparation of doped hierarchically porous graphene (DHPG) electrodes using three precursors sulfonated polystyrene (S-PS) sphères, graphene oxide (GO) and poly(vinyl pyrrolidone) (PVP) *via* facile in situ constructing technique in nickel foam. As the preparation involves no use of binder and the porous graphene is directly grown on the nickel foam, high electronic conductivity is obtained. Even at high current densities, DHPG exhibits good stability and very high capacity. This novel electrode offers116 kW/ kg of power density while maintaining a high energy density of 322Wh / kg at an ultra-high current density of 80 A/g, corresponding to a charge time of 10s. Such électrodes are very useful to fulfil the need for electrochemical storage and provide proper energy density to the battery and power density to the supercapacitor. Due to the combined effect of doping and structure, these électrodes provide long cycle life with no loss of capacity up to 3000 cycles and high capacities for a wide range of temperatures from 20 to 55° C [30]. Hence this carbon-based anode has delivered an outstanding performance so far.

Further development of advanced carbon nanomaterials is also hampered by the high cost of production, the production of waste products from fossil fuels used in production, and the use of toxic and unsafe materials. These problems and shortcomings remains with the material. Combining conventional and advanced materials is one of the possible solutions to this challenge. [24]. To decrease the cost of production waste products can be used as a precursor. Also, the production routes can be optimised to decrease the cost and energy consumption. Nevertheless, nowadays carbon materials are the main and dominant-negative electrode materials in LIBs and have interesting prospects for further development.

CONCLUDING REMARKS

With the boom in electronic devices and demand for hybrid electric vehicles, developing an efficient lithium-ion battery has become a prerequisite. Hence, a significant amount of research endeavours are being made to enhance the LIBs' performance and develop batteries with high specific capacity, minimal price, relatively good cycle life and better safety. For the required outcome in the context of LIBs performance, it is important to develop its electrode materials. The classification of anode or cathode material is based on its energy storage mechanism. The storage of energy in cathode material takes place *via* intercalation or conversion reaction mechanism whereas in the anodes energy is stored by intercalation, conversion reaction and alloying/dealloying. The fast development of lithium insertion compounds has made the commercialisation of lithium-ion batteries a reality. The transition-metal oxides crystallize in rock salt-based layer, spinel and olivine structures such as Li_xMO_2 (M = Co, Ni), LiM'_2O_4 and $LiFePO_4$. Several factors such as specific capacity, cyclability, thermal and chemical stability, electrode voltage and most importantly structural stability need to be considered for developing suitable electrode material for commercial LIBs. For the long service life of LIBs, it is required to eliminate unwanted chemical reactions between electrolyte and electrode. To make LIBs safer the use of non-flammable electrolytes and structurally stable electrodes is of utmost importance. It has been discovered that lowering the size of the particles or designing the architecture of the electrode material to the nanoscale level is one of the most effective techniques for enhancing the electrochemical performance of batteries.

As a final point, in spite of the fact that vast resources have been spent on developing high-performance batteries in the past few decades, there are still numerous challenges pertaining to these batteries that need to be overcome.

CONSENT FOR PUBLICATION

Not applicable.

CONFLICT OF INTEREST

The author declares no conflict of interest, financial or otherwise.

ACKNOWLEDGEMENTS

The authors sincerely acknowledge the financial support received in the form of a research initiation grant (No.- RIG/975/2018) from TEQIP (III), Madan Mohan Malaviya University of Technology. My colleagues' useful recommendations are also acknowledged.

REFERENCES

[1]   Review, A. *Ultrafine Grainrd Nanostructured Mater.,* **2018**, *51*(1), 1-12. elham kamali heidari, A. kamyabi. Electrode Materials for Lithium Ion Batteries:
[http://dx.doi.org/10.22059/JUFGNSM.2018.01.01]

[2]   Zhao, W. A forum on batteries: from lithium-ion to the next generation. *Natl. Sci. Rev.,* **2020**, *7*(7), 1263-1268.
[http://dx.doi.org/10.1093/nsr/nwaa068] [PMID: 34692152]

[3]   Ohzuku, T.; Ueda, A.; Nagayama, M.; Iwahoshi, Y. H. K. Comparative study of LiCoOz, LiNi, ZCol, ZOz and LiNiOz for 4 volt secondary. *Electrochim. Acta,* **1993**, *38*(9), 1159-1167.
[http://dx.doi.org/10.1016/0013-4686(93)80046-3]

[4]   Liu, W.; Song, M.S.; Kong, B.; Cui, Y. Flexible and Stretchable Energy Storage: Recent Advances and Future Perspectives. *Adv. Mater.,* **2017**, *29*(1), 1603436.
[http://dx.doi.org/10.1002/adma.201603436] [PMID: 28042889]

[5]   Lu, X.; Li, G.; Tong, Y. A review of negative electrode materials for electrochemical supercapacitors. *Sci. China Technol. Sci.,* **2015**, *58*(11), 1799-1808.
[http://dx.doi.org/10.1007/s11431-015-5931-z]

[6]   Tarascon, J-M. *Better Electrode Materials for Energy Storage Applications through Chemistry*; Mater. Lithium-Ion Batter, **2000**, pp. 75-103.
[http://dx.doi.org/10.1007/978-94-011-4333-2_4]

[7]   Tai, Z. *Development of Advanced Electrode Materials for Alkali Metal Ions Batteries.,* **2016**.

[8]   Zhang, Y.; Liu, N. Nanostructured Electrode Materials for High-Energy Rechargeable Li, Na and Zn Batteries. *Chem. Mater.,* **2017**, *29*(22), 9589-9604.
[http://dx.doi.org/10.1021/acs.chemmater.7b03839]

[9]   Julien, C.; Mauger, A.; Zaghib, K.; Groult, H. Comparative Issues of Cathode Materials for Li-Ion Batteries. *Inorganics (Basel),* **2014**, *2*(1), 132-154.
[http://dx.doi.org/10.3390/inorganics2010132]

[10]   Zhang, L.; Chen, C. Electrode Materials for Lithium Ion Battery. *Huaxue Jinzhan,* **2011**, *23*(2–3),

275-283.
[http://dx.doi.org/10.1016/j.mset.2018.08.001]

[11]   Xu, B.; Qian, D.; Wang, Z.; Meng, Y.S. Recent progress in cathode materials research for advanced lithium ion batteries. *Mater. Sci. Eng. Rep.,* **2012**, *73*(5-6), 51-65.
[http://dx.doi.org/10.1016/j.mser.2012.05.003]

[12]   Croguennec, L.; Palacin, M.R. Recent achievements on inorganic electrode materials for lithium-ion batteries. *J. Am. Chem. Soc.,* **2015**, *137*(9), 3140-3156.
[http://dx.doi.org/10.1021/ja507828x] [PMID: 25679823]

[13]   Julien, C.; Mauger, A.; Vijh, A.; Zaghib, K. *Lithium Batteries: Science and Technology,* **2015**.
[http://dx.doi.org/10.1007/978-3-319-19108-9]

[14]   Amatucci, G.; Tarascon, J.M.; Klein, L.C. Cobalt dissolution in LiCoO2-based non-aqueous rechargeable batteries. *Solid State Ion.,* **1996**, *83*(1-2), 167-173.
[http://dx.doi.org/10.1016/0167-2738(95)00231-6]

[15]   Chen, Z.; Dahn, J.R. Effect of a ZrO[sub 2] Coating on the Structure and Electrochemistry of Li[sub x]CoO[sub 2] When Cycled to 4.5 V. *Electrochem. Solid-State Lett.,* **2002**, *5*(10), A213.
[http://dx.doi.org/10.1149/1.1503202]

[16]   Jo, M.; Noh, M.; Oh, P.; Kim, Y.; Cho, J. A New High Power LiNi $_{0.81}$ Co $_{0.1}$ Al $_{0.09}$ O $_2$ Cathode Material for Lithium-Ion Batteries. *Adv. Energy Mater.,* **2014**, *4*(13), 1301583.
[http://dx.doi.org/10.1002/aenm.201301583]

[17]   Weaving, J.S.; Coowar, F.; Teagle, D.A.; Cullen, J.; Dass, V.; Bindin, P.; Green, R.; Macklin, W.J. Development of high energy density Li-ion batteries based on LiNi1−x−yCoxAlyO2. *J. Power Sources,* **2001**, *97-98*, 733-735.
[http://dx.doi.org/10.1016/S0378-7753(01)00700-5]

[18]   Thackeray, M.M.; Kang, S.H.; Johnson, C.S.; Vaughey, J.T.; Benedek, R.; Hackney, S.A. Li2MnO3-stabilized LiMO2 (M = Mn, Ni, Co) electrodes for lithium-ion batteries. *J. Mater. Chem.,* **2007**, *17*(30), 3112-3125.
[http://dx.doi.org/10.1039/b702425h]

[19]   Thackeray, M.M.; David, W.I.F.; Bruce, P.G.; Goodenough, J.B. Lithium insertion into manganese spinels. *Mater. Res. Bull.,* **1983**, *18*(4), 461-472.
[http://dx.doi.org/10.1016/0025-5408(83)90138-1]

[20]   Zhang, H.; Zhao, H.; Khan, M.A.; Zou, W.; Xu, J.; Zhang, L.; Zhang, J. Recent progress in advanced electrode materials, separators and electrolytes for lithium batteries. *J. Mater. Chem. A Mater. Energy Sustain.,* **2018**, *6*(42), 20564-20620.
[http://dx.doi.org/10.1039/C8TA05336G]

[21]   Masquelier, C.; Croguennec, L. Polyanionic (phosphates, silicates, sulfates) frameworks as electrode materials for rechargeable Li (or Na) batteries. *Chem. Rev.,* **2013**, *113*(8), 6552-6591.
[http://dx.doi.org/10.1021/cr3001862] [PMID: 23742145]

[22]   Sole, C.; Drewett, N.E.; Hardwick, L.J. *In situ* Raman study of lithium-ion intercalation into microcrystalline graphite. *Faraday Discuss.,* **2014**, *172*(0), 223-237.
[http://dx.doi.org/10.1039/C4FD00079J] [PMID: 25427224]

[23]   Fujimoto, H.; Tokumitsu, K.; Mabuchi, A.; Chinnasamy, N.; Kasuh, T. The anode performance of the hard carbon for the lithium ion battery derived from the oxygen-containing aromatic precursors. *J. Power Sources,* **2010**, *195*(21), 7452-7456.
[http://dx.doi.org/10.1016/j.jpowsour.2010.05.041]

[24]   Marom, R.; Amalraj, S.F.; Leifer, N.; Jacob, D.; Aurbach, D. A review of advanced and practical lithium battery materials. *J. Mater. Chem.,* **2011**, *21*(27), 9938-9954.
[http://dx.doi.org/10.1039/c0jm04225k]

[25]   Emmerich, F.G. Young's modulus, thermal conductivity, electrical resistivity and coefficient of

thermal expansion of mesophase pitch-based carbon fibers. *Carbon,* **2014**, *79*(1), 274-293.
[http://dx.doi.org/10.1016/j.carbon.2014.07.068]

[26]   Ye, L.; Liang, Q.; Huang, Z.H.; Lei, Y.; Zhan, C.; Bai, Y.; Li, H.; Kang, F.; Yang, Q.H. A supercapacitor constructed with a partially graphitized porous carbon and its performance over a wide working temperature range. *J. Mater. Chem. A Mater. Energy Sustain.,* **2015**, *3*(37), 18860-18866.
[http://dx.doi.org/10.1039/C5TA04581A]

[27]   Sun, J.; Du, S. Application of Graphene Derivatives and Their Nanocomposites in Tribology and Lubrication: A Review. *RSC Advances. Royal Society of Chemistry,* **2019**, 40642-40661.
[http://dx.doi.org/10.1039/C9RA05679C]

[28]   Oktaviano, H.S.; Yamada, K.; Waki, K. Nano-drilled multiwalled carbon nanotubes: characterizations and application for LIB anode materials. *J. Mater. Chem.,* **2012**, *22*(48), 25167-25173.
[http://dx.doi.org/10.1039/c2jm34684b]

[29]   Bhardwaj, T.; Antic, A.; Pavan, B.; Barone, V.; Fahlman, B.D. Enhanced electrochemical lithium storage by graphene nanoribbons. *J. Am. Chem. Soc.,* **2010**, *132*(36), 12556-12558.
[http://dx.doi.org/10.1021/ja106162f] [PMID: 20731378]

[30]   Wang, Z.L.; Xu, D.; Wang, H.G.; Wu, Z.; Zhang, X.B. In situ fabrication of porous graphene electrodes for high-performance energy storage. *ACS Nano,* **2013**, *7*(3), 2422-2430.
[http://dx.doi.org/10.1021/nn3057388] [PMID: 23383862]

CHAPTER 4

Design of Supported Catalysts for Nitrogen Reduction Reaction: A Continuous Challenge

Deeksha R.[1] and **Deepak Kumar**[1,*]

[1] *Department of Chemistry, Faculty of Mathematical and Physical Sciences, M.S. Ramaiah University of Applied Sciences, Bengaluru-560058, India*

Abstract: The production of ammonia is facilitated by the nitrogen reduction reaction (NRR), where the inert di-nitrogen molecule is converted to ammonia. Along with being a major carrier of hydrogen, ammonia holds authority in the fertilizer realm. Therefore, it is inevitable to develop a viable and eco-friendly method of production that is cost-effective and resource-efficient. The primary challenge of nitrogen reduction is the cleavage of the particularly stable nitrogen bond. The most popular Haber-Bosch process for ammonia production, although efficient, is highly energy-intensive, and the need for maintaining exceptionally high temperature and pressure conditions is an environmental concern. As an alternative, the direct conversion of nitrogen has been carried out by photocatalysis and electrocatalysis. However, this strategy falls short of achieving superior conversion efficiencies. Consequently, it is conceivable that a fitting catalyst can be the solution for the difficulties associated with NRR. Over the years, several attempts have been made at formulating the best catalyst, including chromium oxynitride nanoparticles, niobium dioxide, various metal (Ru, Al, Rh, Ga) clusters, single-atom catalysts supported on different surfaces, and double-atom catalysts. Recently, perovskites have emerged into the spotlight as excellent catalysts for NRR. In this chapter, we discuss the challenges faced by researchers to formulate righteous catalysts for the sustainable reduction of nitrogen by studying each of these types with a few examples. We also review the recent advancements in the experimental domain of NRR using different electrochemical cells.

Keywords: Ammonia, Density functional theory, Double-atom catalysts, Haber-Bosch process, Heterogeneous catalysis, Nitrogen reduction reaction, Perovskites, Single-atom catalyst.

INTRODUCTION

The major obstacle concerning nitrogen reduction reaction (NRR) is the weakening of the dinitrogen bond due to its high $N\equiv N$ bond energy (940.95

---

* **Corresponding author Deepak Kumar**: Department of Chemistry, Faculty of Mathematical and Physical Sciences, M.S. Ramaiah University of Applied Sciences, Bengaluru-560058, India; Tel: +91 80-49065555; Fax:080 45366677
E-mail: deepakkumar.cy.mp@msruas.ac.in

kJ/mol) [1], electron affinity (-1.9 eV), and the large gap between its HOMO and LUMO (10.82eV) [2, 3], which make electron transfer difficult. Furthermore, the bond length associated with a molecule of nitrogen is very short, 1.09Å [4], indicating that the cleavage would be strenuous. The resulting product of interest from NRR, ammonia, is crucial in various ways. While ammonia is primarily known for its significance in the synthesis of fertilizers, it has several other roles; a few of which are: (i) ecological energy carrier, (ii) transportation fuel, (iii) in agriculture, pharmaceutical, plastic, and textile industries. There are three well-known paths formulated for the synthesis of ammonia; biological nitrogen fixation by *Azotobacter*, the Haber-Bosch process, and electrochemical catalysis. However, the common drawback shared by these methods is the low rate of reaction. *Azotobacter*, Gram-negative aerobic bacteria, reside in the soil. They use the nitrogen available in the atmosphere to synthesize cellular proteins, which are subsequently mineralized upon the death of the bacteria, thereby contributing to the availability of nitrogen in the soil. The Haber-Bosch process has been employed for the synthesis of ammonia since its invention. The typical conditions necessary to carry out the reaction in the presence of Fe/Ru catalysts are temperatures ranging from 300°C-500°C, and pressures ranging from 150-200 atm [4]. In addition to requiring extreme temperature and pressure conditions, the reaction also contributes to global warming by emitting 300 million metric tons of CO_2 every year [4]. Therefore, it is obvious that a clean and sustainable strategy for nitrogen reduction is inevitable. In this regard, electrochemical catalytic methods have gained popularity in the recent years. Although Hydrogen Evolution Reaction (HER), Oxygen Reduction Reaction (ORR), Oxygen Evolution Reaction (OER), and CO_2 Reduction Reaction (CO_2RR) are extensively studied, the electrochemical catalytic reduction reaction of nitrogen to yield ammonia using dinitrogen and water is not thoroughly investigated.

Electrocatalysts can be categorized based on whether they are associated with (i) noble metals, (ii) non-noble metals, or are (iii) metal-free. The activity of the catalyst depends on its properties, including morphology, electronic structure, active sites, and crystallinity. A few ways to augment the catalytic activity are introducing structural defects, doping, increasing the available surface area, and revealing more active sites. Let us look at some of the catalysts that have been explored.

i. Noble metals: Noble metals such as Au, Pt, Ru, Rh have been studied for their catalytic activity in NRR. Yan *et al.* used a binary surfactant mixture to develop tetrahedral Au nanorods and found that the yield of ammonia was low [5]. The low yield was accredited to the surfactant synthesis molecules staying on the catalytic surface and arresting the active sites. Lan *et al.* [6] reported Pt anchored on carbon black for the synthesis of ammonia from air and water under a moderate environment. The yield was found to be 69.8 µg/h/cm². However, Ru catalysts are known to favour the adsorption of hydrogen, which inhibits NRR. Ultimately, the cost involved with noble metal-based processes is too high to be normalized.

ii. Non-noble metals: These are mainly transition metals and are cost-effective. Sun *et al.* [7] proposed an array of MoS_2 nanosheets on carbon cloth for the production of ammonia. The yield was found to be 4.94 µg/h/cm². The key factor involved in the activation of nitrogen is the positive charge present around Mo. The charge is shifted from nitrogen to the edges of Mo, leading to the development of N-Mo bonds and thus, weakening the N≡N. As with several other examples, the drawback here is, the edges of MoS_2 nanosheets are active sites for HER.

iii. Metal-free: Although transition metals are cost-effective, their binding to nitrogen is weak and often fails at activating the molecule. Additionally, their d-orbitals tend to form bonds with hydrogen, which brings on HER. Main group elements, unlike transition metals, do not possess the combination of free and occupied orbitals. Yu *et al.* [8] reported graphene doped with Boron; the positive B atoms preferred nitrogen adsorption and, therefore, ammonia production. The yield was found to be 9.8 µg/h/cm².

Theoretically, the electrochemical reduction reaction of nitrogen giving ammonia as the product is feasible at room temperature and atmospheric pressure, provided that abundant voltage is applied externally. However, practically, very few catalysts can offer high yields.

NITROGEN REDUCTION REACTION – THE MECHANISM

The pathway of electrochemical reduction of nitrogen to ammonia can be outlined as [4]:

$$N_2 + 6H^+ + + 6e^- - \leftrightarrow 2NH_3(g)\ E° = 0.148V\ vs.RHE \tag{1}$$

$$2H^+ + + 2e^- \leftrightarrow H_2\ E° = 0V\ vs.SHE \tag{2}$$

$$N_2 + 6H_2O + 6e^- \leftrightarrow 2NH_3 + 6OH^-\ E° = -0.763V\ vs.SHE\ at\ pH\ 14 \tag{3}$$

$$2H_2O(l) + 2e^- \leftrightarrow H_2(g) + 2OH^-\ E° = -0.828V\ vs.SHE\ at\ pH\ 14 \tag{4}$$

As it can be observed, the equilibrium potentials (E°) of NRR and HER are similar; this indicates that HER is the dominant competing reaction in the process. Hence, it is important that while employing a catalyst, it should show selectivity for NRR. In addition to preferring NRR, the catalyst should also suppress the HER, possess surface properties that enable strong binding with adsorbed nitrogen and hydrogen atoms, show appreciable electrical conductivity and stability.

The electrochemical NRR proceeds as follows [4]:

$$N_2 + H^+ + e^- \leftrightarrow N_2H \quad E° = -3.2V \; vs. \, RHE \tag{5}$$

$$N_2 + 2H^+ + 2e^- \leftrightarrow N_2H_2(g) \; E° = -1.10V \; vs. \, RHE \tag{6}$$

$$N_2 + 4H^+ + 4e^- \leftrightarrow N_2H_4(g) \; E° = -0.36V \; vs. \, RHE \tag{7}$$

$$N_2 + 4H^+ + 4e^- \leftrightarrow N_2H_4(g) \; E° = -0.36V \; vs. \, RHE \tag{8}$$

$$N_2 + e^- \leftrightarrow N_2^- \, (aq) \; E° = -3.37V \; vs. \, RHE \tag{9}$$

The reaction process can follow either of the three routes; associative, dissociative, and enzymatic.

In the associative route, the nitrogen molecule is first adsorbed onto the catalyst surface. Upon adsorption, hydrogenation occurs. This process of hydrogenation can be further broken down into two approaches; distal and alternating. In the distal approach (Fig. **1c**), the nitrogen atom that is located far away from the surface of the catalyst is preferred for hydrogenation. Once the first molecule of ammonia is produced and released, the atom of adsorbed nitrogen that is on the catalyst surface undergoes hydrogenation to produce the next molecule of ammonia. In the alternating approach (Fig. **1b**), as the name suggests, two nitrogen molecules are hydrogenated alternately, and the ammonia molecules are released one after another, closely.

In the dissociative route Fig. (**1a**), the dinitrogen bond (N≡N) is broken first, and the two nitrogen atoms are simultaneously adsorbed onto the surface of the catalyst. Here, they independently undergo the hybrid process.

In the enzymatic route, the hydrogenation of the two nitrogen atoms is carried out alternatively by the proton-electron duos, and the second molecule of ammonia formed is released successively after the first one [9].

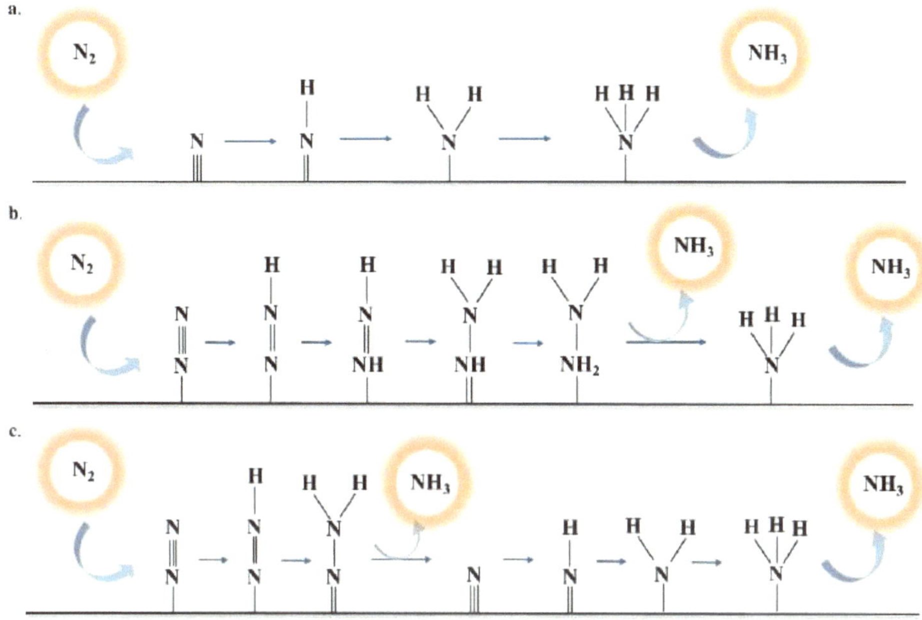

Fig. (1). The different mechanisms of NRR; **(a)** Dissociative **(b)** Associative alternating **(c)** Associative distal

DENSITY FUNCTIONAL THEORY

In the modern-day, several methods are used for computational calculations; Hartree-Fock (HF), Moller-Plesset (MP) perturbation theory, Configuration Interaction (CI), and Density Functional Theory (DFT). DFT is a kind of electronic structure calculation, that has been chosen over the other methods due to its cost-efficient and accurate nature. The main complication concerning all the calculational strategies regarding the electronic structure is the appropriate treatment of the relations between electrons. While HF, MP, and CI are built on the calculation of a wave function, DFT does not need a wave function in its operation and employs total electron density. However, practically, a wave function is involved in some computational areas.

The development of DFT was driven by the shortcomings of the other theories. HF theory could not be applied for most of the chemical reactions, and while MP and CI theories provide highly accurate results for minuscule atoms and molecules, they require significantly large basis sets while computing for larger atoms and molecules and prove to be extremely expensive. DFT offers an advantage over these methods by being both cost-effective and accurate in its results.

Schrodinger proposed a partial differential equation depicting how a quantum system changes with time in 1926 [10]. This equation revolved around the idea that all the information one might need about any quantum system can be given by its wave function.

The simplest form of the equation, the time-independent form, can be expressed as:

$$H\Psi = E\Psi \tag{10}$$

where H is the Hamiltonian operator, and Ψ is the wavefunction that represents the set of solutions for the operator. In a situation involving multiple electrons, the equation takes a more complicated form:

$$\left[\frac{h^2}{2m} \sum_{i=1}^{N} \nabla_i^2 + \sum_{i=1}^{N} V(r)_i + \sum_{i=1}^{N} \sum_{j<i} U(r_i, r_j) \right] \Psi = E\Psi \tag{11}$$

In this form, m represents the mass of the electrons, and the three terms on the LHS of the equation represent the kinetic energy accompanying each electron, the energy associated with the interaction between each electron and nuclei, and the energy associated with the interaction between the electrons themselves. Each electron can be depicted as a function of each of the three spatial coordinates, which is given by Ψ. Since it is obvious that the number of electrons is much greater than the number of nuclei, the equation for a single CO_2 molecule would have to be solved as a 66-dimensional function, considering all the electrons. It is no doubt that this infamous equation introduced a fitting mathematical design to calculate the overall energy of a system, along with other properties, but the Schrodinger equation cannot be used for practical purposes. Furthermore, the wave function that was proposed by Schrodinger had no physical significance; it could not be experimentally observed. Let us now consider another closely related quantity, electron density, n(r) of an electron in a specific location in space. The equation for n(r) can be given as:

$$n(r) = 2 \sum_{i} \Psi_i^*(r)\Psi_i(r) \tag{12}$$

"2" arises from the fact that electrons possess spins and according to Pauli's exclusion principle, each electron wave function can have two separate electrons as long as their spins are opposite. This quantity n(r) is a function of three coordinates only and holds within itself a great deal of information that is also physically observable.

Subsequently, Max Born studied the wave function and interpreted it in a probabilistic fashion. He demonstrated that squaring the absolute value of the amplitude of the wave function gives the probability density, which in turn gives the probability of the occurrence of a quantum event [11].

Later, Thomas and Fermi proposed their calculational findings that did not involve a wave function and considered electron density as the central variable [12, 13]. It was assumed that the total energy of a system, whether it be an atom or a molecule, can be written as a functional of its electron density. Although they did not prove the reality of this parameter, it was realized that electron density, unlike wave function, is experimentally observable. However, this model showed accuracy only within the limits of an infinite nuclear charge and provided inaccurate predictions with real atoms and molecules. Another drawback of the Thomas-Fermi model was its failure to deduce the shell structure of atoms [14]. In 1930, Paul Dirac contributed a term to characteristic exchange energy. Later in 1935, Carl Friedrich von Weizacker contributed to the improvement of the term depicting the kinetic energy of electrons. Despite these corrections, the Thomas-Fermi model continued to yield inaccurate estimations.

In 1950, J.C. Slater conducted dimensional analysis and proposed an exchange functional that was different from that of Dirac's by a mere multiplication constant [15].

The modern DFT is a blessing of three leading figures; Pierre Hohenberg, Walter Kohn, and Lu Jeu Sham. Hohenberg and Kohn proposed that n(r) can be used to define the total electronic energy of any atomic or molecular system. They called it a universal functional and showed that any observable property, along with total electronic energy, can be construed in connection with electron density.

Hohenberg and Kohn put forth two theorems [16], the first of which states that there exists a head-to-head relationship between the electron density and the ground state wave function. The ground state energy can be represented as *E[n(r)]*. They also proved that the global minimum value of the energy functional can be obtained when the exact ground-state electron density is used as the input to the universal functional. Nonetheless, it was unclear as to how the functional should be described, or what its form should be. The second theorem states that the electron density that agrees best with the solution to the Schrodinger equation, is that which minimizes the overall functional. If the real form of the functional is known, it is possible to alter the electron density until the energy is minimized, thereby providing a prescription to find the germane electron density.

$$E[\{\Psi\}] = E_{known}[\{\Psi_i\}] + E_{XC}[\{\Psi_i\}] \tag{13}$$

$$E_{known}[\{\Psi_i\}] = \frac{h^2}{m}\sum_i \int \Psi_i^* \nabla^2 \Psi_i d^3r + \int V(r)n(r)d^3r + \frac{e^2}{2}\int\int \frac{n(r)n(r')}{|r\;r'|}d^3r d^3r' + E_{ion} \tag{14}$$

where the terms on the RHS represent from left to right, the kinetic energies of the electrons, the Columbic interactions between electrons and nuclei, the Columbic interactions between the electrons, and the Columbic interactions between the nuclei. The term E_{XC} is known as the exchange-correlation functional and contains in it the effects that are not included in the E_{known}.

Kohn and Sham [16] simplified the equation and justified that the way to identify the right electron density is by solving a set of equations, each of them involving a single electron.

$$\left[\frac{h^2}{2m}\nabla^2 + V(r) + V_H(r) + V_{XC}(r)\right]\Psi_i(r) = \varepsilon_i\Psi_i(r) \tag{15}$$

The three terms on the LHS denote three types of potentials; the first one represents the interactions between the electron and the nuclei, the second one, known as the Hartree Potential, can be described as $V_H(r) = e^2\int \frac{n(r')}{|r\;r'|}d^3r'$ and represents the Coulombic repulsion interactions between the electron in consideration and all the other electrons in the system. A part of V_H accounts for the self-interactions of the electron, which is corrected by V_{XC}, which can be described as $V_{XC}(r) = \frac{\delta E_{XC}(r)}{\delta n(r)}$. The equation put forth by Kohn and Sham portrayed DFT as a computational tool that could be used practically. Advances in computational methods enable the investigation of efficient catalysts, help in predicting the trends in reaction rates, and provide a better insight into the reaction mechanisms and the interpretation of experimental data. Using computational calculations, the prediction and selection of promising catalysts become easier.

Drawbacks of Density Functional Theory

Despite showing incredible potential in its calculations, DFT does have certain disadvantages to it:

- A deep-seated uncertainty prevails between the energies given by computational calculation with DFT and the true ground state energies of the Schrodinger equation.
- DFT falls short in calculating the excited electronic states.
- Band-gap calculations on semiconducting materials are underestimated in DFT.
- DFT provides significantly inaccurate results in the case of atoms and molecules associated with weak van der Waals forces of attraction.
- The computational expenditure in fixing the problems posed by DFT is immense.

THEORETICAL ADVANCES IN NITROGEN REDUCTION REACTION

By using DFT calculations, several studies have been conducted to investigate the potential catalytic performance of metal surfaces, metal oxides, metal nitrides, metal carbides, single-atom catalysts, and double-atom catalysts (Fig. **2**). Perovskites have recently emerged as electrocatalysts for NRR and could be the future of NRR. Let us look at each one of these with a few examples.

Nitrogen Reduction Reaction on Metal Surfaces

The surfaces of pure metals are one of the modest models for analysis. A variety of transition metals were considered for their surface studies by Skulason *et al.* [17]. Both the flat and stepped forms of the metals were evaluated for their NRR activity. The results showed that Fe and Mo were the most diligent surfaces for the production of ammonia, but they inclined more toward HER. Among the metals that were investigated, Rh, Ir, Ru, Pt, Ni, and Co surfaces were susceptible to be covered by H-adatoms, which led to the lack of active sites for nitrogen adsorption. Sc, Ti, Y, Zr were proposed to adsorb N-adatoms more strongly. These predictions were made at a bias of -1.0 to -1.5V *vs.* NHE. The calculations also showed that an N_2 molecule was able to bind strongly to a stepped surface of Ru (0001).

Ru has gained more attention in its NRR activity because it meets the activation energy barrier of the reaction and has sufficient vacant sites on its surface [18]. In this regard, stepped Ru surfaces were investigated and the computational studies revealed that the activation barrier was found to be low for the associative pathway of NRR. In Ru and Os systems, the $N\equiv N$ bond is debilitated upon its adsorption on the surface. On the other hand, in Rh systems, the weakening of the bond is not sufficient, and therefore, larger activation energy is recorded. It must be noted that the limiting factor for NRR on the surfaces of transition metals is the linear scaling between the energetics of intermediates N_2H and NH_2. Since NRR has a greater overpotential than HER, the latter is preferred. Therefore, it is demanding to develop a catalyst that is selective towards NRR by either stabilizing N_2H or destabilizing NH_2. The use of metal alloys is one strategy to help favour NRR [8, 19]. The alloys can be multifunctional, interacting with the adsorbed intermediates or facilitating the active sites. As both the intermediates bind employing a single nitrogen atom, it is necessary to design appropriate active site configurations to encourage adsorption. For instance, it is favourable to devise an active site that can destabilize one intermediate without influencing the activity of the other. In a study, the thermal conversion of nitrogen to ammonia was explored using a surface single-cluster catalyst consisting of singly dissipated bimetallic sites (M_1A_n) [20]. *Ab initio* simulations and DFT calculations revealed that an isolated $RhCo_3$ cluster on the surface of CoO (011) followed the associative, alternating pathway with both metal sites facilitating hydrogen activation and producing active species. This conclusion can be extended to NRR.

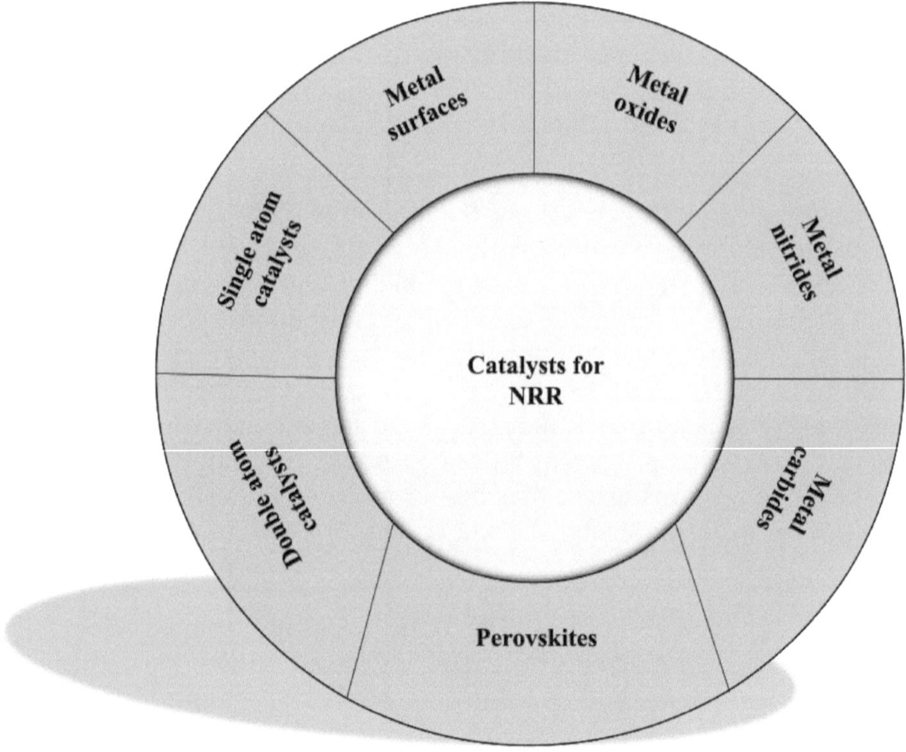

Fig. (2). Different types of catalysts used in NRR.

Nitrogen Reduction Reaction on Metal Oxides/Metal Nitride/Metal Carbide/Metal Sulphide Surfaces

Iron oxide has attracted the spotlight as a potential catalyst for the electrochemical reduction of nitrogen under benign conditions. Licht *et al.* [21] analyzed the synthesis of ammonia from nitrogen in the presence of steam in nano-Fe_2O_3 molten hydroxide suspensions. Subsequently, DFT calculations were performed by a different team to check the reactivity of haematite (0001) surface with $Fe-O_3-Fe-$ and $Fe-Fe-O_3-$ termination [22]. It was observed that NRR followed the associative pathway, and the most taxing step was found to be the addition of the first proton to the adsorbed nitrogen molecule. The dissociative mechanism is not favoured because of the high reaction energy in the initial step of the process. The overall process required an applied potential of -1.14V and -1.84V for $Fe-Fe-O_3-$ and $Fe-O_3-Fe-$ surfaces, respectively. These predicted potentials coincided with Licht's reports. It was also realized that the reactivity of $Fe-Fe-O_3$ was greater than that of $Fe-O_3-Fe$ since the binding energy of N_2H^* species is larger on the former system. The stimulation of the $N{\equiv}N$ bond results from the electron transfer from $Fe-Fe-O_3$ to the adsorbed nitrogen molecule.

Metal nitrides have been proven to be more dynamic for NRR than their pure metal counterparts. Abghoui *et al.* [23 - 26] carried out a series of DFT calculations on several transition metal nitride surfaces at benign conditions. Zincblende (ZB) and rock salt (RS) structure forms of mono-nitrides were investigated. VN, ZrN, NbN, CrN among RS (100) forms displayed low onset potential, as low as -0.5V (VN), capable of avoiding both HER and decomposition of the catalyst. Among the ZB structures, CrN, RuN, WN were found to be effective at low onset potentials (-0.23 to -0.55V). The N-vacancy created on the surface of the catalyst has a tendency to migrate to the bulk form to attain thermodynamic stability, and this makes it necessary that the vacancy be maintained. Fortunately, the activation barrier for migration is quite high, and this leads to stability of the N-vacancy. A good catalyst should have excellent stability and the ability to regenerate itself after the completion of the reaction. NbN alone in the RS forms showed the property of self-regeneration, and it was also reasonably stable. Among the ZB forms, all the catalysts could be regenerated, and they could sustain the catalytic cycles. It is worth noting that the rate-determining step can be different for different nitrides.

Azofra *et al.* [27] explored 2D transition metal carbides as potential candidates for electrochemical NRR. MXenes belonging to d^2, d^3, d^4 transition series were investigated (general formula M_3C_2). An overpotential of 0.64V *vs.* SHE was reported with V_3C_2. The computational results showed that the nitrogen chemisorbed on the MXene nanosheets were capable of elongating and thereby weakening the $N{\equiv}N$ bond, boosting catalytic NRR. The d^3 Ta_3C_2 and d^2 Ti_3C_2 showed large binding energy values, implying that the adsorption of N_2 on the catalytic surface is strong. The substantial extension of the $N{\equiv}N$ bond activates the molecule and drives the reaction.

Transition metal dichalcogenides (TMDs) in their monolayer forms have attracted the spotlight due to their characteristic properties such as availability of abundant surface area for potential catalytically active sites, room for functionalization, good conductivity, provision to create sulfur vacancies to further modify the surface by doping, and others. In this regard, MoS_2 has recently been extensively investigated as a surface to facilitate NRR. Jain *et al.* [28] investigated Fe doped MoS_2 for its NRR activity and found that modifying the surface of MoS_2 leads to the creation of a single sulfur vacancy and the edges of the monolayer sheets serve as the active sites for catalysis. The group also reported that the system showed selectivity for NRR over HER, and the enzymatic pathway was followed. Zhang *et al.* [29] studied the catalytic activity of Fe_2 clusters supported on MoS_2 for NRR and compared the results to that of a single Fe atom supported on MoS_2 surface. It was observed that as opposed to the single Fe atom system, the presence of a second Fe atom helps with the side-on attachment of the N_2 molecule. The $Fe_2@MoS_2$ system recorded a low overpotential of 0.21 eV and showed superior stability at temperatures up to 1000 K. It was concluded that the depletion of electrons on Fe in the presence of MoS_2 makes available the vacant orbitals that subsequently accept the lone pair of electrons from the N_2 molecule, leading to its adsorption. Further, the elongation of the N_2 triple bond was found to be 1.18 Å, and the process of nitrogen reduction was seen to follow the enzymatic route. The most important step in NRR on a metal catalyst surface is the formation of the metal-nitrogen bond. The tighter the bond, the stronger the adsorption. Recently, Lin *et al.* [30] reported that the metastable bulk MoS_2 polymorphs showed a significantly higher NRR activity, owing to the internal localization of electrons surrounding the metal chains in the 1T' phase, which boosts the charge transfer from N_2 to the metal. The yield of ammonia was reported to be 9 times greater than that with $2H\text{-}MoS_2$.

Nitrogen Reduction Reaction on Single-Atom Catalysts

Single-atom catalysts (SACs) have risen as promising candidates for NRR. They are comprised of isolated metal atoms that are attached to a support. In 2014, Tian *et al.* [31] assessed Mo-graphene-based catalyst where the graphene was doped with nitrogen for its catalytic activity towards NRR. In this system, it is understood that Mo/N serves as the active center, the graphene periphery acts as an electron reservoir, and the graphene body functions as an electron bridge for transportation. Nitrogen-doped graphene embedded with FeN_3 was studied for its ability to promote adsorption and subsequent activation of nitrogen due to its spin polarised center [32]. The system was formed by creating a single vacancy in graphene and inserting a single Fe atom in the vacancy (Fig. **3**). It was observed that the symbiotic effect of graphene and FeN_3 allowed the production of ammonia from nitrogen at ambient temperature. Similar to the previous case, graphene plays the role of the electron reservoir, and FeN_3 acts as the electron transmitter and the active site for nitrogen reduction.

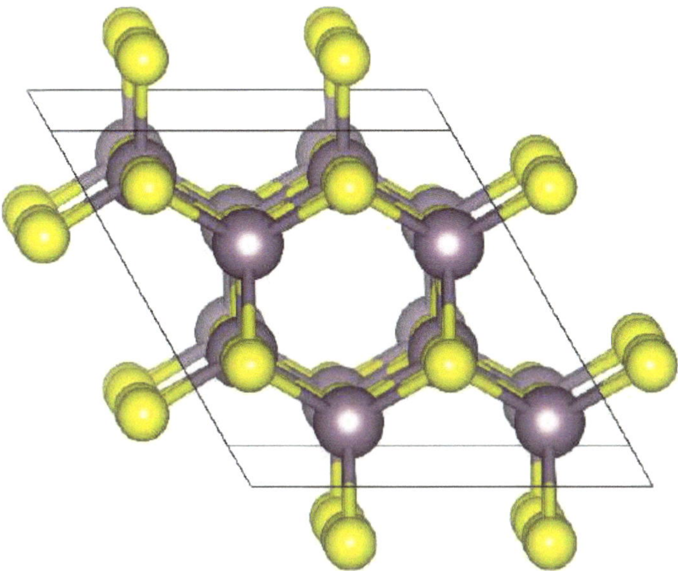

Fig. (3). MoS_2 structure in 2H phase; purple and yellow represent Mo and S atoms respectively.

Zhao and Chen [33] proposed individual atoms of transition metals anchored on a defective boron nitride monolayer as a potential catalyst for NRR at room temperature. The results showed that an individual Mo atom on a defective boron nitride nanosheet with a significantly reduced overpotential of 0.19V was the most active candidate. In the process of adsorption of N_2 on the monolayer, there is a transfer of electrons between Mo in the monolayer to the adsorbed N_2, which brings about the elongation of the N≡N bond. The system showed excellent stability through the large binding energy between Mo atoms and the BN monolayer. Its high efficacy was credited to the synergistic role of the monolayer, and the MoN_3 transmitter component, where the former served as an electron reservoir, and the latter was responsible for the charge transfer between the monolayer and the adsorbed species. In a similar study, Zhou *et al.* [9] proposed a single Mo atom supported on graphene sheets which could be modified by doping with heteroatoms of non-metals such as B, P, N, Se, S, and the like. Through DFT calculations, it was learned that Se (Mo/SeG) was the best dopant to attain the ideal electronic structure of Mo for NRR. It showed a very low potential of 0.41V *vs*. RHE. It was also found to be the most stable for the N_2H intermediate. It was further observed that the reaction occurred via a mixed pathway of alternating and distal routes. Moreover, the system showed excellent selectivity for NRR. The efficiency of Fe-embedded 2D MoS_2 nanosheets was investigated by Azofra *et al.* [34] in aqueous media under ambient conditions. Based on the calculations performed, the system was predicted to automatically arrest nitrogen selectively. The limiting step was anticipated to be the transfer of the proton-electron pair for the N_2H intermediate radical, which demands an activation barrier of 1.02 eV. The system also showed regenerating capacity. Qian *et al.* [35] investigated d-block transition metal atoms supported on C_2N as both single- and double-atom catalysts for their NRR activity. It was reported that among the SACs, Ti, Zr, and Hf showed low overpotential and superior catalytic activity. Although the Cr-Cr catalyst showed promising results as a double-atom system, it was observed that the SACs were better candidates due to their selectivity towards NRR and the conventional donation and back-donation strategies of σ and π electrons respectively.

Nitrogen Reduction Reaction on Double-Atom Catalysts

As the name suggests, double-atom catalysts (DACs) are composed of two different atoms that are attached to a surface. Ma *et al.* [36] investigated heteronuclear dimers of the transition metal; FeM and NiM (M- Ti, Cr, V, Fe, Mn, Co, Cu, Ni) anchored on graphdiyne (GDY) monolayers for support for their catalytic activity with NRR. It was found that FeCo@GDY (potential -0.44V) and NiCo@GDY (potential -0.36V) were the best performing candidates with high catalytic activity, ability to overpower HER, and favourable stability. Through calculations, it was concluded that both catalysts preferred the distal mechanism. In 2019, the same team studied transition metal dimers supported on graphdiyne [37]. Co-dimer (Co_2@GDY) showed the best activity and lowest onset potential of -0.43V. It also displayed enhanced sensitivity for NRR. In 2020, Yang *et al.* [38] studied 8 varieties of Fe-based DACs supported on a graphene substrate; Fe-M/GS (M- Ti, Cr, V, Fe, Mn, Co, Cu, Ni. While FeTi/Gs showed the highest catalytic activity, it was realized that the rate-limiting step was the transfer of the first proton-electron pair. They also concluded that the electron transfer was the factor governing the adsorption of nitrogen, and the presence of the second metal atom (M) serves by enhancing the catalytic activity by promoting the transfer of electrons between nitrogen and the graphene substrate. Graphdiyne is widely investigated and can be considered as one of the most suitable substrates for DACs owing to its advantage of a variation of free energy of the rate-determining step. It also possesses appreciable mechanical and chemical stability, high conjugation of π-electrons, exceptional conductivity, adjustable electronic properties, and wide surface area. A critical drawback concerning DACs is the lack of definitive control over the process of aggregation, which may lead to the formation of metal nanoparticles that have a broad size distribution. Due to this, it becomes necessary to carefully choose the optimal substrate for catalysis. In 2018, a study was conducted to develop a series of SACs and DACs comprising transition metals anchored on a C_2N substrate (M-C_2N and M_2-C_2N, respectively) [39].

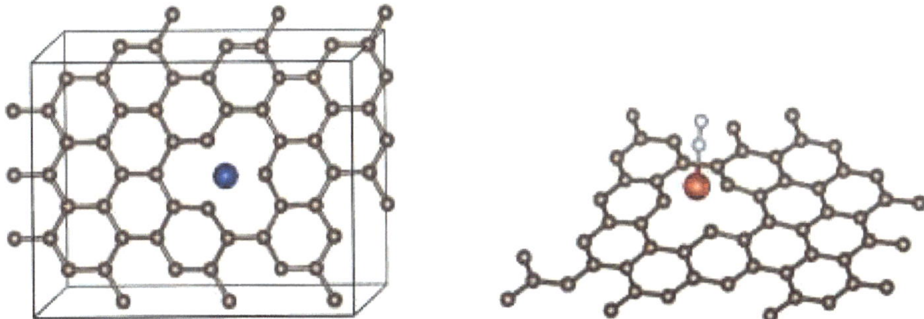

Fig. (4). (L) Unit cell representation of SAC on graphene (R) Adsorption of nitrogen molecule on the catalytic graphene surface.

The presence of strong coupling between the metal atom and the adsorbed nitrogen, along with the porous nature of the substrate, was responsible for stabilizing the catalyst and initiation of electron redistribution to boost the NRR. Mn_2C_2N was reported to have the maximum catalytic activity with a low potential of -0.23V. It was also observed that DACs showed lower potential than SACs, making them better candidates for NRR. The reasoning behind this advantage lies in the additional electron transfer that occurs from the metal d-orbitals to nitrogen in DACs. These electrons are responsible for the elongation of the N≡N bond, leading to its activation. The catalyst also showed a greater preference for NRR than HER, and the enzymatic pathway was found to be optimal. Very recently, Yongkang Xu *et al.* [40] reported Fe-based DACs with transition metals supported on graphdiyne as potential catalysts for NRR. Among the metals investigated, FeNi-GDY, FeCr-GDY, and FeMo-GDY showed the strongest suppression toward HER. Although DACs have only recently surfaced, they may take over SACs due to their superior performance ability. Wei *et al.* [41] reported two hetero-double-atom catalysts, Fe_2 and V_2, on a C_2N surface for NRR. The co-doped C_2N system showed a low limiting potential of -0.17V and facilitated nitrogen reduction via the enzymatic route. The system also successfully suppressed the competing HER. In 2019, Cao *et al.* [42] investigated the catalytic activity of a boron-doped C_2N surface towards NRR. It was found that the DAC $B_2@C_2N$ showed an overpotential of 0.19eV, which was lower than the SAC $B@C_2N$ (0.29eV), and $Ru@C_2N$ (0.80eV). It was also reported that the DAC was more selective towards NRR, as the hydrogenation of N_2 was deemed to be the thermodynamically preferred choice.

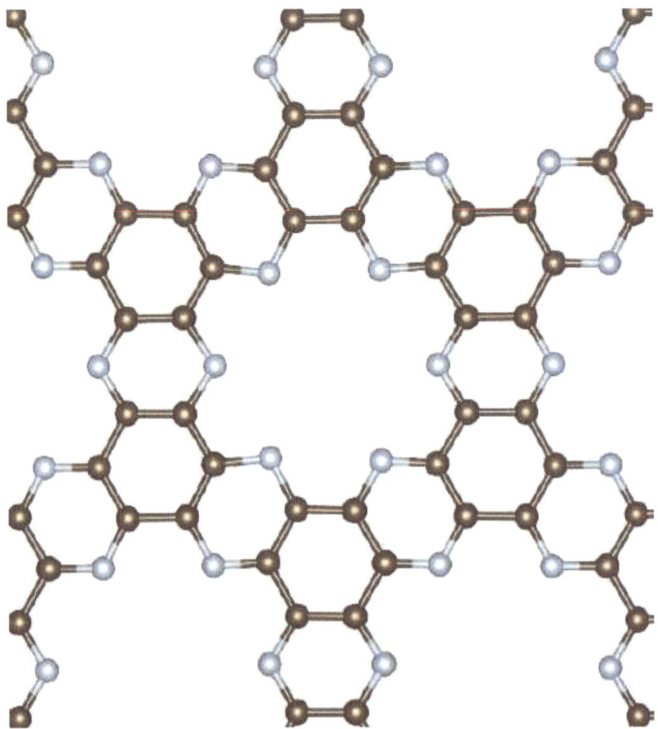

Fig. (5). Structure of C2N (carbon nitride) surface.

EXPERIMENTAL ADVANCES IN NITROGEN REDUCTION REACTION

Experimental evidence of electrocatalytic NRR dates back to 1983, when Sclafani *et al.* examined the feasibility of nitrogen reduction over an iron cathode in an ambient environment in an aqueous electrolyte of 0.6M KOH [43]. The results showed that the rate of production of ammonia relied on the temperature conditions and the potential at the cathode. The peak value of the rate of production was found to be 0.5 mol/h/cm^{-2} at -1.07V *vs.* SCE. Gas diffusion electrodes were implemented in the electrocatalytic reduction of nitrogen to ammonia by Furuya and Yoshiba [44]. The electrodes were modified with Fe-phthalocyanine in different electrolytes at 25°C and 1 atm pressure. In this study, it was seen that the rate of product formation slowed down after 10 minutes. Later in 1990, NRR was tested over fine Fe powder, metal sulfides, metal oxides, ZnSe, SiC, and TiB [45]. It was concluded that metal sulfides showed better catalytic activity than metal oxides. Still, these catalysts displayed very low selectivity for NRR and were not stable.

In the coming of the 21st century, heterogeneous catalysis under ambient conditions became very popular and widely studied. In this regard, we look at four different classes, varying based on the kind of cell used for electrochemical catalysis.

Back-To-Back Type Cell

In this type of cell, solid-state electrolytes are used for the synthesis of ammonia over temperatures extending from 25-800°C. The fundamental constituents of the cell are two permeable electrodes and a Nafion membrane that separates them. In 2009, Xu *et al.* demonstrated NRR from dry nitrogen and wet hydrogen at temperatures ranging from 25-100°C using $SmFe_{0.7}Cu_{0.1}Ni_{0.2}O_3$, a perovskite oxide as the cathode, and $Ni-Ce_{0.8}Sm_{0.2}O_{2-\delta}$ as the anode [46]. The peak rate of production of ammonia was found to be $1.13*10^{-8}$ mol/s/cm^2 at a potential of 2V, at 80°C. Using air and water, Tao *et al.* [6] examined the use of Pt anchored on carbon black (Pt/C) as a catalyst for NRR. The Pt/C electrode served as both the cathode and the anode. The maximum production rate, in this case, was found to be $1.14*10^{-9}$ mol/s/com^2 at a potential of 1.6V. In another study involving Fe, Ni, and FeNi nanoparticles as cathodes, it was seen that the initial effectiveness of Fe materials was 41%, and it instantaneously dropped down to 1.6% within hours. On the contrary, Ni materials showed an initial efficiency of 3%, which progressively increased with time. The rate of ammonia production was recorded as $(1.33-3.80)*10^{-12}$ mol/s/cm^2 at 50°C and 1.2V [47]. With a back-to-back cell, a liquid electrolyte can also be employed, which extends the choice and selection of catalysts as well as electrolytes to a broader domain. The production of ammonia using nano-γ-Fe_2O_3 catalysts was integrated into anion exchange membrane electrodes [48]. The resultant weight-normalized activity of the electrode in KOH measured in half-reaction was found to be 12.5 nmol/h/mg at 0V. The weight-normalized activity measured with γ-Fe_2O_3 nanoparticles coated on the surface of porous carbon paper was found to be 55.9 nmol/h/mg. Zhang *et al.* [49] investigated the surface of MoS_2 nanosheets for their catalytic activity towards NRR. In the double-compartment cell, the MoS_2 nanosheets grown hydrothermally on carbon cloth served as the cathodic catalyst. Optimum production of ammonia was recorded as $8.08 * 10^{-4}$ mol/s/cm at -0.5V, beyond which the performance only slightly declined, due to the competing HER. The activity of the MoS_2 surface was comparable to that of extreme temperature and pressure catalysts. It was also reported that the catalyst showed excellent stability for up to 10 recycle test rounds.

Polymer Electrolyte Membrane (PEM) Cell

In 2000, Kordali *et al.* [50] investigated the catalysis of NRR from gaseous nitrogen and water using Ru supported on carbon felt (Ru/C) as the catalyst in a PEM cell. A 2.0M aqueous KOH electrolyte was used, along with a Nafion membrane for the separation of the electrodes. The peak production value was found to be 1.30 µg/h/cm^2 at 90°C and -1.02V *vs.* Ag/AgCl. When the temperature was reduced to 20°C, a much lower production rate of 0.21 µg/h/cm^2 was recorded. In a similar study, Chen *et al.* [51] investigated iron oxide carried on carbon nanotubes (Fe$_2$O$_3$-CNT) as the catalyst (cathode) in an aqueous NaHCO$_3$ electrolyte (anode) under ambient conditions for the direct production of ammonia from nitrogen and water. A maximum yield of 0.22 µg/h/cm^2 at -2.0V was recorded *vs.* Ag/AgCl. The same group studied the effects of the electrolyte, the iron content in the catalyst, and the applied voltage on the reaction. It was observed that the selection of electrolytes did not have a major effect on the rate of formation of ammonia but influenced the selectivity of the reaction. Optimum performance was recorded on 30% wt. Fe$_2$O$_3$-CNT with a formation rate of 1.06*10^{-11} mol/s/cm^2. The catalyst showed significant stability at -1.0V *vs.* Ag/AgCl. It was further noticed that ammonia was discovered in both the chambers, implying a crossover across the Nafion membrane. In order to limit this crossover, a new gas diffusion layer was introduced between the catalyst and the membrane.

H-Type Cell in Liquid Electrolyte

The reduction of nitrogen and the subsequent production of ammonia was investigated using gold nanorods in an H-type cell with double chambers and a Nafion membrane separating the two chambers. 0.1.M KOH was used as the electrolyte, and a high yield of 1.648 µg/h/cm^2 at -0.2V was recorded *vs.* RHE [5]. In the interest of employing transition metals as catalysts for NRR, recently, Rh nano assemblies were studied in an H-type cell. An excellent yield of 23.88 µg/h/cm^2 was recorded at 0.2V *vs.* RHE [52].

Unlike PEM-type cells, where the reference electrode is situated in the anode chamber, in H-type cells, the working electrode and the reference electrode are both placed in the cathode chamber. This is an advantage as it allows for more accurate measurement of the applied potential. However, similar to PEM cells, H-type cells are also subjected to the crossover effect. Hence, it is preferred to consider both cathode and anode electrolytes to calculate the total amount of ammonia produced.

Single-Chamber Cell in Liquid Electrolyte

In 2006, in the first report of applying a metal-free catalyst for NRR, Koleli and Ropke proposed a polyaniline electrode in a single-chamber cell, capable of serving at high pressures [53]. The polyaniline was coated as a film on a Pt plate in a methanol/LiClO$_4$/H$_2$SO$_4$ electrolyte, and the reaction was carried out under ambient conditions. The production rate at -0.12V and 25°C was found to be

$2.25*10^{-7}$ mol/h/cm^2*vs.* NHE. Subsequently, in 2015, Wessling *et al.* [54] demonstrated the use of Rh and Ru galvanically deposited on Ti felts in 0.50M H_2SO_4 electrolyte in a cell containing a single-chamber. The reaction was performed at 30°C, and the rate of synthesis of ammonia turned out to be eight times greater for Ru ($1.2*10^{-10}$ mol/s/cm^2) than Rh ($1.5*10^{-11}$ mol/s/cm^2). As it is understood that HER is the main competitor, in an attempt to overcome this challenge, Kim *et al.* tested mixtures of 2-propanol and water in the ratio 9:1 as catalysts for NRR [55]. A production rate of $1.54*10^{-11}$ mol/s/cm^2 was reported. A shortcoming associated with single-chamber cells is the possibility of the gaseous products being oxidized at the anode. In the case of double-chamber cells, the presence of an ion-exchange membrane helps to avoid the oxidation of products by allowing the selective transfer of ions. Furthermore, in double-chamber cells, different electrolytes can be used, reducing the influence of cathodic reactions on the anode.

PEROVSKITES FOR NITROGEN REDUCTION REACTION – THE FUTURE

Perovskites are structure complexes described by the general formula ABX_3, where X represents the anion, B represents the small transition metal ion that is a part of the octahedral corners, and A represents the larger cation connected to X by 12-fold coordination.

Perovskites are known for their flexible electronic structure, which can be tuned to suit desired catalytic reactions. Owing to their tuneable properties, perovskites can be exploited in several chemical reactions (Fig. **4**). Their distinct bulk properties, along with their ability to withstand the strain, make them a potential candidate for green electrocatalytic reduction of nitrogen to produce ammonia. In a study by Yangsen Xu *et al.* [56], Sr was doped onto $LaFeO_3$ to generate oxygen vacancies to create $La_{0.5}Sr_{0.5}FeO_{3-\delta}$, a ceramic oxide. The yield of ammonia was found to be 11.51 µg/h/mg$_{cat}$ at -0.6V *vs.* RHE, which is higher than that offered by $LaFeO_3$ nanoparticles. DFT and *ab initio* calculations were conducted to investigate the mechanism and it was seen that the reaction free energy barriers varied appreciably on introducing oxygen vacancies, which stepped up the reaction speed. In a similar study, $LaCrO_3$ was synthesized and analyzed for its catalytic activity in 0.1M Na_2SO_4 in an H-type cell, involving a Nafion membrane separating the two chambers [57]. A high yield of 24.8 µg/h/mg$_{cat}$ was reported at -0.8V *vs.* RHE. DFT calculations were performed to realize the mechanism, and it was revealed that the reaction followed the associative distal pathway. In a study conducted by Yu *et al.*, $La_2Ti_2O_7$ in its nanosheet form was proposed as an impressive catalyst for NRR [58]. The yield provided by the catalyst was found to be 25.15 µg/h/mg$_{cat}$ at -0.55V *vs.* RHE. The catalyst showed high stability, and it was hypothesized that the La atoms behaved as the active sites for the adsorption of N_2. Furthermore, the O vacancies empowered the adsorption of N_2 and promoted catalytic activity. The nanosheets provide a better surface area for the reaction when compared to their bulk counterparts. Very recently, an attempt was made to use $Ce_{1/3}NbO_3$ as a catalyst for NRR [59]. Ce and Nb were chosen for their ability to favour and accelerate NRR; Ce shows selectivity towards NRR owing to its flexible electronic properties, and Nb is capable of back-donating its d-electrons to nitrogen and thereby activating the N≡N bond. Perovskites are yet to be extensively studied for their catalytic activity towards NRR.

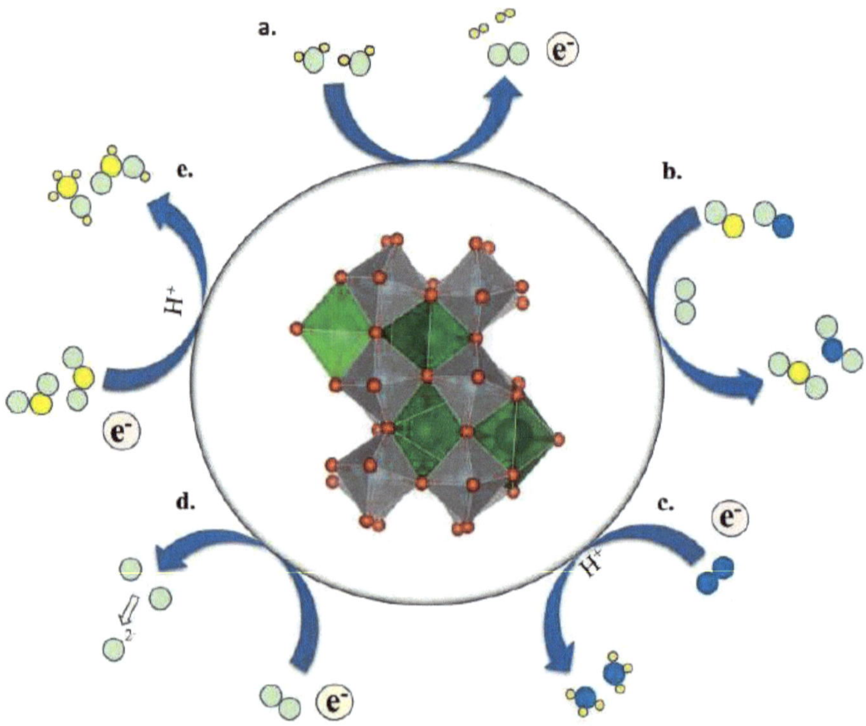

Fig. (6). - Application of perovskites in **(a)** Electrolysis of water, **(b)** Purification of air, **(c)** Reduction of Nitrogen, **(d)** Fuel cell, **(e)** Reduction of Carbon dioxide.

CONCLUSION

The importance of NRR is well established, and it is an ever-growing need for a green method that has led to the development of several catalysts. While, in theory, the conversion of nitrogen to ammonia is feasible under ambient conditions, the practical concept is tricky due to the lack of competent catalysts. An ideal catalyst must not only display high catalytic activity but should also show an excellent preference for NRR and suppress the competing HER. Furthermore, the catalyst should be exceptionally stable [60]. It is possible that after continued use, the stability can decrease, or the catalyst can decompose or deactivate itself. In this regard, it is crucial to understand the active sites present on the catalyst to formulate one that can be self-regenerating. Using theoretical calculations before experimenting reduces the time needed to devise an efficient catalyst. Theoretical tactics aid in understanding in detail the underlying mechanism of the reaction and the factors that can be modified to achieve the best performance. However, these computational calculations are generally conducted in a vacuum. Due to this, experimental conditions such as temperature, pressure, and pH are dilapidated. Therefore, it is also necessary to employ in-situ operational approaches, along with DFT, to produce an effective catalyst. Ranging from simple metal surfaces to single- and double-atom catalysts and perovskites, the search for optimum catalysts has constantly been expanding. Despite the consistent efforts at synthesizing a powerful catalyst, the task remains a challenge due to the poor performance of some when tested experimentally.

CONSENT FOR PUBLICATION

Not applicable.

CONFLICT OF INTEREST

The author declares no conflict of interest, financial or otherwise.

ACKNOWLEDGEMENTS

D. R would like to acknowledge DST-SERB, INDIA, for providing junior research fellowship. D. K also acknowledges DST-SERB, INDIA, for providing financial support (Grant No: CRG/2020/003882).

REFERENCES

[1]   Honkala, K.; Hellman, A.; Remediakis, I.N.; Logadottir, A.; Carlsson, A.; Dahl, S.; Christensen, C.H.; Nørskov, J.K. Ammonia synthesis from first-principles calculations. *Science,* **2005**, *307*(5709), 555-558.
[http://dx.doi.org/10.1126/science.1106435] [PMID: 15681379]

[2]   Jia, H.P.; Quadrelli, E.A. Mechanistic aspects of dinitrogen cleavage and hydrogenation to produce ammonia in catalysis and organometallic chemistry: relevance of metal hydride bonds and dihydrogen. *Chem. Soc. Rev.,* **2014**, *43*(2), 547-564.
[http://dx.doi.org/10.1039/C3CS60206K] [PMID: 24108246]

[3]   Zhan, C.G.; Nichols, J.A.; Dixon, D.A. Ionization Potential, Electron Affinity, Electronegativity, Hardness, and Electron Excitation Energy: Molecular Properties from Density Functional Theory Orbital Energies. *J. Phys. Chem. A,* **2003**, *107*(20), 4184-4195.
[http://dx.doi.org/10.1021/jp0225774]

[4]   Guo, X.; Du, H.; Qu, F.; Li, J. Recent progress in electrocatalytic nitrogen reduction. *J. Mater. Chem.*

A Mater. Energy Sustain., **2019**, *7*(8), 3531-3543.
[http://dx.doi.org/10.1039/C8TA11201K]

[5]   Bao, D.; Zhang, Q.; Meng, F.L.; Zhong, H.X.; Shi, M.M.; Zhang, Y.; Yan, J.M.; Jiang, Q.; Zhang, X.B. Electrochemical Reduction of N_2 under Ambient Conditions for Artificial N_2 Fixation and Renewable Energy Storage Using N_2 /NH_3 Cycle. *Adv. Mater.,* **2017**, *29*(3), 1604799.
[http://dx.doi.org/10.1002/adma.201604799] [PMID: 27859722]

[6]   Lan, R.; Irvine, J.T.S.; Tao, S. Synthesis of ammonia directly from air and water at ambient temperature and pressure. *Sci. Rep.,* **2013**, *3*(1), 1145.
[http://dx.doi.org/10.1038/srep01145] [PMID: 23362454]

[7]   Zhang, L.; Ji, X.; Ren, X.; Ma, Y.; Shi, X.; Tian, Z.; Asiri, A.M.; Chen, L.; Tang, B.; Sun, X. Electrochemical Ammonia Synthesis via Nitrogen Reduction Reaction on a MoS_2 Catalyst: Theoretical and Experimental Studies. *Adv. Mater.,* **2018**, *30*(28), 1800191.
[http://dx.doi.org/10.1002/adma.201800191] [PMID: 29808517]

[8]   Yu, X.; Han, P.; Wei, Z.; Huang, L.; Gu, Z.; Peng, S.; Ma, J.; Zheng, G. Boron-Doped Graphene for Electrocatalytic N2 Reduction. *Joule,* **2018**, *2*(8), 1610-1622.
[http://dx.doi.org/10.1016/j.joule.2018.06.007]

[9]   Zhou, H.Y.; Li, J.C.; Wen, Z.; Jiang, Q. Tuning the catalytic activity of a single Mo atom supported on graphene for nitrogen reduction *via* Se atom doping. *Phys. Chem. Chem. Phys.,* **2019**, *21*(27), 14583-14588.
[http://dx.doi.org/10.1039/C9CP02733E] [PMID: 31241647]

[10]   Schrödinger, E. An Undulatory Theory of the Mechanics of Atoms and Molecules. *Phys. Rev.,* **1926**, *28*(6), 1049-1070.
[http://dx.doi.org/10.1103/PhysRev.28.1049]

[11]   Born, M. Zur Quantenmechanik der Sto☐vorg☐nge. *Eur. Phys. J. A,* **1926**, *37*(12), 863-867.
[http://dx.doi.org/10.1007/BF01397477]

[12]   Fermi, E. Eine statistische Methode zur Bestimmung einiger Eigenschaften des Atoms und ihre Anwendung auf die Theorie des periodischen Systems der Elemente. *Eur. Phys. J. A,* **1928**, *48*(1-2), 73-79.
[http://dx.doi.org/10.1007/BF01351576]

[13]   Thomas, L.H. The calculation of atomic fields. *Math. Proc. Camb. Philos. Soc.,* **1927**, *23*(5), 542-548.
[http://dx.doi.org/10.1017/S0305004100011683]

[14]   Dkhissi, A.; Beljonne, D.; Lazzaroni, R.; Louwet, F.; Groenendaal, L.; Brédas, J. L. **2003**.
[http://dx.doi.org/10.1002/qua.10446]

[15]   Slater, J.C. A Simplification of the Hartree-Fock Method. *Phys. Rev.,* **1951**, *81*(3), 385-390.
[http://dx.doi.org/10.1103/PhysRev.81.385]

[16]   Scholl, D.; Steckel, J. *Density Functional Theory- A Practical Introduction*; John Wiley & Sons, Inc.: Hoboken, New Jersey, **2009**.
[http://dx.doi.org/10.1002/9780470447710]

[17]   Skúlason, E.; Bligaard, T.; Gudmundsdóttir, S.; Studt, F.; Rossmeisl, J.; Abild-Pedersen, F.; Vegge, T.; Jónsson, H.; Nørskov, J.K. A theoretical evaluation of possible transition metal electro-catalysts for N_2 reduction. *Phys. Chem. Chem. Phys.,* **2012**, *14*(3), 1235-1245.
[http://dx.doi.org/10.1039/C1CP22271F] [PMID: 22146855]

[18]   Ishikawa, A.; Doi, T.; Nakai, H. Catalytic performance of Ru, Os, and Rh nanoparticles for ammonia synthesis: A density functional theory analysis. *J. Catal.,* **2018**, *357*, 213-222.
[http://dx.doi.org/10.1016/j.jcat.2017.11.018]

[19]   Fajín, J.L.C.; Cordeiro, M.N.D.S.; Gomes, J.R.B. Water dissociation on multimetallic catalysts. *Appl. Catal. B,* **2017**, *218*, 199-207.
[http://dx.doi.org/10.1016/j.apcatb.2017.06.050]

[20]   Ma, X.L.; Liu, J.C.; Xiao, H.; Li, J. Surface Single-Cluster Catalyst for N_2-to-NH_3 Thermal Conversion. *J. Am. Chem. Soc.,* **2018**, *140*(1), 46-49.
[http://dx.doi.org/10.1021/jacs.7b10354] [PMID: 29244491]

[21]   Licht, S.; Cui, B.; Wang, B.; Li, F.F.; Lau, J.; Liu, S. Ammonia synthesis by N_2 and steam electrolysis in molten hydroxide suspensions of nanoscale Fe_2O_3. *Science,* **2014**, *345*(6197), 637-640.
[http://dx.doi.org/10.1126/science.1254234] [PMID: 25104378]

[22]   Nguyen, M.T.; Seriani, N.; Gebauer, R. Nitrogen electrochemically reduced to ammonia with hematite: density-functional insights. *Phys. Chem. Chem. Phys.,* **2015**, *17*(22), 14317-14322.
[http://dx.doi.org/10.1039/C4CP04308A] [PMID: 25482262]

[23]   Abghoui, Y.; Garden, A.L.; Howalt, J.G.; Vegge, T.; Skúlason, E. Electroreduction of N_2 to Ammonia at Ambient Conditions on Mononitrides of Zr, Nb, Cr, and V: A DFT Guide for Experiments. *ACS Catal.,* **2016**, *6*(2), 635-646.
[http://dx.doi.org/10.1021/acscatal.5b01918]

[24]   Abghoui, Y.; Skúlasson, E. Transition Metal Nitride Catalysts for Electrochemical Reduction of Nitrogen to Ammonia at Ambient Conditions. *Procedia Comput. Sci.,* **2015**, *51*(1), 1897-1906.
[http://dx.doi.org/10.1016/j.procs.2015.05.433]

[25]   Abghoui, Y.; Skúlason, E. Computational Predictions of Catalytic Activity of Zincblende (110) Surfaces of Metal Nitrides for Electrochemical Ammonia Synthesis. *J. Phys. Chem. C,* **2017**, *121*(11), 6141-6151.
[http://dx.doi.org/10.1021/acs.jpcc.7b00196]

[26]   Abghoui, Y.; Garden, A.L.; Hlynsson, V.F.; Björgvinsdóttir, S.; Ólafsdóttir, H.; Skúlason, E. Enabling electrochemical reduction of nitrogen to ammonia at ambient conditions through rational catalyst design. *Phys. Chem. Chem. Phys.,* **2015**, *17*(7), 4909-4918.
[http://dx.doi.org/10.1039/C4CP04838E] [PMID: 25446373]

[27]   Azofra, L.M.; Li, N.; MacFarlane, D.R.; Sun, C. Promising prospects for 2D d^2–d^4 M_3C_2 transition metal carbides (MXenes) in N_2 capture and conversion into ammonia. *Energy Environ. Sci.,* **2016**, *9*(8), 2545-2549.
[http://dx.doi.org/10.1039/C6EE01800A]

[28]   Jain, A.; Bar Sadan, M.; Ramasubramaniam, A. Identifying a New Pathway for Nitrogen Reduction Reaction on Fe-Doped MoS_2 by the Coadsorption of Hydrogen and N_2. *J. Phys. Chem. C,* **2021**, *125*(36), 19980-19990.
[http://dx.doi.org/10.1021/acs.jpcc.1c04499]

[29]   Zhang, H.; Cui, C.; Luo, Z. MoS_2-Supported Fe_2 Clusters Catalyzing Nitrogen Reduction Reaction to Produce Ammonia. *J. Phys. Chem. C,* **2020**, *124*(11), 6260-6266.
[http://dx.doi.org/10.1021/acs.jpcc.0c00486]

[30]   Lin, G.; Ju, Q.; Guo, X.; Zhao, W.; Adimi, S.; Ye, J.; Bi, Q.; Wang, J.; Yang, M.; Huang, F. Intrinsic Electron Localization of Metastable MoS_2 Boosts Electrocatalytic Nitrogen Reduction to Ammonia. *Adv. Mater.,* **2021**, *33*(32), 2007509.
[http://dx.doi.org/10.1002/adma.202007509] [PMID: 34219276]

[31]   Le, Y.Q.; Gu, J.; Tian, W.Q. Nitrogen-fixation catalyst based on graphene: every part counts. *Chem. Commun. (Camb.),* **2014**, *50*(87), 13319-13322.
[http://dx.doi.org/10.1039/C4CC01950D] [PMID: 24934900]

[32]   Li, X.F.; Li, Q.K.; Cheng, J.; Liu, L.; Yan, Q.; Wu, Y.; Zhang, X.H.; Wang, Z.Y.; Qiu, Q.; Luo, Y. Conversion of Dinitrogen to Ammonia by FeN_3-Embedded Graphene. *J. Am. Chem. Soc.,* **2016**, *138*(28), 8706-8709.
[http://dx.doi.org/10.1021/jacs.6b04778] [PMID: 27383680]

[33]   Zhao, J.; Chen, Z. Single Mo Atom Supported on Defective Boron Nitride Monolayer as an Efficient Electrocatalyst for Nitrogen Fixation: A Computational Study. *J. Am. Chem. Soc.,* **2017**, *139*(36),

12480-12487.
[http://dx.doi.org/10.1021/jacs.7b05213] [PMID: 28800702]

[34]   Azofra, L.M.; Sun, C.; Cavallo, L.; MacFarlane, D.R. Feasibility of N_2 Binding and Reduction to Ammonia on Fe□Deposited MoS_2 2D Sheets: A DFT Study. *Chemistry,* **2017**, *23*(34), 8275-8279.
[http://dx.doi.org/10.1002/chem.201701113] [PMID: 28524268]

[35]   Qian, Y.; Liu, Y.; Zhao, Y.; Zhang, X.; Yu, G. Single *vs* double atom catalyst for N_2 activation in nitrogen reduction reaction: A DFT perspective. *EcoMat,* **2020**, *2*(1)
[http://dx.doi.org/10.1002/eom2.12014]

[36]   Ma, D.; Zeng, Z.; Liu, L.; Jia, Y. Theoretical screening of the transition metal heteronuclear dimer anchored graphdiyne for electrocatalytic nitrogen reduction. *Journal of Energy Chemistry,* **2021**, *54*, 501-509.
[http://dx.doi.org/10.1016/j.jechem.2020.06.032]

[37]   Ma, D.; Zeng, Z.; Liu, L.; Huang, X.; Jia, Y. Computational Evaluation of Electrocatalytic Nitrogen Reduction on TM Single-, Double-, and Triple-Atom Catalysts (TM = Mn, Fe, Co, Ni) Based on Graphdiyne Monolayers. *J. Phys. Chem. C,* **2019**, *123*(31), 19066-19076.
[http://dx.doi.org/10.1021/acs.jpcc.9b05250]

[38]   Yang, W.; Huang, H.; Ding, X.; Ding, Z.; Wu, C.; Gates, I.D.; Gao, Z. Theoretical study on double-atom catalysts supported with graphene for electroreduction of nitrogen into ammonia. *Electrochim. Acta,* **2020**, *335*, 135667.
[http://dx.doi.org/10.1016/j.electacta.2020.135667]

[39]   Chen, Z.W.; Yan, J.M.; Jiang, Q. Single or Double: Which Is the Altar of Atomic Catalysts for Nitrogen Reduction Reaction? *Small Methods,* **2019**, *3*(6), 1800291.
[http://dx.doi.org/10.1002/smtd.201800291]

[40]   Xu, Y.; Cai, Z.; Du, P.; Zhou, J.; Pan, Y.; Wu, P.; Cai, C. Taming the challenges of activity and selectivity in the electrochemical nitrogen reduction reaction using graphdiyne-supported double-atom catalysts. *J. Mater. Chem. A Mater. Energy Sustain.,* **2021**, *9*(13), 8489-8500.
[http://dx.doi.org/10.1039/D1TA00262G]

[41]   Wei, Z.; He, J.; Yang, Y.; Xia, Z.; Feng, Y.; Ma, J. Fe, V-co-doped C2N for electrocatalytic N2-t--NH3 conversion. *Journal of Energy Chemistry,* **2021**, *53*, 303-308.
[http://dx.doi.org/10.1016/j.jechem.2020.04.014]

[42]   Cao, Y.; Deng, S.; Fang, Q.; Sun, X.; Zhao, C.; Zheng, J.; Gao, Y.; Zhuo, H.; Li, Y.; Yao, Z.; Wei, Z.; Zhong, X.; Zhuang, G.; Wang, J. Single and double boron atoms doped nanoporous $C_2N–h$ 2D electrocatalysts for highly efficient N_2 reduction reaction: a density functional theory study. *Nanotechnology,* **2019**, *30*(33), 335403.
[http://dx.doi.org/10.1088/1361-6528/ab1d01] [PMID: 31026848]

[43]   Sclafani, A.; Augugliaro, V.; Schiavello, M. Dinitrogen Electrochemical Reduction to Ammonia over Iron Cathode in Aqueous Medium. *J. Electrochem. Soc.,* **1983**, *130*(3), 734-736.
[http://dx.doi.org/10.1149/1.2119794]

[44]   Furuya, N.; Yoshiba, H. Electroreduction of nitrogen to ammonia on gas-diffusion electrodes modified by Fe-phthalocyanine. *J. Electroanal. Chem. Interfacial Electrochem.,* **1989**, *263*(1), 171-174.
[http://dx.doi.org/10.1016/0022-0728(89)80134-2]

[45]   Furuya, N.; Yoshiba, H. Electroreduction of nitrogen to ammonia on gas-diffusion electrodes loaded with inorganic catalyst. *J. Electroanal. Chem. Interfacial Electrochem.,* **1990**, *291*(1-2), 269-272.
[http://dx.doi.org/10.1016/0022-0728(90)87195-P]

[46]   Cui, X.; Tang, C.; Zhang, Q. A Review of Electrocatalytic Reduction of Dinitrogen to Ammonia under Ambient Conditions. *Adv. Energy Mater.,* **2018**, *8*(22), 1800369.
[http://dx.doi.org/10.1002/aenm.201800369]

[47]   Renner, J.N.; Greenlee, L.F.; Ayres, K.E.; Herring, A.M. Electrochemical Synthesis of Ammonia: A

Low Pressure, Low Temperature Approach. *Electrochem. Soc. Interface,* **2015**, *24*(2), 51-57.
[http://dx.doi.org/10.1149/2.F04152if]

[48]   Kong, J.; Lim, A.; Yoon, C.; Jang, J.H.; Ham, H.C.; Han, J.; Nam, S.; Kim, D.; Sung, Y.E.; Choi, J.;
       Park, H.S. Electrochemical Synthesis of NH$_3$ at Low Temperature and Atmospheric Pressure Using a
       γ-Fe$_2$O$_3$ Catalyst. *ACS Sustain. Chem.& Eng.,* **2017**, *5*(11), 10986-10995.
       [http://dx.doi.org/10.1021/acssuschemeng.7b02890]

[49]   Zhang, L.; Ji, X.; Ren, X.; Ma, Y.; Shi, X.; Tian, Z.; Asiri, A.M.; Chen, L.; Tang, B.; Sun, X.
       Electrochemical Ammonia Synthesis via Nitrogen Reduction Reaction on a MoS$_2$ Catalyst:
       Theoretical and Experimental Studies. *Adv. Mater.,* **2018**, *30*(28), 1800191.
       [http://dx.doi.org/10.1002/adma.201800191] [PMID: 29808517]

[50]   Kordali, V.; Kyriacou, G.; Lambrou, C. Electrochemical synthesis of ammonia at atmospheric pressure
       and low temperature in a solid polymer electrolyte cell. *Chem. Commun. (Camb.),* **2000**, (17), 1673-
       1674.
       [http://dx.doi.org/10.1039/b004885m]

[51]   Chen, S.; Perathoner, S.; Ampelli, C.; Mebrahtu, C.; Su, D.; Centi, G. Electrocatalytic Synthesis of
       Ammonia at Room Temperature and Atmospheric Pressure from Water and Nitrogen on a Carbon-
       Nanotube-Based Electrocatalyst. *Angew. Chem. Int. Ed.,* **2017**, *56*(10), 2699-2703.
       [http://dx.doi.org/10.1002/anie.201609533] [PMID: 28128489]

[52]   Liu, H.M.; Han, S.H.; Zhao, Y.; Zhu, Y.Y.; Tian, X.L.; Zeng, J.H.; Jiang, J.X.; Xia, B.Y.; Chen, Y.
       Surfactant-free atomically ultrathin rhodium nanosheet nanoassemblies for efficient nitrogen
       electroreduction. *J. Mater. Chem. A Mater. Energy Sustain.,* **2018**, *6*(7), 3211-3217.
       [http://dx.doi.org/10.1039/C7TA10866D]

[53]   Köleli, F.; Röpke, D.; Aydin, R.; Röpke, T. Investigation of N2-fixation on polyaniline electrodes in
       methanol by electrochemical impedance spectroscopy. *J. Appl. Electrochem.,* **2011**, *41*(4), 405-413.
       [http://dx.doi.org/10.1007/s10800-010-0250-3]

[54]   Kugler, K.; Luhn, M.; Schramm, J.A.; Rahimi, K.; Wessling, M. Galvanic deposition of Rh and Ru on
       randomly structured Ti felts for the electrochemical NH$_3$ synthesis. *Phys. Chem. Chem. Phys.,* **2015**,
       17(5), 3768-3782.
       [http://dx.doi.org/10.1039/C4CP05501B] [PMID: 25556769]

[55]   Kim, K.; Yoo, C.Y.; Kim, J.N.; Yoon, H.C.; Han, J.I. Electrochemical Synthesis of Ammonia from
       Water and Nitrogen in Ethylenediamine under Ambient Temperature and Pressure. *J. Electrochem.
       Soc.,* **2016**, *163*(14), F1523-F1526.
       [http://dx.doi.org/10.1149/2.0741614jes]

[56]   Xu, Y.; Xu, X.; Cao, N.; Wang, X.; Liu, X.; Fronzi, M.; Bi, L. Perovskite ceramic oxide as an efficient
       electrocatalyst for nitrogen fixation. *Int. J. Hydrogen Energy,* **2021**, *46*(17), 10293-10302.
       [http://dx.doi.org/10.1016/j.ijhydene.2020.12.147]

[57]   Ohrelius, M.; Guo, H.; Xian, H.; Yu, G.; Alshehri, A.A.; Alzahrani, K.A.; Li, T.; Andersson, M.
       Electrochemical Synthesis of Ammonia Based on a Perovskite LaCrO$_3$ Catalyst. *ChemCatChem,* **2020**,
       12(3), 731-735.
       [http://dx.doi.org/10.1002/cctc.201901818]

[58]   Yu, J.; Li, C.; Li, B.; Zhu, X.; Zhang, R.; Ji, L.; Tang, D.; Asiri, A.M.; Sun, X.; Li, Q.; Liu, S.; Luo, Y.
       A perovskite La$_2$Ti$_2$O$_7$ nanosheet as an efficient electrocatalyst for artificial N$_2$ fixation to NH$_3$ in
       acidic media. *Chem. Commun. (Camb.),* **2019**, *55*(45), 6401-6404.
       [http://dx.doi.org/10.1039/C9CC02310K] [PMID: 31094366]

[59]   Hu, X.; Sun, Y.; Guo, S.; Sun, J.; Fu, Y.; Chen, S.; Zhang, S.; Zhu, J. Identifying electrocatalytic activity and mechanism of Ce1/3NbO$_3$ perovskite for nitrogen reduction to ammonia at ambient conditions. *Appl. Catal. B,* **2021,** *280*(280), 119419.
[http://dx.doi.org/10.1016/j.apcatb.2020.119419]

[60]   Rasool, A.; Anis, I.; Dixit, M.; Maibam, A.; Hassan, A.; Krishnamurty, S.; Dar, M.A. Tantalum Based Single, Double, and Triple Atom Catalysts Supported on g-C2N Monolayer for Effective Nitrogen Reduction Reaction: A Comparative DFT Investigation. *Catal. Sci. Technol.,* **2022,** *12,* 310-319.
[http://dx.doi.org/10.1039/D1CY01292D]

CHAPTER 5

Role of Nanocomposites in Environmental Remediation: Recent Advances and Challenges

Premanjali Rai[1,*]

[1] *Trace Organics Laboratory, Central Pollution Control Board, Parivesh Bhavan, East Arjun Nagar, Delhi-110032, India*

Abstract: Nanocomposites offer an exclusive advantage over bulk materials in terms of efficiency on account of their greater surface area, higher reactivity, ease of modification, good dispersion, and hence, multi-faceted applications. The various forms of nanocomposites derived from low-cost resources, especially carbon-based materials, are of unique interest. Activated carbons offer the unique advantage as the matrix for nanocomposites synthesis due to their graphite structure, thereby providing strength and the ease of modification on the surface of nanocomposites while introducing desired functional groups. Apart from this, they are widely popular for their large surface area and porosity. Therefore, carbon-based nanocomposites offer vivid applications in various fields, such as environmental remediation as adsorbents, suitable sorbents in the analytical determination of organics, targeted drug delivery, diagnostic agents, fuel cells and sensors, to name a few. Amongst these, the role of nanocomposites as sensors and environmental remediation tools has been studied extensively. The varied modes of action include adsorption, nano-catalysis, membrane filtration, *etc* ., for pollutants ranging from inorganic ions, heavy metals, pesticides, dyes, anti-bacterials, oil spills, and many more. However, there are constraints in their stability, cost, storage and disposal triggered by varying environmental conditions.This chapter presents a review of the synthesis, application and challenges of nanostructured composite materials in environmental remediation.

Keywords: Carbon nanotubes, Environmental decontamination, Nanocomposites, Synthesis.

INTRODUCTION

Composite materials are prepared by the amalgamation of two or more raw materials embedded in a matrix, each having its own unique properties, to create a new material having characteristic features of greater advantage. These are

[*] **Corresponding author Premanjali Rai**: Trace Organics Laboratory, Central Pollution Control Board, Parivesh Bhavan, East Arjun Nagar, Delhi-110032, India; Tel: +919559570905; E-mail:premanjalirai@yahoo.com

broadly categorized into polymer matrix-based composites, carbon matrix-based composites, metal matrix-based composites, and ceramic matrix-based composites. When the size dimensions are reduced to a nanometric scale, they are known as nanocomposites. Nanocomposites are those entities that exist in the nanometric-sized phase, *i.e.*, those particles having a diameter in the range of 10^{-9} m. The reduced size and greater surface area thereof provide them an edge over other conventional microcomposites. Reported as the material of the 21st century, nanocomposites find versatile applications in an array of studies. The very first advent of nanocomposites is believed to have started with clay/polymer composite. In 1990, Toyota company used clay and nylon-6 nanocomposites to manufacture timing belt covers in cars. Ever since then, the application of nanocomposites has proliferated in every sector. Reportedly, around 13420 papers have been published on nanocomposites in the last several years, specifically between the years 1988-2008.

Today, the application of nanocomposites is found in every field, starting from fuel cells, aerospace engineering, biomedical applications, sensors, environmental remediation tools, and many more (Fig. **1**). The reduction in size or the change in the size of the particle size induces a change in the behavior of the particle properties. In a review study [1], the "critical size" or the feature size at which such changes in properties are expected to occur in the nanocomposites has been enlisted (Table **1**).

Table 1. Critical size limits in nanocomposites beyond which changes in properties occur [1].

Properties	Reported Critical Size (nm)
Catalytic activity	<5
Hard magnetic materials turn soft	<20
Change in Refractive Index	<50
Super paramagnetism observed	<100
Change in strength and toughness	<100
Change in hardness and plasticity	<100

The ceramic matrix nanocomposites (CMNC) mainly have aluminium oxide (Al_2O_3) and silicon dioxide (SiO_2) as the matrix (examples include Al_2O_3/SiO_2, Al_2O_3/TiO_2, Al_2O_3/SiC, and Al_2O_3/CNT) and as the reinforcing material. The CMNC find applications in aerospace designing, engine exhausts, hypersonic vehicles, nuclear power industry, *etc* ., owing to their high thermal resistance, creep resistance, and inertness properties, to name a few.

The metal matrix nanocomposites (MMNC) consist of an alloy matrix in which nanosized materials are reinforced (examples include Fe-Cr/Al$_2$O$_3$, Ni/Al$_2$O$_3$, Co/Cr, Fe/MgO, Mg/CNT, and Al/CNT). The MMNCs have good ductility, elastic modulus, fracture toughness and strength imbibed from the metal and ceramic constituents, and they find great applications in high-strength structured materials in aerospace and automobile material manufacturing. One of the drawbacks of MMNCs is the high surface area which leads to agglomeration of nanoparticles on the surface, thereby reducing the strengthening capacity of the nanocomposites. However, this nanocomposite is a newly developed material, and more research studies are focusing on overcoming this challenge by modifying preparation routes.

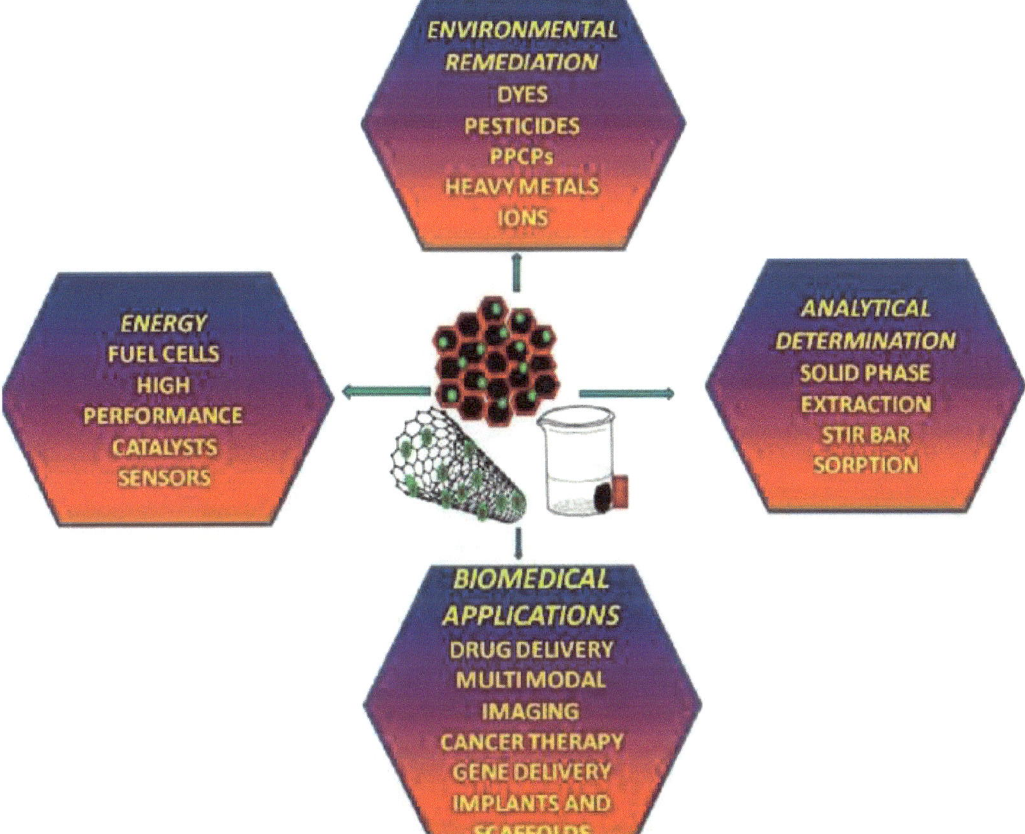

Fig. (1). Functional perspectives of nanocomposites.

The Polymer Matrix Nanocomposites (PMNC) consist of the polymeric matrix such as the thermostat or thermoplastic embedded with nanosized particles. Another classification based on the number of nanoscale dimensions they have is zero-dimensionality particles (such as bimetallic or trimetallic nanoparticles), one dimensional materials (such as the carbon nanotubes and nanorods), and two dimensional materials (such as the nanomembranes). The most promising candidate amongst the composite material recently being investigated for environmental applications is the carbon matrix impregnated with metals in nanometric dimensions. The carbon is modified by intercalation or impregnation of magnetic materials such as magnetite (Fe_3O_4), ferrite (Fe_2O_3), hematite (α-Fe_2O_3), maghemite (γ-Fe_2O_3) and metals, alloys and compounds containing Fe, Ni and Co. As an enhancement in their functional perspective, magnetic metal oxides are impregnated on the carbon matrix for enhanced maneuverability. The activated carbon renders a solid support owing to its large specific surface area and the porous structure holds the metallic nanoparticles within it. Symbiotically, the magnetic metal oxides in the nanometric sizes offer magnetic moments to the composite, which enhances or facilitates its navigation at the time of application. The nanoscale dimensions are an influential parameter as the particles elucidate the best magnetic behavior at a critical diameter of around 10-20 nm [2]. In such cases, each nanoparticle becomes a single magnetic domain with a large constant magnetic moment and behaves like a giant paramagnetic atom (superparamagnetism). It shows a quick response to the applied magnetic field with negligible remanence (residual magnetism) and coercivity (the field required to bring the magnetization to zero).

Extensive studies have been devoted to the application of carbon nanocomposites (magnetic and nonmagnetic) in the field of environmental remediation. Regarding their applications, they render utmost utility in fields pertaining to environmental remediation, catalysis, electrochemical sensing, extraction, magnetic recording devices, and recognition of biomolecules, pharmaceutical delivery and disease diagnosis [3]. Since the present study is focused on environmental decontamination, a short review has been presented on the routes of synthesis and application of carbon metal composites.

SYNTHESIS OF CARBON NANOCOMPOSITES

The microstructure of the carbon constitutes parallel stacks of polyaromatic graphene sheets and interstices within them. The "active sites" located along the edges of the carbon layers, contain high densities of unpaired electrons. They allow good sorption/chemisorption of heteroatoms such as oxygen, hydrogen, nitrogen, sulfur, *etc* ., allowing the formation of a stable surface [4]. Such heteroatoms arise from the spontaneous reaction of carbon on exposure to air or may be developed from the activating agents which oxidize the surface such as H_2O_2, HNO_3, H_2SO_4 [5]. Magnetic carbon nanocomposites comprise carbon surface modified by intercalation or impregnation of magnetic materials such as magnetite (Fe_3O_4), ferrite (Fe_2O_3), hematite (α-Fe_2O_3), maghemite (γ-Fe_2O_3) and metals, alloys and compounds containing Fe, Ni and Co. They are composed of ferromagnetic nanoparticles dispersed in a non-magnetic matrix [6]. The routes for the synthesis of carbon nanoparticles have been vast, ranging from simple precipitation by alkaline hydrolysis at room temperature followed by reduction using external reducing agents to one step carbothermal reduction (reduction of oxides is caused due to gaseous intermediates) in order to gain more control over the shape and size of the magnetic nanoparticles with high stability and desired dispersion. The metal nanoparticle is introduced into the matrix by impregnation with an aqueous solution of the precursor salt, followed by reduction treatments (thermal or external reducing agent, *e.g.*, borohydride), which enable dispersion of metal nanoparticles on and within the matrix. Studies on the synthesis of metallic nanoparticles without solid support have reported certain disadvantages. Firstly, they have the tendency to undergo surface oxidation owing to the high surface reactivity accountable to their large surface area. Hence, they need to be stored always under passive conditions. Secondly, they are unstable at lower pH conditions due to the possibility of leaching out in the aqueous phase. Thirdly, due to the high surface energy, they have the tendency to agglomerate which reduces the number of reactive sites [7]. Such drawbacks of these highly efficient nano scaled particles are ruled out in the presence of a solid support like carbon. It not only prevents the oxidation of the metal nanoparticles to a certain extent but also provides stability and controls their nucleation, thereby, providing a narrow size distribution [8]. These heteroatoms impart surface functionalities to the carbon which may be broadly divided into acidic and basic functional groups. The acidic character of activated carbons is primarily attributed to surface oxygen groups such as carboxylic acid, lactone, phenol, and lastly groups while the basic character arises from other oxygen containing functional groups such as pyrones, quinones and nitrogen containing functional groups and other inorganic impurities. Apart from this the high density delocalized π electrons on the polyaromatic basal plane sheets which act as electron rich Lewis base sites for forming π-complexes with cations such as (H_3O^+- π) contribute to the basic nature of activated carbons. There are two major routes for the synthesis of carbon nanocomposites with metal nanoparticles, *i.e.*, the wet or solvent phase synthesis and the dry or calcination technique.

WET SYNTHESIS

The procedure adopted during wet synthesis is relatively simple, easily controllable and less time-consuming. It is also the most extensively adopted method for the synthesis of carbon metal nanocomposites. Zarei *et al.* [9]

prepared the carbon nanocomposite using the wet chemical precipitation method. The activated carbon was first made hydrophilic by acid treatment. Then, the modified carbon surface was impregnated with iron ($FeCl_3.6H_2O$ and $FeSO_4.7H_2O$) under alkaline condition (pH11) and stirred for 2 hours followed by drying in oven at 100°C. The iron was well dispersed over the carbon matrix under nanometric dimension and showed magnetic behavior. In another process, a hydrothermal route of synthesis was adopted [10]. Carbon nanotubes having COOH groups were ultrasonicated with hexadecyltrimethylammonium chloride in deionized water (DI). The mixture was magnetically stirred at 40°C for 20 min. After the reaction was complete, the product was washed with ethanol and DI. It was again dispersed in $CuCl_2$ solution and stirred at room temperature while adding ascorbic acid until complete dissolution. Thereafter, NaOH solution was added under vigorous stirring for 10 min. The mixed solution was autoclaved at 120 °C for 2 h and vacuum dried later for 24 h. Characterization results showed that ball shaped Cu was uniformly deposited on the surface of the carboxylated CNTs to give the Cu-CNT-COOH nanocomposites. Carbon nanofibres (CNFs) were also used to prepare Ag-Cu-CNF composites [11]. The CNFs were firstly acidified for functionalization and dispersed in a solution of $CuNO_3$, $AgNO_3$ and NaOH was then added dropwise for precipitation of the metal ions. Thereafter, the resultant product was dried and calcined at 300°C. Characterization results showed that the composites were spongy globular agglomerated mass where the needle shaped carbon nanofibre was decorated with Au and Cu nanoparticles within an average size of 30-50 nm. The bimetallic nanocomposites Ag-Cu/CNF showed superior adsorptive affinity compared to single metal CNF composite towards Rhodamine B dye and 2,4-D pesticide from water at adsorption capacity of 201 mg/g and 112 mg/g, respectively.

Rai and Singh [12] developed a magnetic nanocomposite out of PET waste and studied its application in the removal of pharmaceuticals from water. The schematic illustration of the steps involved in synthesis has been given in Fig. (**2**). The PET bottles and containers were initially charred and carbonized into activated carbon in N_2 and CO_2 flow. Thereafter, magnetic modification of the activated carbon was done following the classical wet impregnation method. In order to prevent agglomeration and aid better dispersion of the resultant adsorbent particles, the solvent polarity was reduced by adding ethanol (60 ml) as a cosolvent to the ferric chloride solution. The pre-weighed activated carbon was added to the mixture solution and stirred continuously for 2 h. This enabled the adsorption of iron onto the activated carbon. Then, alkaline solution of 3M NaOH was added to raise the pH to 8. At the this time, hydrolysis of Fe (III) took place and reddish brown precipitates of iron (III) hydroxide were seen. The optimum pH of precipitation was chosen on account of the pH_{zpc} (point of zero charge) of the activated carbon determined to be 7. With the addition of the alkali, electrostatic attraction between the positively charged ferric hydroxide sol and negatively charged functional groups on the carbon surface (dissociated carboxylate groups) took place. Also, the self hydrolysis of $NaBH_4$ which is undesired for effective reduction is also avoided by NaOH stabilization. Fe (III) was then reduced *in situ* by $NaBH_4$ (1M) to finally obtain black precipitates of iron oxides. The reduced black iron sol on carbon particles was filtered, washed thoroughly with ethanolic water, dried and stored until further use. Characterization studies revealed good dispersion of the magnetic nanoparticles of Fe_3O_4 having a high saturation magnetization value of 35 emu/g (Fig. **3**). The microporosity and surface area were reduced after the modification and Raman analysis revealed the graphitic nature of the PET-derived carbon.

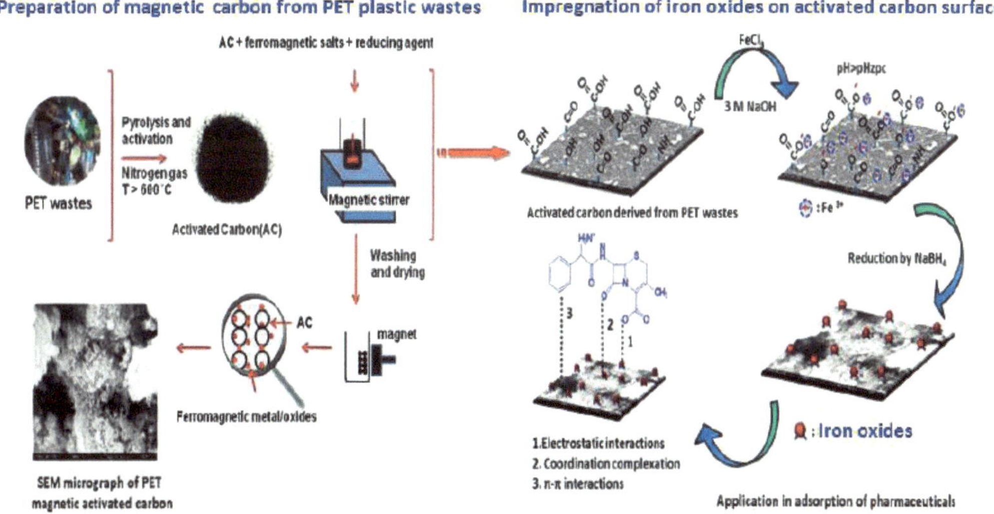

Fig. (2). Schematic illustration of the development of magnetic carbon nanocomposite prepared from PET wastes [12].

A solvothermal route of synthesis involves reduction reactions between $FeCl_3$ and ethylene glycol in closed reactor vessels under high temperatures and pressures. In one study [13], graphite oxide (GO) and CNTs were chosen as the matrix and were mixed with ethylene glycol in ultrasonicator bath for 2 h. Thereafter, the solution was filtered and washed thoroughly. The solution was prepared by adding $FeCl_3.6H_2O$ and NaAc into the GO-CNT mixture under stirring conditions. The entire mixture was autoclaved at 200°C for 6 h. The obtained product was dried under a vacuum. Characterization results revealed that the Fe_3O_4 NPs was successfully impregnated onto GO-CNT nanocomposite.

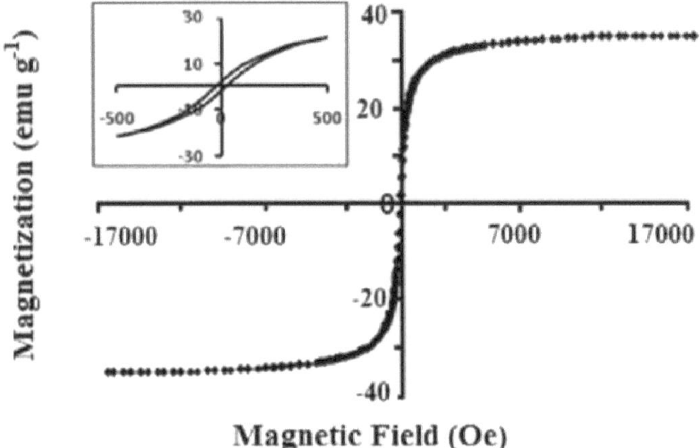

Fig. (3). VSM hysteresis curve of PET magnetic activated carbon composites [12]

In one more study [14], graphene oxide (GO) was used as the carbon source and CuO nanoparticles were dispersed on the surface. The GO was initially dispersed in methanol and Cu (acac)$_2$ was added to the mixture until evaporation of methanol under reflux conditions. The slurry was mixed well on a mortar pestle and calcinated at 350° C under an inert atmosphere. The obtained composite after characterization by TEM showed the CuO nanoparticles to be in the particle size of 12-35 nm. The Cu-C bonding was revealed to be a strong covalent bond through Raman and XPS studies.

A hydrothermal route of synthesis was reported [10] where a composite of functional CNTs and copper was prepared. In this method, COOH functionalized CNTs and hexadecyltrimethylammonium chloride were mixed in DI water and ultrasonicated for 1 h. The mixture was magnetically stirred at 40°C for 20 min and then the derived product was washed with ethanol and DI water after centrifugation. CuCl$_2$ was added to the above mixture and stirred at room temperature. This was followed by the ascorbic acid addition and after the ascorbic acid was completely dissolved, 0.5 mmol/L NaOH was rapidly added to the mixture dropwise. The reagents were all vigorously shaken and the mixture solution was autoclaved at 120° C for 2 h. The finally obtained composites were cooled, and washed with ethanol and DI water. The final products were dried in a vacuum for 24 h. The Cu nanocomposites were ball shaped and deposited over the surface of the CNT-COOH, with visible agglomeration.

DRY SYNTHESIS

The dry method for the synthesis of nanocomposites has been seen to offer more advantages compared to the wet chemical process. The foremost advantages of this method are the simplicity of the steps involved, ease of control of the parameters, better anchorage of the metal nanoparticles, *etc* . However, its limitation stems from the fact that large scale production is not possible using the dry route of synthesis [15]. A very simple method of preparation of activated carbon nanocomposites is the deposition of the carbon matrix into the metal solution and calcination at high temperatures under inert conditions. Sovizi *et al.* [16] took activated carbon, made hydrophillic with nitric acid and soaked it in $Fe(NO_3)_3 \cdot 9H_2O$ solution for 45 min under stirring conditions for complete adsorption. Thereafter, the mixture was filtered and calcined under an Argon atmosphere at 750°C for 3 h to complete the formation of magnetic nanoparticles of Fe_3O_4 inside the pores of the activated carbon. The magnetic iron oxide activated carbon composite was stable for several months under environmental conditions. TEM and VSM analysis showed that the iron oxides within the composite were 40-80 nm in size while the saturation magnetization value was 7.28 emu/g.

In another study [17], direct biomass (maize straw) was chosen as the suitable carbon precursor and immersed in a ferric chloride solution for impregnation for a contact time of 30 min. The biomass was filtered and dried for 24 h. The biomass was then carbonized in presence of nitrogen while the ferric chloride was converted into Fe_3O_4. The prepared adsorbent was applied in studying its sorptive behaviour towards aflatoxin and was characterized for the surface morphology, elemental analysis and particle size. It was found that the agglomerated Fe_3O_4 nanoparticles were somewhat cubical and the composite size ranged from 80-300 nm calculated from the XRD Debye-Scherrer crystallite size (Fig. **4**). Surface morphological characterization results showed aggregation of particles and Fe_3O_4 structures to be somewhat cubical. The surface area was 70.50 m^2/g while the total pore volume was 102.86 cm^3/g. Infrared spectroscopy results showed the presence of hydroxyl and absorbed water along with Fe-O str *etc* h vibrations. Thermogravimetric analysis showed that there was a loss in mass which could be attributed to the water of hydration of the nanoparticles.

A ruthenium decorated on graphene nanosheet composites was developed for application as nanocatalyst by a dry method [18]. The graphene nanosheets (GNs) were prepared from graphene nanoplatelets by an exfoliation method. Thereafter, the obtained GNs were functionalized (oxidized) in the presence of sulfuric and nitric acid. Once done, the functionalized GNs were mixed well with Ru(III)acetylacetonate and calcined at 300°C under argon atmosphere for a soaking time of 3 h. Characterization of the composite revealed ultrafine Ru nanoparticles (particle diameter below 3 nm) on the surface of the GNs.

Fig. (4). SEM images of magnetic carbon nanocomposites at different magnification [17].

APPLICATION IN THE FIELD OF ENVIRONMENTAL DECONTAMINATION

The use of nanocomposites can be seen to be maximum in case of their use as adsorbents for the decontamination of water. Zarei *et al.* [9] reported efficient adsorption of the highly carcoinogenic and mutagenic chemical, the unsymetrical dimethyl hydrazine, a dialkyl hydrazine, known as UDMH, used as rocket fuel, chemical industries for the synthesis of pharamceutiacls, dyes and other chemicals, by the prepared magnetic carbon-iron oxide nanocomposite. It was observed that a high removal efficiency of greater than 90% was achieved with an adsorbent dose of 20 mg for an initial UJDMH concentration of 10 mg/l and a contact time of 30 min. Magnetic nanocomposite prepared from iron oxides and activated carbon derived from scrap tires was used as an adsorbent in the removal of a dye, the reactive blue 19 (RB19) from an aqueous solution [19]. A monolayer sorption capacity of 119.05 mg/g was found for the adsorbent in experimental studies. Further regeneration studies in the presence of 0.1 M NaOH indicated that the removal efficiency decreased from 93.22% in the first cycle to 71.36% in the fifth cycle of its reuse, suggesting good regeneration potential. Rai and Singh [12] reported another waste derived carbon nanocomposite showing magnetic behavior and used it to study the adsorptive capacity of an antibiotic in water. They reported enhanced adsorption capacity of the composite compared with the parent carbon derived from waste PET bottles. An enhanced adsorptive capacity of 71.42 mg/g was reported for the prepared magnetic composite as compared to the parent carbon which showed reduced sorption capacity of 21.27 mg/g towards cephalexin, a beta-lactam antibiotic. The interactions between the antibiotic molecule and the developed adsorbents were elaborately comprehended from the spectra obtained by IR analysis of the adsorbents post adsorption. The FTIR results clearly demonstrated changes in the spectra of the spent adsorbents in the form of sorption induced shifts and the emergence of new characteristic peaks. The intensity of the broad band observed at 3396-3418 cm^{-1} was severely reduced after adsorption suggesting adsorption of the adsorbate molecules onto the adsorbents *via* displacement of the physisorbed water molecules. New peaks emerged at 1384 cm^{-1} and 1523 cm^{-1} which was attributed to the symmetric str *etc* hed vibrations of COO and bending vibration of NH groups in cepahlexin. The Fe-O band shifted from 613 cm^{-1} to 626 cm^{-1} in cepahlexin loaded adsorbent indicating an auxiliary role played by iron oxides in chemisorption of cephalexin by forming complexes with the reactive ligands.

In the field of analytical determination of organics in environmental samples, the sample preparation step seems to be very crucial. It involves a series of steps where the analyte needs to be extracted from the matrix followed by proper clean-up *i.e.*, along with enrichment there should be minimal loss as much as possible. This is generally done by solid phase extraction process where the sample is loaded onto the cartridge and suitable solvents are used for the elution. This process can be time consuming and costly in the case of commercial sorbents used in the cartridge. The novel carbon based composites have gained a popular alternative to the commercial sorbents in this case. The carbonaceous sorbents have good stability, ruggedness and also selectivity in the case of organic pollutants. A magnetic SPE composite material using CNTs, Graphene and Fe_3O_4 was also developed [13]. Due to intrinsic Van der Waals forces between the graphene layers, agglomeration of the graphene sheets takes place. The presence of CNTs was seen to reduce this effect. The composite was much efficient in adsorptive determination of organophosphorus pesticides from water and real wastewater compared to Fe_3O_4, G-Fe_3O_4 and CNT-Fe_3O_4 alone. The recoveries for the pesticides ranged from 75.6% to 102.5% in river water samples with a relative standard deviation between 3.2% and 8.4%. The limit of detection ranged from 1.4 to 10 pg/ml for the pesticides after extraction by the above method and determination by HPLC analysis. The technique of extracting analyte molecules from the liquid phase by stirring in the presence of the magnetic nanocomposite is known as a stir bar sorptive-dispersive microextraction. The magnetic nanocomposite is held strongly to the magnetic bar on account of the magnetic forces. Once stirring starts, the rotational speed surpasses magnetism and the composite particles are dispersed throughout the sample solution. As the stirring speed slows down, the magnetic composite particles attract again and settle on the magnetic stirrer bar. Later, desorption takes place and the analyte of interest is measured. This sample preparation technique is non-invasive, requires very less reaction time, and is convenient in handling. This technique was reported by Madej *et al.* [20], where they utilized magnetically modified graphene as the sorbent to extract a mixture of pesticides from spiked water. The composite gave good response and an average extraction yield was found to be 20-75% of the mixture of pesticides (boscalid, chlorpyrifos, deltamethrin, dimethenamid-P, dimoxystrobin, metazachlor and tebuconazol). The optimized extraction conditions were 20 mg of the prepared sorbent at sample pH of 6 for a contact time of 12 min in presence of 1 g NaCl and 4.5 ml acetone as the desorbing solvent. The stirbar sorptive technique was also found to be quick and convenient due to the non invasive technique of collection of the spent sorbents and also efficient in case of larger sample volumes.

Fuel cells are environmentally friendly appliances for producing electrical energy from chemical energy. They have high energy conversion efficiency and low emissions. The use of the carbon based materials in fuel cells has been seen to reduce the content of the metal catalyst such as platinum, which is expensive in itself, while retaining the catalytic activity. The carbon based support materials offers higher surface area, high electrical conductivity, strong mechanical properties and they include activated carbon, grapheme nanosheets, carbon nanotubes (CNT), and other such forms of carbon. The polymer electrolyte membrane fuel cell (PEMFC) which operates on the hydrogen fuel is seen as a promising candidate as a clean energy generation source. However, the commercialized application is limited due to the high cost of the platinum catalyst. Jha *et al.* [21] devised a single walled carbon nanotube based PEMFC loaded with Platinum at 0.06 mg/cm^2 against the value of 0.125 mg/cm^2 set as the US DOE 2017 technical target for platinum group metal loading. With reduced Pt load, it was still able to achieve good cell potential under optimized conditions.

The nanocomposites have found immense application potential in biomedical field for targeted drug delivery, repair and implantation, diagnosis of diseases such as in magnetic resonance imaging, immunoassay, *etc* . Amongst these, application in selective drug delivery seems to be quite challenging. Yang *et al.* [22] used magnetite functionalized multiwalled CNTs composite material in delivering chemotherapeutic agents into the lymph nodes using an external magnet. The magnetic CNTs slowly released the chemotherapeutic drugs into the targeted cells which helped in selective killing of the tumour cells. Magnetic CNTs were also used in cell tracking which contributed to comprehending regenerative medicine such as in haematopoiesis. Gul *et al.* [23] developed magnetic SWCNTs with Ni and Y at the tips of the composite structure. The CNTs were first oxidized and reacted with thionyl chloride. This was followed by conjugation with 2'-(ethylenedioxy)-bis-(ethylamine) which gave amino group functionalized CNTs. These amino groups were then reacted with fluorescein-isothiocyanate (FITC), a fluorescent dye to give the fluorescent labelled CNTs. For studying further, umbilical cord blood or peripheral blood leukapheresis CD34+ cells were isolated and seeded in poly-L-lysine culture dishes. The fluorescent labelled functionalized magnetic SWCNTs were put in contact with the cells at varying concentrations. High uptake efficiencies of 83, 90 and 100% were noted after a contact period of 6 h at the nanocomposite concentrations of 10, 20 and 40 µg/ml, respectively. The study suggested that the prepared nanosystem had no impairment on the cell viability and can well be used in other cellular systems too.

Biopolymers have also been actively researched in biomedical applications as the polymeric material used in orthopedic treatment needs to be biocompatible along with superior mechanical strength and stability. CNTs having high mechanical integrity, low density, high flexibility and high surface area make them ideal as incorporation fillers for biopolymers. At the same time, carbon is a natural constituent of the body. Taking advantage of these facts, a group of researchers synthesized modified multi-walled carbon nanotubes with hydroxyapatite nanorods reinforced into polypropylene to obtain biocomposites [24]. The nanocomposites were seen to be suitably fabricated with the melt processing approach. The characterization analysis showed good mechanical strength with y-ray diffraction pattern proving no structural change in polypropylene from the addition of the CNTs and hydroxyapatite at different compositions. The hybrid nanocomposites have been claimed to be a suitable cast as implants in various orthopedic applications.

CHALLENGES

Although there are many benefits arising from the application of nanocomposites during ease and improvement in the quality of life, there are certain challenges during the preparation and storage of nanocomposites. One of the major challenges faced in the preparation of the nanocomposite material is its stability. The metallic nanoparticles grafted on the surface of the carbon matrix might undergo oxidation and hence suffer reactivity issues under varying experimental conditions. In such cases, storage conditions need to be closely monitored and also mass production of the composite is limited. In addition to this, poor dispersion of the metallic nanoparticles has also been observed to take place. Such limitations may be overcome by using a suitable surfactant as the matrix material of choice.

The other challenges faced are the high cost of the commercial carbon source, commercial activated carbons, CNTs, graphene sheets, *etc* . which require considerable production costs. However, low cost alternatives may be adopted such as the carbon rich waste products such as scrap tires, municipal wastes, biomass, agricultural by-products, *etc* . Reusing these waste products will significantly lower high production costs and waive the logistic burden in waste handling. Lastly, the main challenge lies in implementation of the nanocomposite in real field applications. Designing such systems requires a lot of time and study in optimization of the controlling parameters.

CONCLUSION

Nanocomposites can be synthesized from a wide variety of raw materials. The raw materials range from simple activated carbon to 3D multi-walled carbon nanotubes. Due to their ease of modification, the impregnation of functional groups mostly follows a simple process of acid treatment. The functionalized carbon structure acts as a good base for the dispersion of magnetic metallic nanoparticles. The route for dispersion can be any suitable method of synthesis, such as the calcination dry method or the wet chemical impregnation method. The prepared nanocomposites upon characterization show promising results in terms of their stability, maneuverability, and application efficacy. They are versatile in their applications and pose to be a promising candidate in the field of environmental remediation. A lot of research has focused on the adsorptive capacity of nanocomposites. However, research needs to be extended in the field of real environmental samples.

CONSENT FOR PUBLICATION

Not applicable.

CONFLICT OF INTEREST

The author declares no conflict of interest, financial or otherwise.

ACKNOWLEDGEMENTS

Declared none.

REFERENCES

[1]   Camargo, P.H.C.; Satyanarayana, K.G.; Wypych, F. Nanocomposites: synthesis, structure, properties and new application opportunities. *Mater. Res.,* **2009**, *12*(1), 1-39.
[http://dx.doi.org/10.1590/S1516-14392009000100002]

[2]   Leslie-Pelecky, D.L.; Rieke, R.D. Magnetic properties of nanostructured materials. *Chem. Mater.,* **1996**, *8*(8), 1770-1783.
[http://dx.doi.org/10.1021/cm960077f]

[3]   Zhu, M.; Diao, G. Review on the progress in synthesis and application of magnetic carbon nanocomposites. *Nanoscale,* **2011**, *3*(7), 2748-2767.
[http://dx.doi.org/10.1039/c1nr10165j] [PMID: 21611651]

[4]   Bandosz, T.J.; Ania, C.O. *Surface chemistry of activated carbons and its characterization in Activated Carbon Surfaces in Environmental Remediation, Interface Science and Technology*; Bandosz, T.J., Ed.; Elsevier, **2006**, Vol. 7, pp. 159-229.
[http://dx.doi.org/10.1016/S1573-4285(06)80013-X]

[5]   Knappe, D.R.U. *Surface chemistry effects in activated carbon adsorption of industrial pollutants, Interface Science and Technology in Drinking Water Treatment*; Newcombe, G.; Dixon, D., Eds.; Elsevier, **2006**, Vol. 10, pp. 155-177.

[6]   Nayak, B.B. Magnetic nanocomposite materials. In: *Thesis, Indian Institute of Technology*; Bombay, India, **2006**.

[7]   Huber, D.L. Synthesis, properties, and applications of iron nanoparticles. *Small,* **2005**, *1*(5), 482-501.
[http://dx.doi.org/10.1002/smll.200500006] [PMID: 17193474]

[8]   Wang, Y.; Huang, Q.; Xian, Q.; Sun, C. Preparation of activated carbon fiber supported nanoscale Fe0 for simultaneous adsorption and dechlorination of chloroform in water. *Adv. Mat. Res.,* **2011**, *399-401*, 1386-1391.

[9]   Zarei, A.R.; Pedram, A.; Rezaeivahidian, H. Adsorption of 1,1-dimethylhydrazine (UDMH) from aqueous solution using magnetic carbon nanocomposite: kinetic and thermodynamic study. *Desalination Water Treat.,* **2016**, *57*(40), 18906-18914.
[http://dx.doi.org/10.1080/19443994.2015.1092891]

[10]   Liu, J.; Feng, Y.; Yin, J.; Jiao, T. Preparation of Functional CNT-COOH-Cu Nanocomposites Using Carbon Nanotubes and Application for Reduction of p-Nitrophenol. *Integr. Ferroelectr.,* **2020**, *208*(1), 97-103.
[http://dx.doi.org/10.1080/10584587.2020.1728721]

[11]   Saharan, P.; Kumar, V.; Mittal, J.; Sharma, V.; Sharma, A.K. Efficient ultrasonic assisted adsorption of organic pollutants employing bimetallic-carbon nanocomposites. *Sep. Sci. Technol.,* **2021**, *56*(17), 2895-2908.
[http://dx.doi.org/10.1080/01496395.2020.1866608]

[12]   Rai, P.; Singh, K.P. Valorization of Poly (ethylene) terephthalate (PET) wastes into magnetic carbon for adsorption of antibiotic from water: Characterization and application. *J. Environ. Manage.,* **2018**, *207*, 249-261.
[http://dx.doi.org/10.1016/j.jenvman.2017.11.047] [PMID: 29179114]

[13]   Razmi, H.; Jabbari, M. Development of graphene-carbon nanotube-coated magnetic nanocomposite as an efficient sorbent for HPLC determination of organophosphorus pesticides in environmental water samples. *Int. J. Environ. Anal. Chem.,* **2015**, *95*(14), 1353-1369.
[http://dx.doi.org/10.1080/03067319.2015.1090567]

[14]   M., Gopiraman; D., Deng; S.G., Babu; T., Hayashi; R., Karvembu; I.S., Kim Sustainable and versatile CuO/GNS nanocatalyst for highly efficient base free coupling reactions. *Sustainable Chem. Eng, 3*, 2478-2488.**2015**,

[15]   Gopiraman, M.; Kim, I.S. Carbon Nanocomposites: Preparation and Its Application in Catalytic Organic Transformations in Nanocomposites - Recent Evolutions. *Intechopen.,* **2018**, (November) [http://dx.doi.org/10.5772/intechopen.81109]

[16]   Sovizi, M.R.; Eskandarpour, M.; Afshari, M. Synthesis, characterization, and application of magnetic-activated carbon nanocomposite (m-Fe_3O_4 @ACCs) as a new low-cost magnetic adsorbent for removal of Pb(II) from industrial wastewaters. *Desalination Water Treat.,* **2016**, *57*(59), 28887-28899. [http://dx.doi.org/10.1080/19443994.2016.1193062]

[17]   Zahoor, M.; Ali Khan, F. Aflatoxin B1 detoxification by magnetic carbon nanostructures prepared from maize straw. *Desalination Water Treat.,* **2016**, *57*(25), 11893-11903. [http://dx.doi.org/10.1080/19443994.2015.1046147]

[18]   Gopiraman, M.; Ganesh Babu, S.; Khatri, Z.; Kai, W.; Kim, Y.A.; Endo, M.; Karvembu, R.; Kim, I.S. Dry Synthesis of Easily Tunable Nano Ruthenium Supported on Graphene: Novel Nanocatalysts for Aerial Oxidation of Alcohols and Transfer Hydrogenation of Ketones. *J. Phys. Chem. C,* **2013**, *117*(45), 23582-23596. [http://dx.doi.org/10.1021/jp402978q]

[19]   Abdollahzadeh, H.; Fazlzadeh, M.; Afshin, S.; Arfaeinia, H.; Feizizadeh, A.; Poureshgh, Y.; Rashtbari, Y. Efficiency of activated carbon prepared from scrap tires magnetized by Fe_3O_4 nanoparticles: characterisation and its application for removal of reactive blue19 from aquatic solutions. *Int. J. Environ. Anal. Chem.,* **2020**, (April), 1-15. [http://dx.doi.org/10.1080/03067319.2020.1745199]

[20]   Madej, K.; Jonda, A.; Borcuch, A.; Piekoszewski, W.; Chmielarz, L.; Gil, B. A novel stir bar sorptive-dispersive microextraction in combination with magnetically modified graphene for isolation of seven pesticides from water samples. *Microchem. J.,* **2019**, *147*, 962-971. [http://dx.doi.org/10.1016/j.microc.2019.04.002]

[21]   Jha, N.; Ramesh, P.; Bekyarova, E.; Tian, X.; Wang, F.; Itkis, M.E.; Haddon, R.C. Functionalized single-walled carbon nanotube-based fuel cell benchmarked against US DOE 2017 technical targets. *Sci. Rep.,* **2013**, *3*(1), 2257. [http://dx.doi.org/10.1038/srep02257] [PMID: 23877112]

[22]   Yang, F.; Fu, D.L.; Long, J.; Ni, Q.X. Magnetic lymphatic targeting drug delivery system using carbon nanotubes. *Med. Hypotheses,* **2008**, *70*(4), 765-767. [http://dx.doi.org/10.1016/j.mehy.2007.07.045] [PMID: 17910909]

[23]   Gul, H.; Lu, W.; Xu, P.; Xing, J.; Chen, J. Magnetic carbon nanotube labelling for haematopoietic stem/progenitor cell tracking. *Nanotechnology,* **2010**, *21*(15), 155101. [http://dx.doi.org/10.1088/0957-4484/21/15/155101] [PMID: 20299726]

[24]   Khan, F.S.A.; Mubarak, N.M.; Khalid, M.; Walvekar, R.; Abdullah, E.C.; Ahmad, A.; Karri, R.R.; Pakalapati, H. Functionalized multi-walled carbon nanotubes and hydroxyapatite nanorods reinforced with polypropylene for biomedical application. *Sci. Rep.,* **2021**, *11*(1), 843. [http://dx.doi.org/10.1038/s41598-020-80767-3] [PMID: 33437011]

CHAPTER 6

The Excess Refractive Indices of Some Organic and Inorganic Components

Dilbar T. Bozorova[1], Shukur P. Gofurov[2,5], Abdulmutallib M. Kokhkharov[1], Mavlonbek A. Ziyayev[3], Feruza T. Umarova[1] and Oksana B. Ismailova[1,4,5,*]

[1] *Institute of Ion-plasma and laser technologies, Tashkent, Uzbekistan*

[2] *Tsukuba University, Ibaraki-shi, Japan*

[3] *Institute of Bioorganic Chemistry, Tashkent, Uzbekistan*

[4] *Turin Polytechnic University in Tashkent, Uzbekistan*

[5] *Uzbekistan-Japan Innovation Center of Youth, Tashkent, Uzbekistan*

Abstract: To reveal the concentration-dependent optical properties of aqueous, ethanol and toluene binary solutions, the refractometry method was used. The peak of refractive indices corresponds to formed heteromolecular complexes with a certain concentration. The bonds between the molecules formed in mixtures are also directly related to their chemical structure. The bonds that form between polar and non-polar compounds and also between polar protic and polar aprotic compounds are different. All solutions were measured in the concentration range $\sim0\div1$ mole fraction at room temperature. In our work, we have shown that, along with Infrared and Raman spectroscopy, the refractometric method is effective for the determination of formed structures in mixtures.

Keywords: Binary solutions, Dipole-dipole interaction, Excess refractive indices, Hydrogen bonds, Intermolecular interactions, Organic and inorganic mixtures, The refractometric method.

INTRODUCTION

Wide usage of mixed solvents in various nanotechnological processes necessitates establishing the relationship of physical and chemical properties of individual components with the structure and composition of their binary and multicomponent systems. The application of mixed organic systems is due to the possibility of a different variation of their physicochemical features in multicomponent solutions to create a system with predetermined characteristics. It

* **Corresponding author Ismailova Oksana Bakhtiyarovna:** Uzbekistan-Japan Innovation Center of Youth, 100095, 2B Universtiteskaya, Uzbekistan; Tel: (99871) 203 00 22; E-mail: oismailova56@gmail.com.

is known that in polar solvents and their solutions, reactions are accompanied by complex formation and associations of molecules due to the presence of a dipole moment in molecules characterized by a complex set of intermolecular interactions. To establish the formed structures of the associates in solvents and in their binary solutions, as well as to study the role of hydrogen bonds in the formation of various types of associates, the concentration features of the optical characteristics of solutions were studied by the refractometry method. Refractometry is one of the methods that allows to obtaining information about the nature of the interaction between the components and the structure of characteristics of the system. In this chapter, we present experimental measurements of refractive indices of some organic and inorganic mixtures.

Experimental Setup

All pure components presented in this work and their mixture of water, toluene and ethanol were measured using high-sensitivity digital refractometer PAL-BX/RI (ATAGO, Japan). To analyze the refractive index, mixtures were carried out five times over the entire concentration range at a fixed temperature of 25 ± 0.05 °C. The average value has been chosen. To calculate the excess refractive index n^E , the measured refractive indices were used in equation 1 [1]:

$$n^E = n_{solutions} - (xn_1 + (1 - x)n_2) \qquad (1)$$

Refractive Indices of Pure Component

The refractive index is used to determine optical density of the material and calculated excess refractive index can show change in the molecular structure of mixed liquids, and to study physicochemical properties of the solutions. An advantage of the refractometric method quickly and accurately analyze and orient whether the organic/inorganic liquid under study contains impure compounds and whether the liquids obtained are pure for the experiment. Organic liquids are used in many such fields, including the synthesis and optoelectronic properties of sulvanite nanocrystals [2, 3]. Refractive indices of several organic and inorganic compounds used in our experiments are illustreted further in Table **1** to compare with our results.

Table 1. Experimental refractive indices (*n*) values of several pure liquids at the temperature 298 K.

S.NO	Liquid	Molecular Formula	Refractive Index, *n*), at T=298K	Literature
1.	Dimethylformamide	C_3H_7NO	1.4279	[4]
2.	Water	H_2O	1.3610	[4]

(Table 1) cont.....

S.NO	Liquid	Molecular Formula	Refractive Index, *n*), at T=298K	Literature
3.	Ethanol	C_2H_6O	1.3324	[4]
4.	Tetrahydrofuran	C_4H_8O	1.4045	[4]
5.	Acetic acid	$C_2H_4O_2$	1.3699	[5]
6.	Toluene	C_7H_8	1.4908	[7]
9.	Acetonitrile	CH_3CN	1.3402	[8]
10.	Cyclohexane	C_6H_{12}	1.4237	[9]
11.	Benzene	C_6H_6	1.4898	[10]
12.	Hexane	C_6H_{14}	1.3437	[11]

Refractive Indicies of Some Binary Solutions

Aqueous Solutions

Refractive indices of freshly prepared water solutions of ethanol, acetic acid, tetrahydrofuran, and dimethylformamide (DMF) are measured and plots of the changes of the absolute values of the excess refractive index of this solutions are shown in Fig. (**1**) . It can be seen that maximum values of refractive indices are achieved at component concentrations of *C* ~0.2÷0.3 mole fraction. A non-polar hydrophobic part and a polar hydrophilic group of ethanol and acetic acid molecules participate in interactions with water molecules.

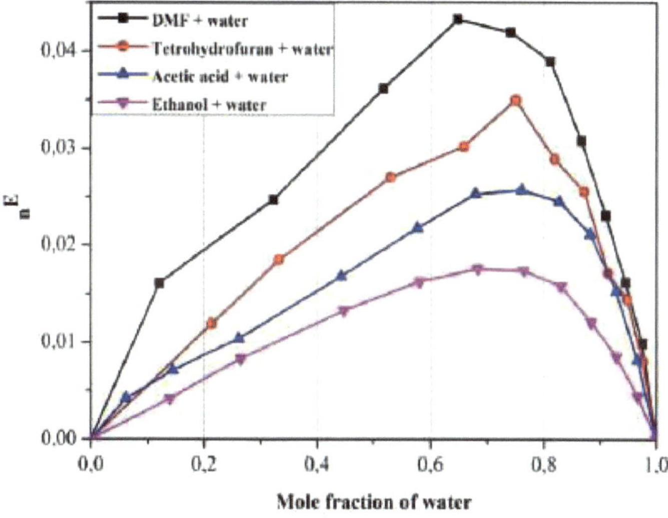

Fig. (1). Changes of the absolute values of excess refractive index of aqueous solutions of ethanol (pink triangle), acetic acid (blue triangle), tetrahydrofuran (red circle) and DMF (black square).

In this state, the maximum number of hydrogen bonds are realized in the systems, which correspond to the heteromolecular structure formation with H-bond in binary mixtures, and with further increasing of ethanol concentration, the hydrogen bonds are broken [11] (see Fig. **2**).

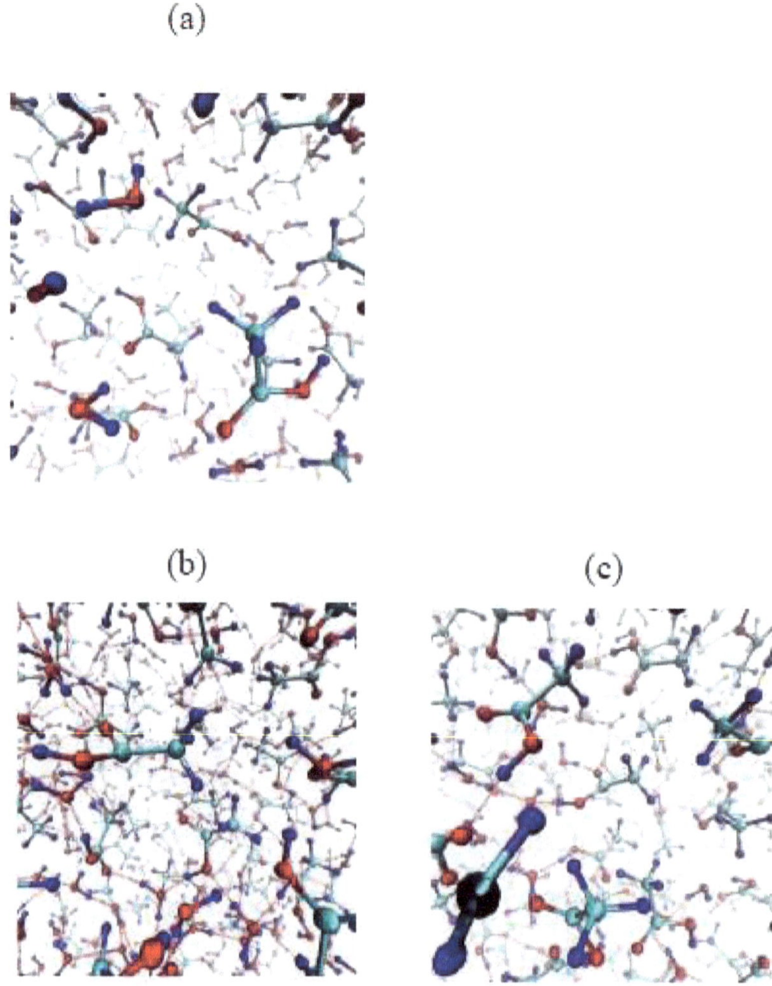

Fig. (2). Molecular dynamic simulation image of several molecules of water and acetic acid: due to the influence of water, hydrophobic CH3 groups are oriented opposite to each other at a low concentration of acetic acid **(a)**. The maximum number of hydrogen bonds is characteristic for formation of heteromolecular structures between water and acetic acid molecules at the concentration of 0.3 mole fraction in solution **(b)**. Acetic acid molecules form parallel structures at a concentration of 0.8 mole fraction **(c)**.

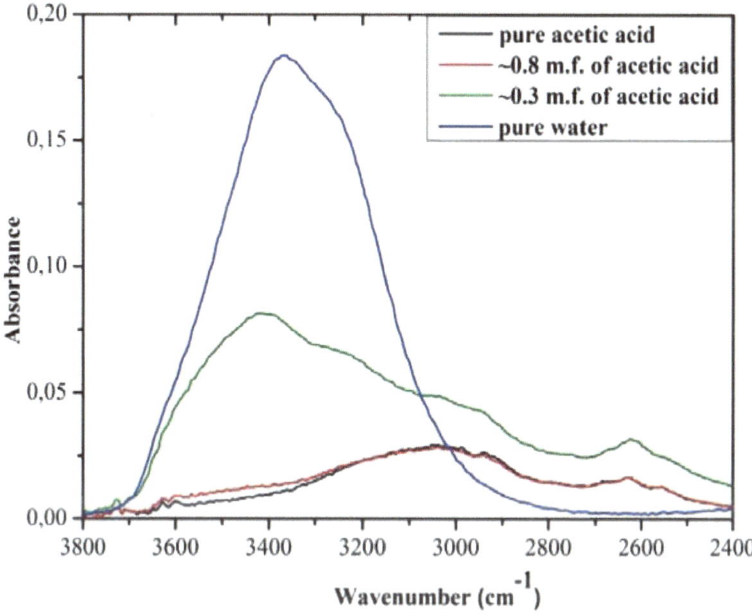

Fig. (3). FTIR spectra of pure components and their mixtures at concentrations of ~0.3 and ~0.8 mole fraction in wavelength range (2400-3800 cm $^{-1}$) [5].

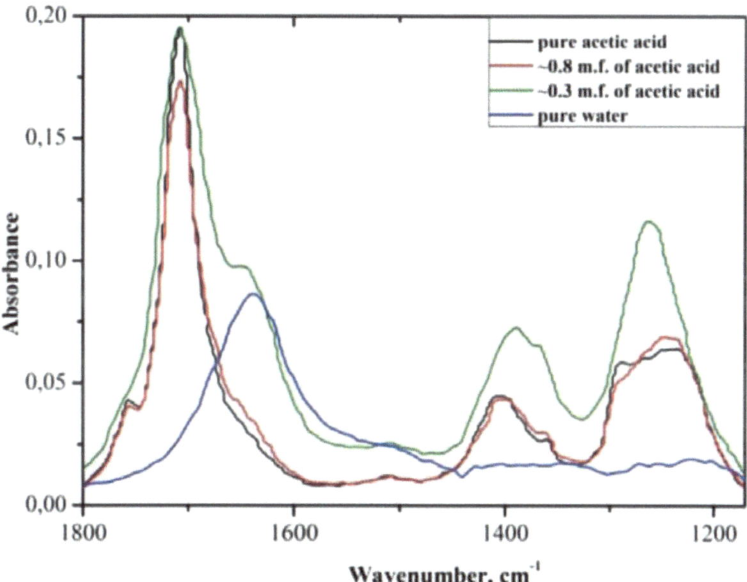

Fig. (4). FTIR fingerprint of pure components and their mixtures at concentrations of ~0.3 and ~0.8 mole fraction in wavelength range (1180-1800 cm $^{-1}$) [5].

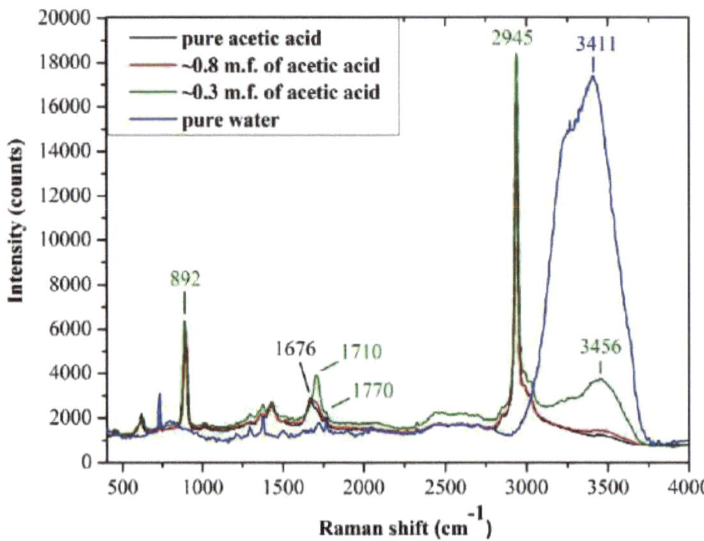

Fig. (5). Raman spectra of pure components and their mixtures at concentrations of ~0.3 and ~0.8 mole fraction in wavelength range (400-4000 cm $^{-1}$) [5].

This fact indicates that at certain concentrations, there is an equilibrium state of intermolecular interactions in the fluid [12 - 15]. All graphs are parabolic and slightly elongated at high concentrations. Although the deviation of parabolic form occurs at a concentration of ~0.8 m.f. for an aqueous solution of acetic acid due to the parallel alignment of the formed structure of acetic acid molecules. Information on structural rearrangments in liquids obtained by refractometry of acetic acid and water correlates with IR and Raman spectroscopy of solutions (see Figs. **3 - 5**). IR spectra of pure water and acetic acid, as well as aqueous solutions of acetic acid has concentrations of ~ 0.3 and ~ 0.8 mole fraction. Peaks with maxima at 3050 cm^{-1} and 2934 cm^{-1} correspond to CH_3 and OH– vibrations for pure acetic acid and its aqueous solutions. The dipole-dipole interaction between acetic acid molecules at a concentration of ~ 0.8 mole fraction is predominant, characterized by the appearance of a maximum at 1379 cm^{-1} (see Fig. **4**)). According to Raman spectra, dipole-dipole interactions between carboxylic groups of acetic acid and OH parts of water molecules appear around ~3450 cm^{-1} at a concentration of ~0.3 mole fraction of acetic acid.

Toluene Solutions

The measured values refractive indices of toluene solutions of tetrahydrofuran, DMF, ethanol, and acetic acid are given in Table **3**. Fig. (**6**) shows plots of data on excess refractive indices of freshly prepared toluene solutions of tetrahydrofuran,

DMF, ethanol and acetic acid over the entire range of solute concentrations. According to experimental data, the maximum value of the excess refractive index is observed at ~0.5 mole of component for toluene solutions, characterizing the heteromolecular structure formation between molecules of toluene and tetrahydrofuran, ethanol, and acetic acid, due to dipole-dipole interaction [7 - 9]. Thus, the maximum value of the refractive index was observed at ~0.4÷0.5 mole fraction toluene in toluene-DMF mixtures.

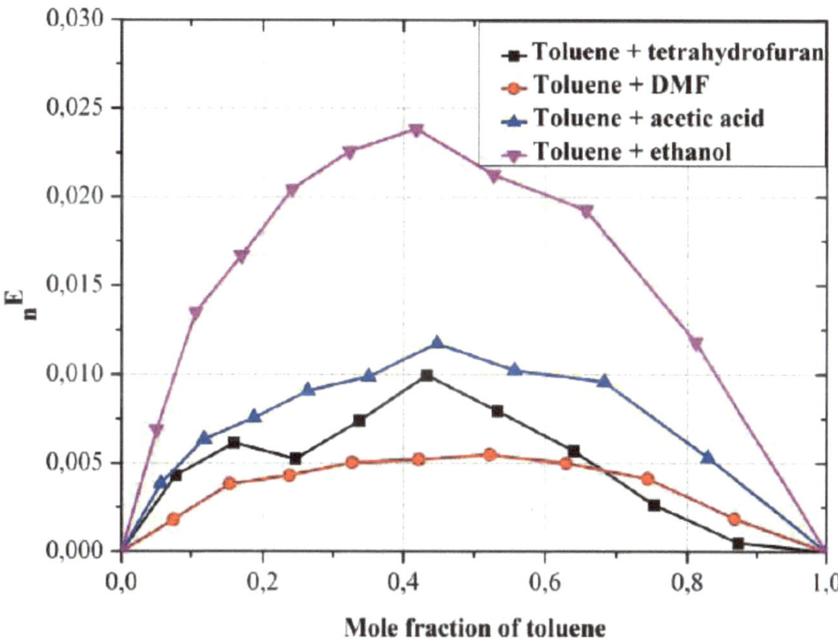

Fig. (6). Changes of the absolute values of excess refractive index of toluene solutions of ethanol (pink triangle) acetic acid (blue triangle), tetrahydrofuran (black square) and DMF (red circle).

Strong dipole-dipole interactions between toluene and tetrahydrofuran are stimulated at ~0.4 mole fraction of toluene and weak interactions are detected at ~0.4 mole fraction toluene. When comparing the excess refractive index of various mixtures in this study, the lowest maximum value of excess refractive index is observed at a ~0.5 mole in the toluene+DMF system. Orientation of molecules of one component in the interstitial positions of the structural network of molecules of the other component. Maximum value n^E of toluene+DMF is lower than that of other mixtures of weak donor-acceptor-type, and a specific interaction occurs in toluene+DMF mixtures. The value of the excess refractive index decreases with a further increase in the mole fraction of toluene, shown in Fig. (**6**).

Ethanol Solutions

Fig. (**7**) illustrates plots of the excess refractive index of ethanol solutions of water, DMF, toluene and cyclohexane over the entire range of component concentrations. The measured refractive indices are given in Tables **2**, **3**, and **4**. It has been shown that heteromolecular complexes in binary solutions of DMF-ethanol, and cyclohexane-ethanol are detected at the concentration of ~0.5 mole fraction of those compounds due to strong H-bonds. Relatively weak interactions are determined at certain concentrations: ~0.2 and ~0.9 mole fraction of DMF and ~0.2 and ~0.8 mole fraction of cyclohexane [9]. At the concentration of ethanol of 0.4÷0.5 mole fraction the dipole-dipole type of interactions occurs in all alcohols, as a result, dipole-dipole interactions are also manifested in toluene-ethanol mixtures. According to Fig. (**4**), dipole-dipole interactions of toluene-ethanol are stronger than those of other mixtures [7]. The maximum value of excess refractive indices is recorded at C ~0.3÷0.4 mole fraction of solute concentration in water-ethanol solution, respectively [9].

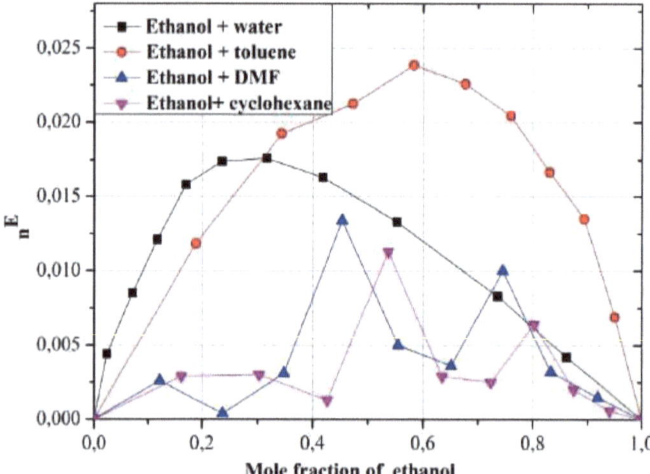

Fig. (7). Changes in absolute values of excess refractive index of ethanol solutions of water (black square), cyclohexane (pink triangle), toluene (red circle) and DMF (blue triangle).

Table 2. Experimental mole fraction x, refractive index n, excess refractive index n^E and excess refractive index for tetrahydrofuran-water, DMF-water, ethanol-water and acetic acid-water binary solutions 298 ± 0.05 K [4 - 6].

x, Tetrahydrofuran	n	n^E	x, DMF	n	n^E
1.00000	1.4045	0.0000	1.00000	1.4279	0.0000
0.80830	1.4014	0.0119	0.8080	1.4249	0.0161
0.66641	1.3990	0.0185	0.67673	1,4218	0.0247

(Table 2) cont.....

x, Tetrahydrofuran	n	n^E	x, DMF	n	n^E
0.47030	1.3933	0.0270	0.48198	1.4147	0.0362
0.34120	1.3872	0.0302	0.35180	1.4093	0.0433
0.24978	1.3854	0.0350	0.25866	1.3991	0.0420
0.18164	1.3744	0.0289	0.18871	1.3895	0.0390
0.12890	1.3673	0.0256	0.13425	1.3761	0.0308
0.08686	1.3558	0.0172	0.09065	1.3642	0.0231
0.05257	1.3506	0.0145	0.05496	1.3539	0.0162
0.02407	1.3421	0.0080	0.02519	1.3448	0.0099
0.00000	1.3324	0.0000	0.00000	1.3324	0.0000
x, ethanol	n	n^E	x, acetic acid	n	n^E
1.000000	1.3610	0.0000	0.000000	1.3325	0.00000
0.861300	1.3612	0.0042	0.33376	1.3419	0.00814
0.735059	1.3618	0.0083	0.07289	1.3505	0.01527
0.552188	1.3615	0.0133	0.11878	1.3581	0.02116
0.418366	1.3607	0.0163	0.17333	1.3635	0.02452
0.316194	1.3591	0.0176	0.23925	1.3672	0.02575
0.235631	1.3566	0.0174	0.32054	1.3698	0.02531
0.170477	1.3531	0.0158	0.42324	1.3701	0.02177
0.116698	1.3479	0.0121	0.55713	1.3702	0.01680
0.071553	1.3430	0.0085	0.73894	1.3705	0.01036
0.02407	1.3421	0,0080	0.85664	1.3716	0.00706
0.00000	1.3324	0.0000	0.93906	1.3718	0.00418
-	-	-	1.00000	1.3699	0.00000

Experimental mole fraction x, refractive index n excess refractive index n^E and excess refractive index for tetrahydrofuran-toluene, DMF-toluene, ethanol-toluene and acetic acid-toluene binary solutions 298 ± 0.05 K [7].

x, Acetic Acid	n	n^E	n^E, Tetrahydrofuran	x	n^E
0.000000000	1.4908	0.00000	0.000000000	1.4908	0.00000
0.171039286	1.4750	0.00532	0.127110106	1.4801	0.00050
0.317052791	1.4612	0.00956	0.246786427	1.4716	0.00262
0.443158592	1.4463	0.01023	0.359662562	1.4647	0.00569
0.553168226	1.4342	0.01172	0.466302093	1.4575	0.00792
0.649978587	1.4204	0.00987	0.567208001	1.4506	0.00994
0.735830762	1.4090	0.00908	0.662830594	1.4396	0.00739

(Table 3) cont.....

x, Acetic Acid	n	n^E	n^E, Tetrahydrofuran	x	n^E
0.812485661	1.3980	0.00754	0.753574227	1.4294	0.00522
0.881346090	1.3883	0.00635	0.839803023	1.4227	0.00614
0.943543374	1.3781	0.00383	0.921845746	1.4136	0.00429
1.000000000	1.3673	0.000000	1.000000000	1.4024	0.00000
x, DMF	n	n^E	x, ethanol	n	n^E
0.000000000	1.4908	0.00000	0.000000000	1.4908	0.00000
0.132397941	1.4838	0.00186	0.188721092	1.4775	0.01182
0.255595045	1.4778	0.00410	0.343573095	1.4643	0.01923
0.370518208	1.4710	0.00499	0.472922776	1.4491	0.02125
0.477973867	1.4643	0.00548	0.582590745	1.4371	0.02384
0.578666959	1.4573	0.00521	0.676751529	1.4233	0.02258
0.673216396	1.4508	0.00504	0.758477043	1.4103	0.02045
0.762167804	1.4441	0.00429	0.830078084	1.3970	0.01668
0.846004055	1.4380	0.00380	0.893326299	1.3854	0.01350
0.925154038	1.4307	0.00179	0.949602767	1.3713	0.00689
1.000000000	1.4239	0.00000	1.000000000	1.3577	0.00000

Table 4. Experimental values of mole fraction x, refractive index n, for ethanol-DMF and ethanol-cyclohexane solutions 298$\pm$0.05 K [9].

x, DMF	n	n^E	x, cyclohexane	n	n^E
0.0000	1.3610	0.0000	0.0000	· 1.3604	0.0000
0.0819	1.3680	0.0015	0.0600	1.3648	0.0006
0.1672	1.3753	0.0032	0.1258	1.3704	0.0020
0.2561	1.3822	0.0100	0.1978	1.3756	0.0064
0.3487	1.3878	0.0036	0.2773	1.3805	0.0025
0.4454	1.3957	0.0050	0.3653	1.3864	0.0029
0.5464	1.3983	0.0134	0.4633	1.3922	0.0113
0.6521	1.4075	0.0031	0.5731	1.3980	0.0013
0.7626	1.4122	0.0004	0.6971	1.4075	0.0030
0.8784	1.4221	0.0026	0.8382	1.4164	0.0029
1.0000	1.4276	0.0000	1.0000	1.4237	0.0000

CONCLUDING REMARKS

In this research, the concentration dependence of different compounds and their mixtures on the formation of hydrogen bonds in solutions was investigated. The structure of the mixtures differs significantly from the structure of pure components, accompanying the local rearrangement of the formed complexes and associates. This is indicated by the fact that the excess refractive index n^E is maximum at some concentration values. Refractometry is one of the most informative methods for studying structures formed in mixtures of the solutions. Based on our experimental studies and analysis, it can be concluded that the heteromolecular structures are formed due to hydrogen bonds or/and Van der Waals interaction. IR and Raman spectra of binary solutions of water and acetic acid confirm structural analysis of their molecular and intermolecular vibrations. Comparison of the data obtained by using FT-IR and Raman spectroscopy confirms that the refractometry method is sensitive to determine the features of structural rearrangements of molecules in mixtures.

CONSENT FOR PUBLICATION

Not applicable.

CONFLICT OF INTEREST

The author declares no conflict of interest, financial or otherwise.

ACKNOWLEDGEMENT

This study has been supported by grant MIRAI, UZB-IND-83 and by a fundamental research of the Uzbekistan Academy of Sciences of the Republic of Uzbekistan.

REFERENCES

[1]   Chen, C.C.; Stone, K.H.; Lai, C.Y.; Dobson, K.D.; Radu, D. Sulvanite (Cu_3VS_4) nanocrystals for printable thin film photovoltaics. *Mater. Lett.,* **2018**, *211*(211), 179-182.
[http://dx.doi.org/10.1016/j.matlet.2017.09.063]

[2]   Liu, M.; Lai, C.Y.; Zhang, M.; Radu, D.R. Cascade synthesis and optoelectronic applications of intermediate bandgap Cu_3VSe_4 nanosheets. *Sci. Rep.,* **2020**, *10*(1), 21679.
[http://dx.doi.org/10.1038/s41598-020-78649-9] [PMID: 33303797]

[3]   Samoc, A. Dispersion of refractive properties of solvents: Chloroform, toluene, benzene, and carbon disulfide in ultraviolet, visible, and near-infrared. *J. Appl. Phys.,* **2003**, *94*(9), 6167-6174.
[http://dx.doi.org/10.1063/1.1615294]

[4]   Gofurov, Sh.; Ismailova, O.; Makhmanov, U.; Kokhkharov, A. Heteromolecular structure formation in aqueous solutions of ethanol, tetrahydrofuran and dimethylformamide. *International Journal of Chemical, Molecular, Nuclear. Materials and Metallurgical Engineering,* **2017**, *11*(4), 300-304.

[5]   Gofurov, S.; Makhmanov, U.; Kokhkharov, A.; Ismailova, O.B. Structural and optical characteristics

of aqueous solutions of acetic acid. *Appl. Spectrosc.,* **2019**, *73*(5), 503-510.
[http://dx.doi.org/10.1177/0003702819831325] [PMID: 30700097]

[6]   Bozorova, D.T.; Gofurov, S.P.; Kokhkharov, A.M.; Ismailova, O.B. Terahertz spectroscopy of aqueous solutions of acetic acid. *J. Appl. Spectrosc.,* **2021**, *88*(4), 719-722.
[http://dx.doi.org/10.1007/s10812-021-01230-3]

[7]   Bozorova, D.T.; Gofurov, Sh.P.; Kokhkharov, A.M.; Ismailova, O.B. Study of heteromolecular structures in toluene solutions of tetrahydrofuran, dimethylformamide, ethanol and acetic acid by refractometric method. **2020.**

[8]   Bakhramov, S.A.; Kokhkharov, A.M.; Gofurov, Sh.P.; Makhmanov, U.K.; Ismailova, O.B. Optical features of tetrahydrofuran+acetonitrile systems (in russian). *Reports of Academy Sciences of the Republic of Uzbekistan,* **2019**, *1*, 19-23.

[9]   Bozorova, D.T.; Gofurov, Sh.P.; Kokhkharov, A.M.; Ismailova, O.B. Study of heteromolecular interaction of nanaqueous binary solutions. *Uzb. J. Phys,* **2020**, *22*(2), 111-114.

[10]   Bozorova, D.; Gofurov, Sh.; Kokhkharov, A.M.; Ismailova, O.B. Study excess refractive indecies of benzene + ethanol binary systems. *Senhri Journal of Multidisciplinary Studies,* **2020**, *5*, 19-23.
[http://dx.doi.org/10.36110/sjms.2020.05.01.002]

[11]   Plastinin, I.V.; Burikov, S.A.; Gofurov, S.P.; Ismailova, O.B.; Mirgorod, Y.A.; Dolenko, T.A. Features of self-organization of sodium dodecyl sulfate in water-ethanol solutions: Theory and vibrational spectroscopy. *J. Mol. Liq.,* **2020**, *298*(112053), 112053.
[http://dx.doi.org/10.1016/j.molliq.2019.112053]

[12]   Antipova, M.L.; Petrenko, V.E.; Odintsova, E.G.; Bogdan, T.V. Study of solvation of substituted propylbenzene in ethanol-water solutions under subcritical conditions by molecular dynamics. *J. Supercrit. Fluids,* **2020**, *155*(104649), 104649.
[http://dx.doi.org/10.1016/j.supflu.2019.104649]

[13]   Egorov, G.I.; Makarov, D.M. Densities and thermal expansions of (water + tetrahydrofuran) mixtures within the temperature range from (274.15 to 333.15) K at atmospheric pressure. *J. Mol. Liq.,* **2020**, *310*(113105), 113105.
[http://dx.doi.org/10.1016/j.molliq.2020.113105]

[14]   Belandria, V.; Pimentel-Rodas, A.; Mohammadi, A.H.; Galicia-Luna, L.A.; Richon, D. Clarification of the volumetric properties of the (tetrahydrofuran + water) systems. *J. Chem. Thermodyn.,* **2009**, *41*, 1386-82.

[15]   Jia, X.Q.; Li, Y.; Zhang, C.X.; Gao, Y.C.; Wu, Y. Supramolecular clusters clarification in ethanol-water mixture by using fluorescence spectroscopy and 2D correlation analysis. *J. Mol. Struct.,* **2020**, *1219*(128569), 128569.
[http://dx.doi.org/10.1016/j.molstruc.2020.128569]

CHAPTER 7

The Fundamental Concepts of Nanotechnology-Based Nanomaterials: Recent Developments and Challenges of Metal Oxide Nanoparticles

Asma Almontasser[1] and **Azra Parveen**[1,*]

[1] *Department of Applied Physics, Z. H. College of Engineering & Technology, Aligarh Muslim University, Aligarh-202002, India*

Abstract: Nanotechnology is one of the most promising new technologies in the recent decade, which involves structures, devices, and systems with novel properties and functions due to their atomic arrangement on scales of 1–100 nm. In general, the materials at the nanoscale have unique physical, mechanical, chemical, and biological properties compared to their bulk materials. Thus, these properties lead to a new novelty and variety of technological applications in the different fields (physics, science and engineering, electrical and computer science, materials, chemistry, biology, and medicine). This chapter provides a brief introduction of nanotechnology, nanomaterials, and their applications while discussing structural, optical, and magnetic properties as well as antibacterial activities. This chapter overviews the current research being carried out on the properties and application of metal oxide nanoparticles, especially chromium oxide (Cr_2O_3) nanoparticles, indium oxide nanoparticles (In_2O_3), and magnesium oxide (MgO) nanoparticles. These materials are considered novel materials for biological and smart applications, such as antimicrobial, drug delivery systems, and cancer therapy. Multi-drug and antibiotic resistance are among the great challenges that confront researchers in designing and developing efficient antimicrobial and biomedical agents. Inspired by the remarkable developments in nanoparticles in recent times, antibacterial metal oxide nanoparticles have been discovered as potential antibiotics to restrict infectious diseases. Thus, the mechanism of anti-microbial activities of metal oxide nanoparticles is discussed here in detail. In this chapter, a brief literature survey related to the present study is also performed. By the end of the chapter, the concluding remarks with views on the recent progress and future challenges of metal oxide nanoparticles are appended.

Keywords: Antimicrobial activities, Chromium oxide, Indium oxide, Metal oxide nanoparticles, Magnesium oxide nanoparticles, Magnetic properties, Nanotechnology, Optical properties.

---

* **Corresponding author Azra Parveen:** Department of Applied Physics, Z. H. College of Engineering & Technology, Aligarh Muslim University, Aligarh-202002, India; Tel: 0091 9411653414; E-mail: azrap2015@gmail.com

Dibya Prakash Rai (Ed.)

INTRODUCTION TO NANOSCIENCE AND NANOTECHNOLOGY

Nanoscience and nanotechnology are progressive studies to create and use structures in the form of very advantageous functional materials, devices, and systems possessing novel properties and functions within the dimensions of less than 100 nanometers (one billionth of a meter) length scale [1]. Nanotechnology is the capability to study and manipulate the behavior and properties of the materials at atomic, molecular, and supramolecular levels and organize them systematically. It implies the creation and utilization of different systems as chemical, physical, biological, and others with structural characteristics between molecules or single atoms to submicron size particles, encompassing consequent nanosize structures to larger units [2, 3]. Nanoscale's uprising in technology and different fields endures many reasons for being significant marvelously, some of which have been listed below [4].

1. Nanosystems possess a high-density texture, in addition to its enhanced conducting behavior of electricity which may create new devices of electronics with advanced functions, developed circuits (smaller and faster), and greatly reduce the energy consumption simultaneously through controlling the interactions and complication of nanostructure [5].
2. Improvement in the surface area to volume ratio at the nanoscale provides the effect of a new quantum mechanical potential. Therefore, the reduction of particles size leads to the significant change in the electronic properties of materials. By manipulating the nanoscale design of materials, there is a possibility for changing their properties, like macroscopic and microscopic, magnetization, charge capacity, and melting temperatures, without altering their chemical composition [6].
3. Chemical properties of nanoscale provide features for catalysis, according to the chemistry of interfaces and surfaces and the large ratios of surface area to volume [7]. Nanostructures are idyllic for energy storage, composite materials, catalysis, and target drug delivery. Additionally, it can be used in medicine due to numerous properties, like increased erosion resistance and the probability of inhibiting bacterial growth.
4. Research and nanotechnology involve the development of engineered particles, and molecular structures orman-made nanoscale objects. When nanoscale objects are incorporated with biotechnology components, nanotechnology becomes significant with large potential applications such as biological sensors, new pharmaceuticals, diagnostic devices, medical imaging, and others [8]. As well as, this technology contributed the possibility to examine the matter macrostructure and microstructure by using molecular self-assembly [9].

There are two principal factors of nanomaterials that cause significant differences in properties and behavior of nanomaterials from the other bulk material:

a. Surface area (the fraction of atoms at the surface and typical smoothly scaling properties due to surface effects). The drastic increase in the ratio of surface area/volume is linked to the protuberances on the surface.
b. Quantum effect known as "Quantum confinement effect" which exhibits discontinuous behavior, according to completion of shells on systems with delocalized electrons.

Both of these factors have an effect on the nanomaterial's properties such as mechanical, optical, structural, electrical, transport, and magnetic properties, also affect the chemical reactivity of materials [10]. Over the last few decades, the development has been particularly cleared in the three areas where size effects are important, *i.e.*, metals (Ag, Au, *etc.*), semiconductors (Ag_2S, CdSe, *etc.*), and magnetic materials (Fe, CO, Ni, *etc.*) [11]. The size of particles affects the materials properties, which is associated with the confinement of electrons in different dimensions (quantum confinement effects). Thus, the properties of nanomaterials differ from their bulk material. The quantum confinement of excitons (electron-hole pairs) describes the electrons in the terms of energy levels, whereas the confinement means to restrict the random motion of the electron which is moving to the particular levels of energy. When the particle's materials have a dimension in nanoscale, the confinement dimensions lead to discrete energy levels, resulting in a broad energy band gap [12]. In the terms of bulk metals, between the valence and conduction bands, there is no band gap (bands overlap). So, the bulk metals differ from nanostructures of metal, whereas in the metal nanostructures two bands are separated by an energy gap. A diagram of the electronic bands in bulk metals, semiconductors, semiconducting nanoparticles and insulators is shown in Fig. (**1**).

Nanotechnology has attracted increasing attention from researchers and scientists to synthesize materials that have extraordinary magnetic, optical, photocatalytic, electrical, and chemical properties. Nanotechnology is contributing to changing the world around us which possesses widespread and pervasive applications across various sectors. Since previous years, nanotechnology has touched the everyday life of people in various forms with myriad applications. Nanotechnology has greatly and decisively contributed to the development and production of innovative materials and processes in numerous fields. Nanotechnology has an extensive range of applications such as biotechnology, biomedical & drug delivery, optical engineering & communication, cosmetics & plants, agriculture & food, textile, defense & security, metallurgy & materials, electronics and energy storage. The applications of nanotechnology in different fields are shown in Fig. (**2**).

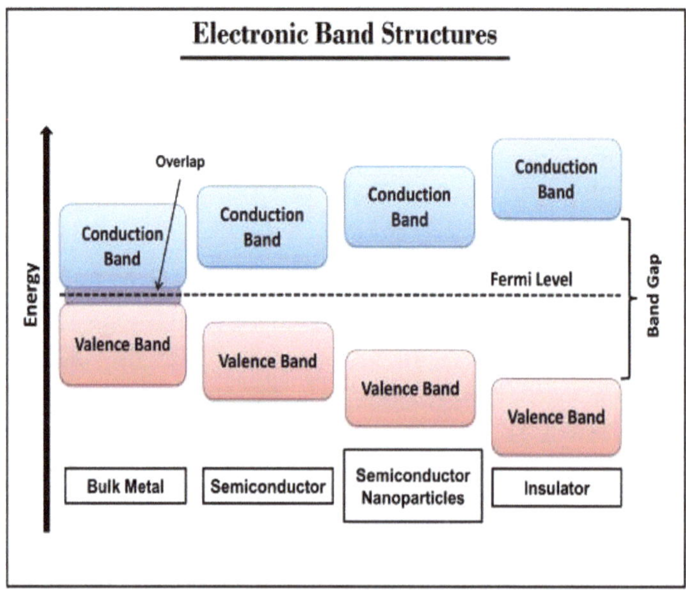

Fig. (1). Scheme of electronic band structures in bulk metal, semiconductor, semiconductor nanoparticles, and insulator.

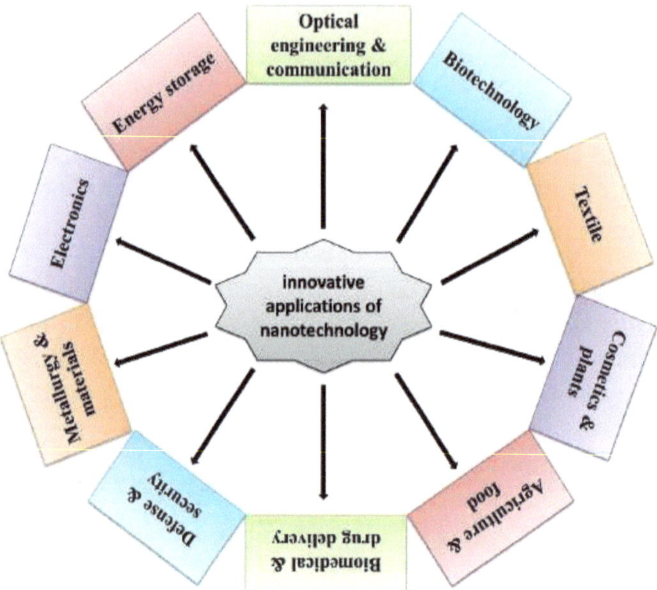

Fig. (2). Applications of nanotechnology.

CLASSIFICATION OF NANOMATERIALS

Due to modern classification, there are different types of nanomaterials classified on their dimensionality, zero, one, two and three dimensional nanomaterials. According to electron confinement in the different dimensions, the properties of nanomaterials change like physical, chemical, electrical and optical. Fig. (**3**) illustrates the different types of nanomaterials.

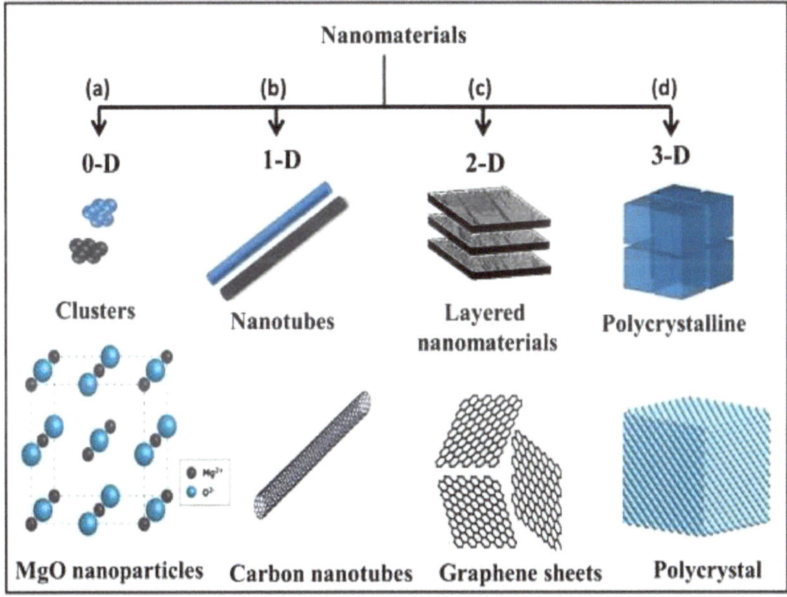

Fig. (**3**). Schematic representation of (**a**) 0-D, (**b**) 1-D, (**c**) 2-D, and (**d**) 3-D nanomaterials.

Zero Dimensional (0-D)

In the 0-D system, the electrons are localized at the point and they are not free to move in any direction. All dimensions in this system are not larger than 100 nm (all dimensions at nanoscale). The most common examples of 0-D are nanoparticles (amorphous, single-crystalline, polycrystalline) [13]. If the dimension is very small and the quantum confinement effect is observed, the nanoparticles are known as quantum dots (QDs). Semiconductor QDs behave as artificial atoms or artificial molecules according to quantum confinement effects and display the discrete energy level controllably [14]. Quantum dots semiconductors strongly confine the electrons or electron holes. There are many unique properties of quantum dots. Furthermore, the quantum dots display interesting phenomena like narrow emission, broad excitation, emission wavelength which depends on size, and the capability to emit light with different colors [15].

One Dimensional (1-D)

1-D in nanoscale range and two dimensions in macroscale range, which means in this classification the electron motion is confined in two directions while the electrons can move easily in one direction only. 1-D nanomaterials include nanofibers, nanorods, nanotubes, and nanowires, *etc*. Some examples of metals (Si, Au, Ag, Si, *etc*), and metal oxides (TiO_2, ZnO, CeO_2, *etc*).

Two Dimensional (2-D)

In 2-D nanomaterials, the electron motion is confined only in one dimension and the electron motion is free in the other two dimensions. 2-D nanomaterials include nanocoatings, graphene, nanolayers, and nanofilms [16].

Three Dimensional (3-D)

In 3-D objects of large size, the electron moves freely in any direction (no confinement), and the electronic structure is not restricted by the dimension of the material. Examples of 3-D are bulk powders, multi-nanolayer, dispersions of nanoparticles, nanotubes, and bundles of nanowires.

The density of states of a system describes the number of different states at the particular energy level, which are to be occupied by the electrons. Fig. (**4**) shows a diagram of quantization of the electronic density of state as a function of energy, which is a result of variation in the material dimensions (0-D, 1-D, 2-D, and 3-D).

PROPERTIES OF NANOMATERIALS

Optical Properties

In general, the nanosized materials have unique features (physical, mechanical, biological and chemical) as compared with bulk materials. Thus, the properties of these materials lead to a new novelty and a variety of technological applications in the different fields (physics, electrical and computer science, science and engineering, materials, chemistry, biology, and medicine). Physical properties of nanomaterials are correlated to thickness, color, boiling with melting point phase, and radioactivity. Mechanical properties are related to optical, electrical, thermal and magnetic parameters. Both properties of nanomaterials (mechanical and physical) depend on size effects. By their chemical reactivity, oxidation, catalytic effects and surface modification ability, the chemical properties of nanomaterials are determined.

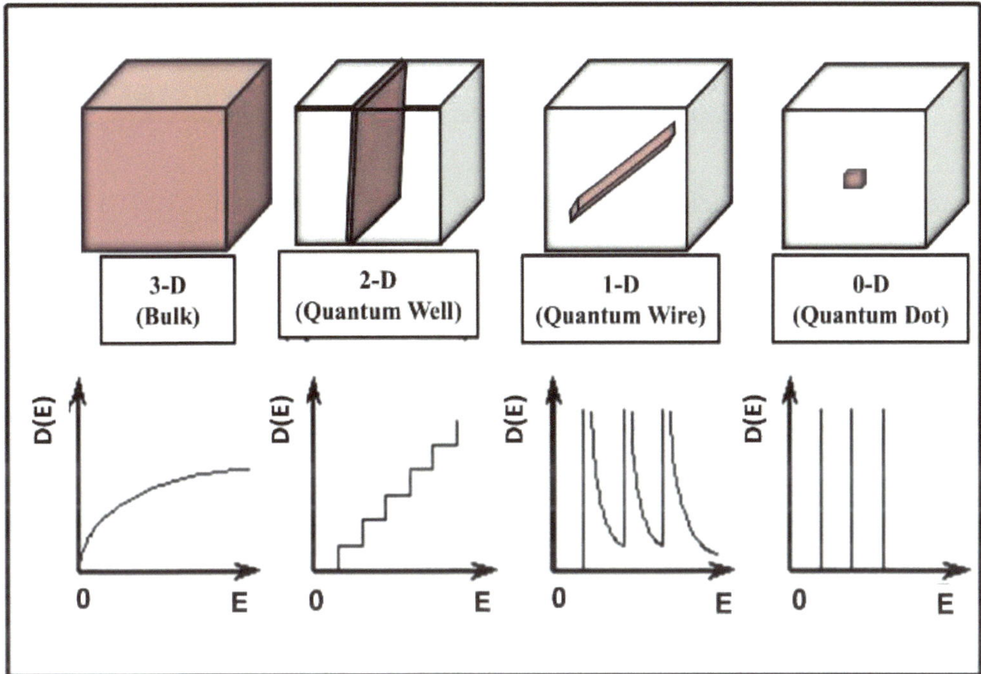

Fig. (4). Schematic representation of the electronic density of states based on dimensionality.

In nanomaterials, the optical properties are considerably one of the most attractive and useful aspects of nanomaterials. There are many applications depending on the optical properties of nanomaterials such as biomedicine, optical detector, sensors, phospholaser, imaging display, photoelectrochemistry, solar cell, and photocatalysis. The optical properties of nanomaterials are based on the different factors: (i) feature size, (ii) surface characteristics, (iii) shape. Additionally, other variables such as doping materials and interaction with other nanostructures or the surrounding environment. In optical properties of nanomaterials, particularly the metallic and semiconductor nanoparticles have different colors, thus the best concert with the applications is correlated imaging. It is essential to find out the reflectance and absorption value of the materials to distinguish which materials can be used for specific applications. Fig. (**5**) shows the size of particles dependent color of materials. The smallest quantum dots emit blue light while larger one emits red light. Furthermore, by controlling the size of the quantum dot nanoparticle gives the ability to absorb and emit the energy in the entire spectrum of visible light, this light emission process is called photoluminescence (PL).

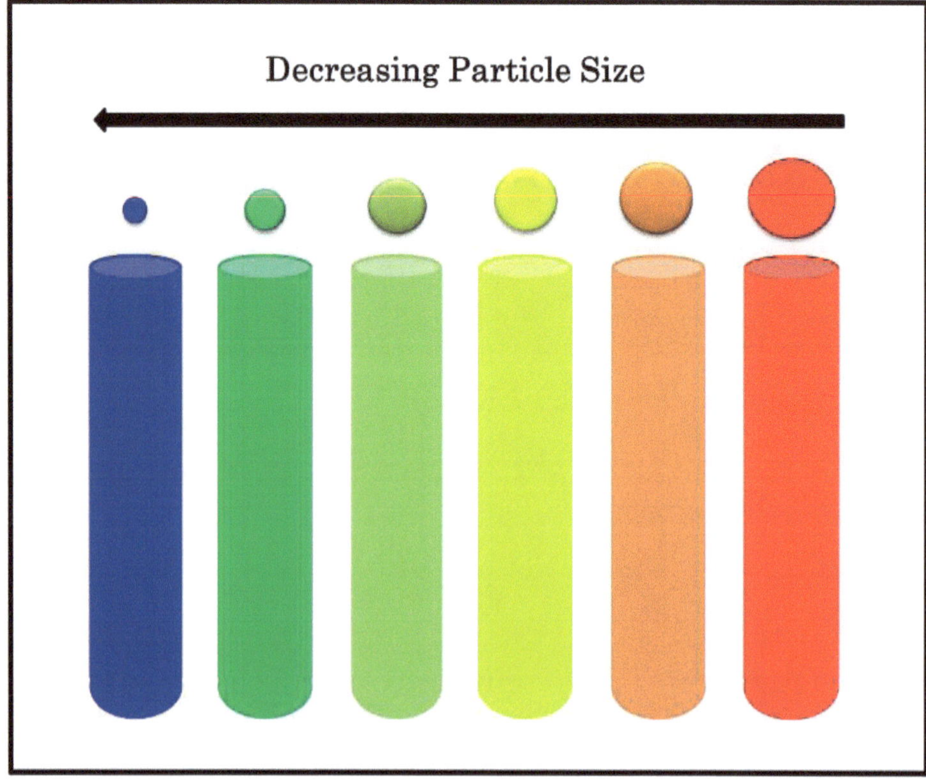

Fig. (5). Colors of materials based on the particle size.

The size-dependent photoluminescence color of the material was exploited in numerous applications such as *in vivo* and *in vitro* imaging and analysis, light emitting diodes, and lasers [17 - 20]. The material's color can be determined by the photon wavelength that is absorbed by materials from the incident light on its surface. The photon absorption occurs when the electron gains energy, and then excites from a lower to a higher energy level. In nanomaterials, the energy band structure and charge carrier density are different from their bulk counterparts [21]. Semiconductor nanomaterials possess discrete energy levels, size of materials inversely proportional to the energy band gap. So, with decreased particle size the energy gap increases, which means a broad band gap presence in nanomaterials compared to their bulk materials [22]. The absorption and emission spectra shift as a function of size. The blue shift effect occurred clearly when the optical absorption edge shifted to a high frequency (decrease in wavelength- increase in energy) with particle size decreases, while the opposite effect indicated the red shift. Fig. (6) (a and b) shows the blue shift and red shift of PL peaks depending upon the quantum dot size.

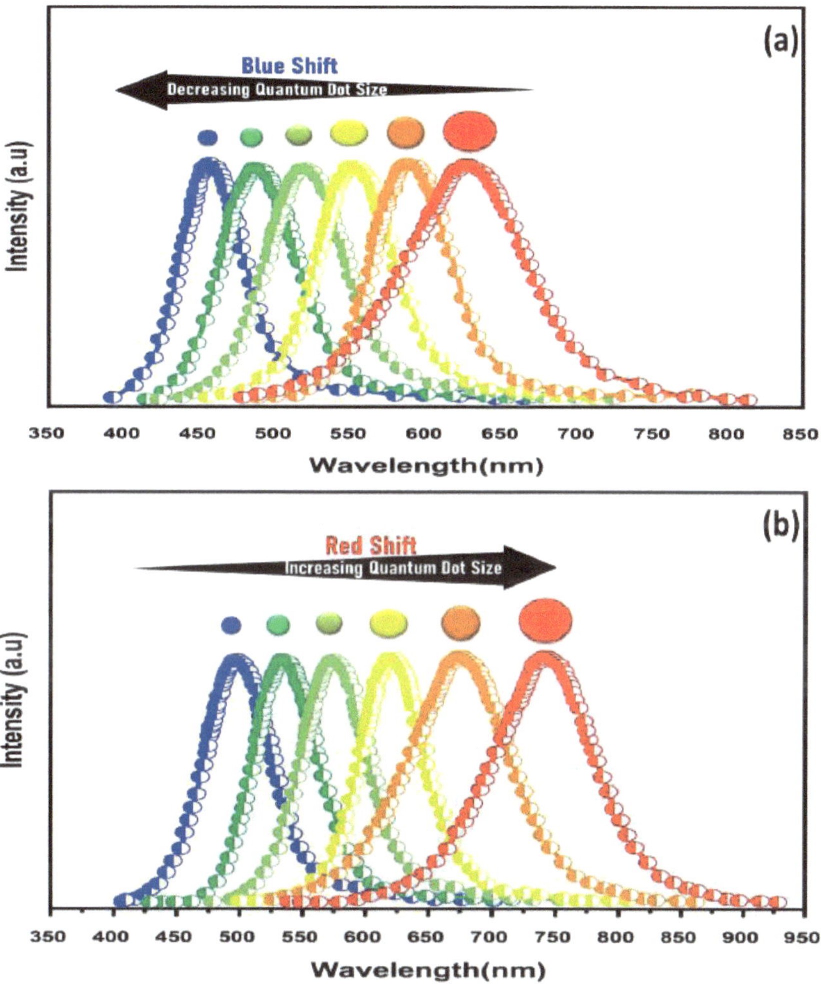

Fig. (6). Schematic of **(a)** Blue shift and **(b)** Red shift of PL peaks depending on the quantum dot size.

Magnetic Properties

The magnetic nature of most convertible systems is a consequence of contributions from both size effects and interaction. The magnetic properties are affected by the confinement of the materials to nanoscale dimension. Magnetic characteristics of NPs are influenced by different parameters such as morphology, size, composition, shape, defects and inter-planer spacing. There is a dramatic change in the magnetic behaviors of magnetic nanomaterials, which is produced when the critical length is in the range of NPs [23]. In addition, the large ratio of the surface area to volume influences the large ratio of atoms which is having a different coupling of magnetic with compared to neighboring atoms, causing the change in magnetic properties. Thus, nanomaterials in the existence of applied magnetic fields have the ability to enhance the magnetic properties by interacting atoms with the nearest neighbors, as the number of atoms increases on the surface. Due to dimensionality reduction, the magnetic behavior was enhanced in the ultrasmall ferromagnetic metal nanoparticles and their clusters [24]. So, the reduction in the size of particles in bulk material displays ferromagnetic properties to a particular dimension, causing superparamagnetic phenomena as a result of the formation of a single magnetic domain. The superparamagnetism (SPM) nanomaterials show zero coercive fields and the magnetic moment can randomly flip direction under the effect of temperature, by applying the external magnetic field the magnetic moment tends to align [25]. The magnetic properties of nanomaterials differ in several substantial respects from their bulk material properties. For example, the large fraction of atoms that exist on the surfaces or interfaces, which have local environments that differ extremely from those of the interior atoms, causes them to be unable to distinguish between interior and exterior properties. In small ferromagnetic particles, the values of magnetic saturation (Ms), remnant magnetization (M_r), coercivity of the material (H_c), and Curie temperature (T_c) might be different from the values of their bulk counterparts in a size-dependent way. Nanometer-sized ferromagnetic particles have a unique phenomenon that differ from their bulk ferromagnetic materials, where usually small ferromagnetic particles form only a single magnetic domain, while the bulk ferromagnetic materials consist of several magnetic domains. The simplest example; Gold (Au) and platinum (Pt) are non-magnetic in their bulk form, while nanosized behave as magnetic particles [26]. Au and Pt nanoparticles exhibit ferromagnetic behavior in nature when they are capped with thiol Carbon, thiol alkane, and thiols. Furthermore, the effects of large surface-to-volume ratios and single-domain behavior can therefore cause new and unusual magnetic properties [27, 28].

MAGNETIC NANOMATERIALS

Magnetic nanomaterials are a class of nanomaterials that can be manipulated by magnetic fields. Magnetic nanomaterials include iron (Fe), cobalt (Co), or nickel (Ni) which produce magnetic fields directly or indirectly [29]. The strength of a magnet is determined by the saturation magnetization (M_s) and magnetic coercivity (H_c) values, in which H_c is the field required for the reduction of the magnetization of the materials to zero after the magnetization is driven to saturation, and applied in a direction reverse to the original saturating field, both values of saturation magnetization and magnetic coercivity increase with a decrease in the grain size, as well as an increase in the surface area per unit volume of the grains.

The magnetic behavior of materials can be categorized into the different types, in Fig. (7) the diagram represents the configurations of the atomic magnetic dipole for a diamagnetic, paramagnetic, antiferromagnetic, ferrimagnetic, and ferromagnetic materials.

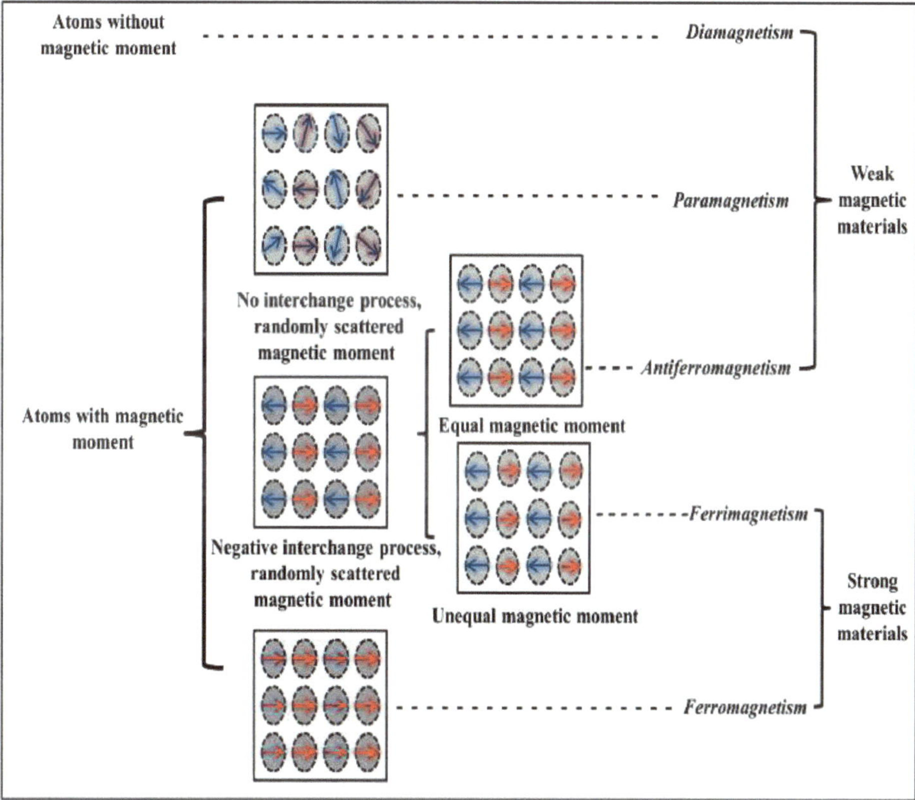

Fig. (7). Diagramatic representation of the magnetic behavior of materials and atomic dipole configuration for all magnetic materials.

Diluted Magnetic Semiconductors (DMSS)

DMSs are a class of materials that have both magnetic and semiconductor properties. The magnetic characteristics of diluted magnetic semiconductor materials have been explained by an indirect exchange interaction between magnetic ions.

Indirect exchange mechanism acts between moments over a comparatively large distance, according to the distance between moments the exchange interaction can be either positive or negative. In this interaction, the mechanism is that a spin polarization arises in its neighboring by the magnetic ions. Other magnetic ions influence the actual spin polarization (P) in the conduction electron in the range. Thus, two moments are coupled, which are located in different positions. This coupling is called indirect exchange coupling (IEC). Additionally, there is super-exchange interaction that is considered a type of indirect exchange, which is acting between two magnetic ions separated by nonmagnetic anions. Diluted magnetic semiconductors are doped with transition metals (TMs), especially 3d (TMs) instead of electronically active elements. Ordinarily, the magnetic moments have arisen from atomic 3d transition metals (TMs) chains [30].

DMS family includes the standard semiconductors, the position of magnetic moments that correlated with the magnetic atoms interacted with the holes in a valence-band and delocalized electrons in a conduction- band, some fraction of cations of host in non-magnetic semiconductor (a sizable fraction of atom) are substituted by magnetic ions (3d ions), leading to the effect of the electronic structure by strong hybridization between 3d orbitals and p orbitals [31]. Consequently, the hybridization originates from the strong magnetic interaction between the charge carriers in the host valence band and localized 3d spins, so localized magnetic moments in the semiconductor matrices are created as illustrated in Fig. (**8**).

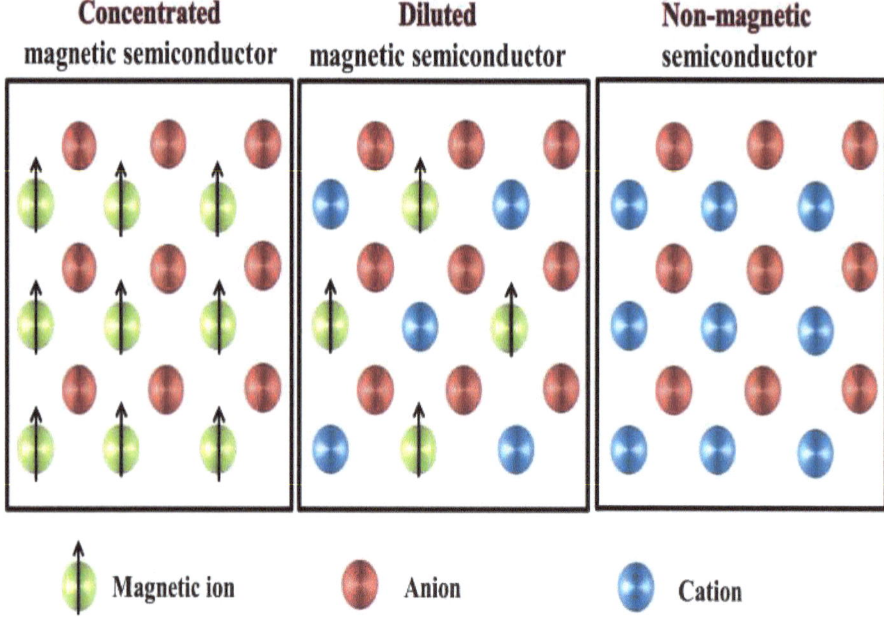

Fig. (8). Schematic of diluted magnetic semiconductor.

The manipulation of the properties and spin injection can be used in non-magnetic semiconductors to contribute to the spintronic applications, where the doping of the transition metal element causes spin polarization or spin order, resulting in the non-magnetic semiconductors becoming magnetic semiconductors [32]. DMS materials are extremely sensitive to preparation methods, preparation conditions (chemical reaction, temperature, impurities and lattice defects), and an external magnetic field. So, during the preparation of DMS special care or a clean room is required.

In the late 1960's, the diluted magnetic semiconductor field originated, where the studies exhibited in some ferromagnetic semiconductors the cohabitation of magnetism and semiconductor, such as europium-chalcogenides [33], and Cr, which are based on the spinels of crystal structure [34]. Ferromagnetic materials and semiconductors were widely studied from the late 1960's to early 1970's, and opened a research field of interaction between the properties of ferromagnetism and semiconductor materials [35]. In the late 1970's to 1980's, the DMS research was mainly focused on II-VI based systems, and adopting II–VI semiconductors as a host material. In this period, the most materials studied are II–VI compounds (ZnSe, CdS, CdTe, and CdSe) doped with transition metal ions, replacing their original cations [36]; as well as it was realized that the doping of magnetic ions (impurity ions) would change the electronic transport, optical properties and simultaneously induce magnetic spins [37]. Subsequently, the studies were expanded toward II-VI host material doped with transition metal ions Fe [38, 39], Co [40, 41], and Cr [42, 43].

In the 1990s, a rapid progression in diluted magnetic semiconductor research arose to a large scope. Initiated from the crystal growth method it was developed far from thermodynamic equilibrium, basically by molecular beam epitaxy (MBE) and laser ablation techniques. DMS researchers have detected the ferromagnetism (FM) in transition metal elements (mainly Mn)-doped III-V semiconductors. Initiated in (In, Mn) As and then in (Ga, Mn) As. The ferromagnetic ordering for Curie temperature was at 35 K and 170 K in (In, Mn) As [44] and (Ga, Mn) As [45], respectively. This type of material is challenging for synthesizing high quality diluted magnetic semiconductor nanomaterials, and their Curie temperature is low, so they found that it was not suitable for practical applications. The long domain ferromagnetic order between positioned moments in Mn-doped III-V group was raised by substitutional Mn through Ruderman–Kittel–Kasuya–Yosida (RKKY) exchange coupling [46], and also through mediated-hole mechanisms. In a wide-band-gap III-V semiconductors, the systems are not yet well understood, there is no consensus regarding the origin of FM in DMS nanomaterials, although a large number of the reports offered the existence of ferromagnetic order at and above room temperature (RT). Recently, dilute magnetic semiconductors (DMS) have become a prime focus of magnetic semiconductor research. Due to their fascinating properties, it has attracted considerable attention from researchers and scientists. Ferromagnetism at or above room temperature in oxides has been explained by a number of studies, such as the double exchange (DE) interaction model, super-exchange model, Zener model, RKKY model, bound magnetic polaron (BMPs) model, *etc.*

The comparative of the major features of magnetic interactions is briefly given in the following: in DMS direct exchange refers to the direct coupling of magnetic ions *via* overlapping of their orbitals whereas indirect exchange is the correlation of spins of two magnetic ions. Additionally, superexchange interaction is that indirect interaction in which magnetic ions interact through coulomb exchange over a large distance. It is antiferromagnetic coupling. On the other hand, by virtual hopping, the magnetic ions couple in different charge states in double exchange. In this case, interaction occurs between ions with different oxidation states and this can be ferromagnetic or antiferromagnetic.

Structure and Properties of Cr_2O_3 Nanoparticles

Chromium (III) oxide (Cr_2O_3) is an interesting material for metal oxide nanostructures. It is also called dichromium trioxide, or chromia, or chromium sesquioxide, and it has a broad band gap around 3.87 eV p-type metal oxide nanostructures. Due to the existence of cation vacancies, the intrinsic nature of Cr_2O_3 is a wide band-gap p-type semiconductors. In this material, two chromium cations (Cr_2) hold a charge of +3 each, totaling +6, while each oxygen ion (O_3) holds a charge of -2, amounting to a total -6. On the other hand, in Cr_2O_3 each chromium ion is surrounded by six ions of oxygen and each ion is surrounded by four chromium ions. Fig. (**9**) shows the rhombohedral structure of Cr_2O_3. Cr_2O_3 appears as a fine light to dark green. It is amphoteric and insoluble in water. Cr_2O_3 nanostructure possesses improved properties (stiffness, hardness, and *etc.*) compared to its bulk phase.

Previous studies reported that Cr_2O_3 nanoparticles possess rhombohedral structure having a space group R-3c with crystal parameters a, b and c (where a = b = 4.953 Å, and c = 13.578 Å) [47 - 49]. Gibot *et al.*. confirmed the rhombohedral structure and sphere-shaped morphology of Cr_2O_3 NPs [50]. Cr_2O_3 transition metal oxide has been used in electronic and optical devices, catalysis, colorants, and wear resistant [51 - 54]. Present article's author synthesized Cr_2O_3 NPs *via* sol-gel route and pointed out that the Cr_2O_3 NPs exhibited a single phase with a rhombohedral structure and a mean crystallite size of about 13.5 nm [55].

Anandan *et al..* confirmed that Cr_2O_3 optical features are related to the size of its particles, where the optical band gap values of Cr_2O_3 nanoparticles are decreased with increasing the particle size [56]. Mohanapandian reported the morphological, structural and optical analysis of Cr_2O_3 NPs doped with Cu [57]. Zekaik *et al..* studied the absorption properties and optical band gap of Cr_2O_3 and Ni-doped Cr_2O_3 thin films [58]. Tian *et al..* confirmed that a higher temperature results in a higher crystallinity of the Cr_2O_3. Consequently, it is also attributed to an increase in the crystallite size. As well as, they reported that FTIR absorption peaks shifted towards the higher wavelength (red shift) due to the reduction of oxygen defects with increasing the particle size [59]. Almontasser *et al..* confirmed the spherical morphology of Cr_2O_3 nanocrystals, and pointed PL peak positions for Cr_2O_3 at 312 nm, 340 nm and 425 nm, which are related to oxygen vacancies, or Cr vacancies, and oxygen interstitials or Cr interstitials [60].

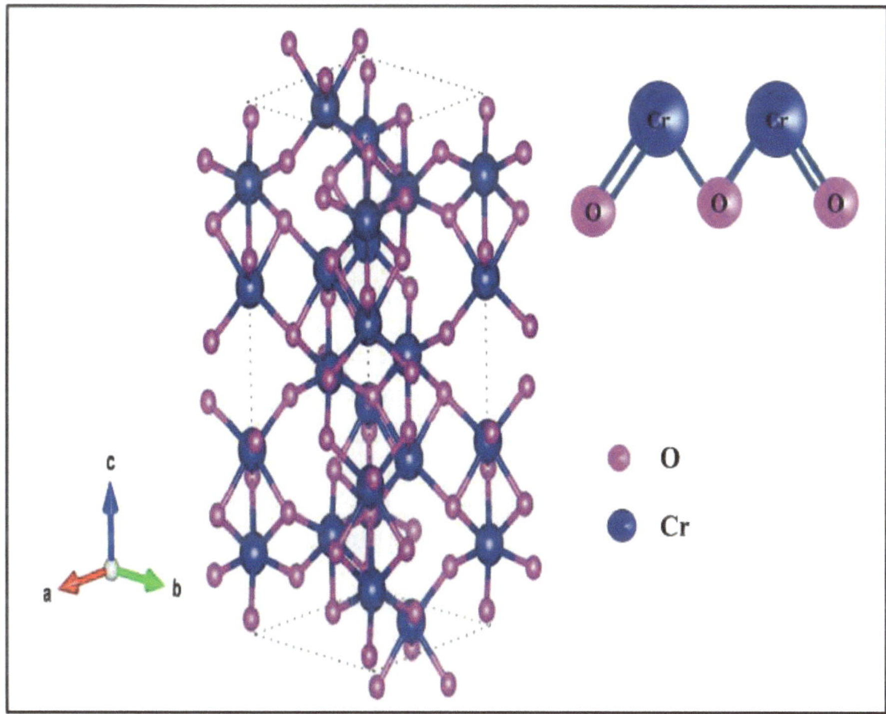

Fig. (9). Rhombohedral crystal structure of Cr_2O_3.

Chen *et al..* proved that NiFe-Cr_2O_3 nanoparticles have a considerably enhanced catalytic activity, they observed that the increasing amount of Fe leads to an increase in the average nanocrystal sizes of NiFe-Cr_2O_3 nanoparticles, where Fe possesses a strong magnetic characteristic and tended to agglomerate [61]. Bulk

Cr_2O_3 has a zero magnetic moment at absolute zero due to its anti-ferromagnetic behavior. The atomic magnetic moments in bulk chromium are aligned in parallel direction in the presence of an external magnetic field, leading to properties such as very high resistance against magnetization and near-zero net magnetization [62]. Opezan *et al.*. reported that Cr_2O_3 nanoparticles possess an antiferromagnetic behavior, and showed the existence of a net magnetic moment on the crystal surface due to the large surface/volume ratio [63]. Ferromagnetism properties can exhibit on the surfaces of chromium thin films and chromium crystals [64 - 67], although bulk chromium has an antiferromagnetic structure. Si *et al.*. reported that Cr_2O_3 nanoparticles show an enhanced saturation magnetization due to their uncompensated spins at the surface [68]. Jin *et al.*. studied the chromium oxides (Cr_2O_3, Cr_2O_5, CrO_2, CrO_3, and Cr_3O_8) and confirmed exhibiting a total of different magnetic properties and structures dependent upon the ratio of Cr/O [69]. Lei *et al.*. confirmed a high coercive field in antiferromagnetic (AFM) Cr_2O_3 [70]. Vázquez *et al.*. studied the evidence of weak ferromagnetism in Cr_2O_3 nanoparticles [71]. Yang *et al.*. studied the magnetic behavior of Cr_2O_3 corundum structure, where results showed superparamagnetic behavior of Cr_2O_3NPs, in which the Cr_2O_3 particles surfaces become superparamagnetic, while the cores particle retain their original antiferromagnetic behavior [72].

Structure and Properties of In_2O_3 Nanoparticles

Indium oxide (In_2O_3) is one of the transparent semiconductor oxides (TSOs) materials, In_2O_3 semiconductor material is n-type material with a broad band gap (3.7 eV). In_2O_3-based oxides have been used as transparent conducting oxides (TCOs) [73, 74]. In_2O_3 is an interesting material for different applications due to its fascination properties such as electrical conductance, strong interaction between molecules of the toxic gas and its surfaces as well as high transparency to visible light high [75, 76]. The cubic crystalline structure of In_2O_3 is shown in Fig. (**10**).

Chen *et al.*. reported that at room temperature the light could not emit from bulk In_2O_3. However, the photoluminescence spectrum of In_2O_3 nanocrystal at room temperature exhibited an emission peak at position 380 nm and excitation at 325 nm, which is ascribed to the presence of oxygen vacancies. Also, they confirmed the existence of a blue shift in the absorption spectrum of In_2O_3 nanocrystal [77, 78]. Maensiri *et al.*. reported the spectrum absorbance of In_2O_3 nanoparticles annealed at 400 °C, 500 °C, and 600 °C for 2 hrs. They also confirmed that photoluminescence emission of In_2O_3 nanoparticles ascribed to an electron occupying oxygen vacancies with a photo-excited hole [79]. Ye *et al.*. reported that the strong UV absorption of In_2O_3 octahedral nanocrystals was exhibited at

300 nm, and also a very strong band-edge UV emission is shown at 367 nm, also confirming that in the visible region the weak photoluminescence of In_2O_3 nanostructures was due to the defects (mainly oxygen vacancies) in the crystal lattice of In_2O_3 nanocrystals [80].

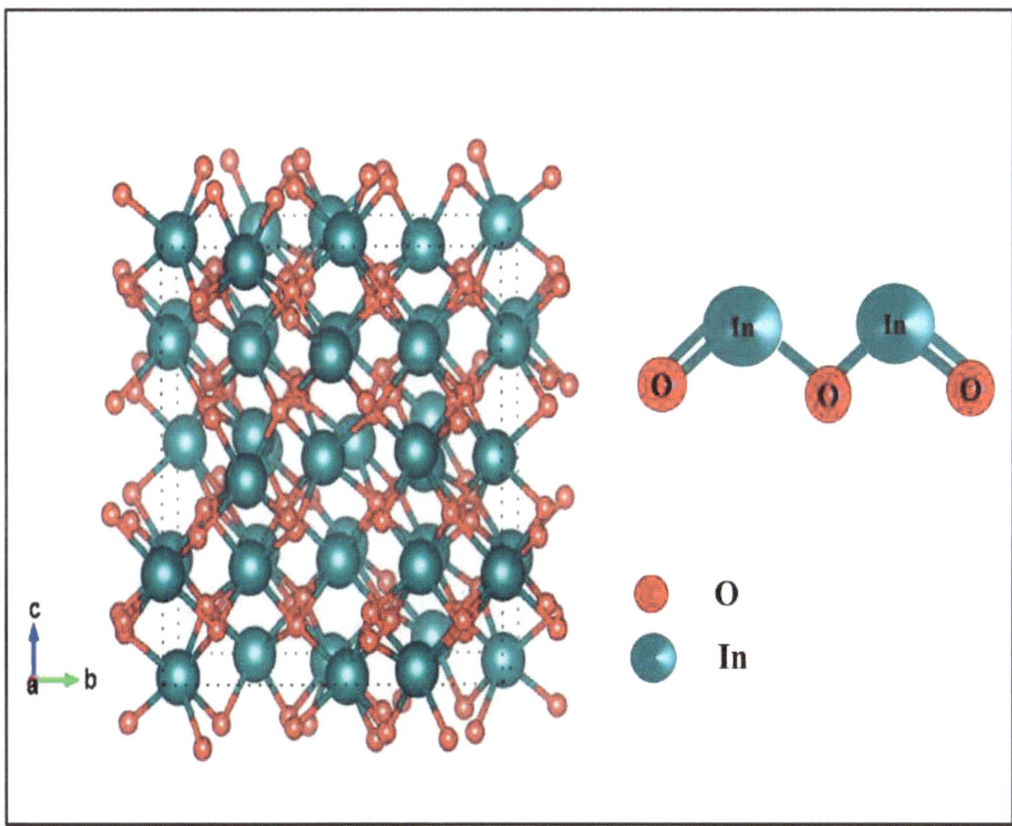

Fig. (10). Cubic crystal structure of In_2O_3.

Farvid *et al..* confirmed that grain size and crystal structure of colloidal In_2O_3 nanocrystalline can be controlled and manipulated by dopant ions, where Mn dopant inhibited the nanocrystal growth of In_2O_3 [81]. The charge carrier transport properties of indium oxide (In_2O_3) nanocrystal are studied by Forsh *et al.* [82]. Gurlo *et al..* reported the relation between the sensing properties with grain size distribution and surface morphology, in which the decrease in the grain sizes of In_2O_3 causes increasing sensor resistance and NO_2 sensitivity [83]. Shen *et al..* discussed the photoluminescence and construction of nanotubes of In_2O_3 [84]. Hosamani *et al..* investigated the morphology and optical properties of In_2O_3, and they found that In_2O_3 nanoparticles possess a wide energy band gap of about

2.9–3.6 eV [85]. Almontasser *et al..* pointed the purity phase of synthesized cubic In_2O_3 nanoparticles, with the energy band gap around 3.6 eV, and reported that the three peak PL emission of In_2O_3 at 307.8, 393.1, and 408.2 nm were due to oxygen vacancies [86].

Prakash *et al..* confirmed that Ni- doped In_2O_3 nanoparticles have a ferromagnetic behavior at RT, while the pure (In_2O_3) nanoparticles show weak ferromagnetism [87]. Wongsaprom *et al..* reported that at RT the pure (In_2O_3) nanoparticles exhibit a diamagnetic behavior, while Fe-doped In_2O_3 nanoparticles exhibit ferromagnetism ascribed to the oxygen vacancy related defects on the surface of samples [88]. Singhal *et al..* confirmed a weak ferromagnetic behavior of In_2O_3 NPs doped with Fe at RT due to the low defect concentration [89]. Meng *et al..* reported that due to the strong d-d coupling of the magnetic ions in Co-doped In_2O_3 nanocrystal, the ferromagnetic nature of this material was observed [90]. Yoo *et al..* and Jiang *et al..* observed the existence of ferromagnetic properties in Fe-doped In_2O_3 films at room temperature with large magnetic moments [91, 92]. Sun *et al..* reported the improvement of magnetic nature of In_2O_3 nanostructure doped by Fe^{3+}, which transformed from PM to SPM nature [93]. Prakash *et al..* confirmed that Co-doped In_2O_3 nanoparticles exhibit room temperature ferromagnetism [94]. Li *et al..* observed the ferromagnetic behavior in Fe- and Ni-co-doped In_2O_3 nanoparticles at room-temperature [95]. Naik *et al..* reported that Co^{2+} incorporated with In_2O_3 causes the redaction in crystal lattice, they also reported the Co-doped In_2O_3 NPs behave as ferromagnetic at room temperature [96].

Structure and Properties of MgO Nanoparticles

MgO is a substantial inorganic material and interesting metal oxide with a wide band-gap of (5.45 eV). Basically, MgO nanoparticles are white powder and nontoxic and have a high hardness and high melting point. MgO is a unique solid of high ionic character, it has crystal structure and simple stoichiometry. MgO nanoparticles possess a cubic structure. Cubic nanoparticles of MgO are called MgO nanocubes or, more strictly, nanocuboids. Fig. (**11**) represents the cubic structure of MgO and the arrangement of Mg and O ions in the unit cell.

Badar *et al..* reported that the optical band gap values (5.89 eV, 5.95 eV) of MgO nanoparticles are considerably smaller than the value expected for bulk MgO (7.8 eV) [97]. Roessler studied the optical properties of single-crystal MgO [98]. Stankic *et al..* confirmed optical properties of MgO nanocubes depending on the particle size [99]. Wobbe *et al..* analyzed the optical properties of cubic MgO NPs, mainly in the ultraviolet absorption spectra, where the ultraviolet light is absorbed by MgO nanoparticles and re-emits it quite efficiently at the visible wavelengths [100]. Hadia *et al.*, examined the structural and optical properties of MgO nanowire, they found that MgO is highly crystallized in cubic structure with the energy band gap of about 4.51 eV [101]. Niedermeier *et al.*. reported that the Ni-doped MgO optical band gap increases marginally to 4.8 eV [102]. Asakijar *et al.*. confirmed the cubic crystal structure of transition metal (Cr^{3+}, and Fe^{3+})-doped MgO [103]. The present article's authors synthesized MgO *via* a precipitation method and confirmed the single crystalline nature of cubic MgO nanostructure, with the optical band gap of around 4.8eV [104].

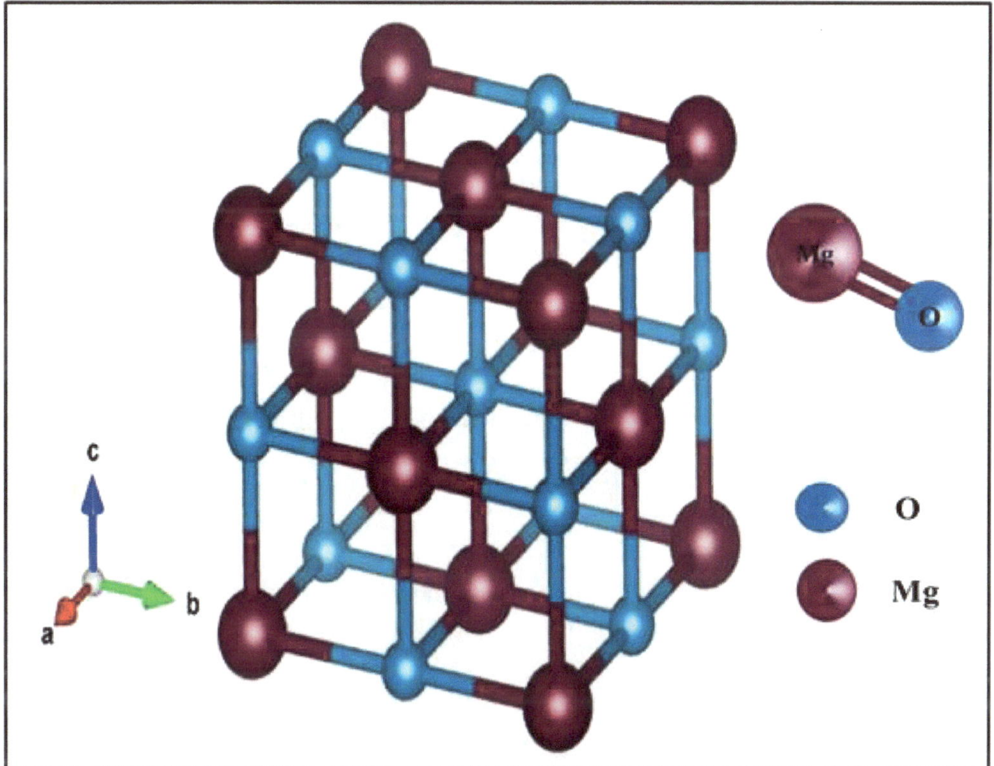

Fig. (11). Cubic crystal structure of MgO.

By doping ions into the MgO lattice, many different properties of MgO can be improved. Various ions have been used as dopant materials (such as Al, Li, N, Cr, Cu, Fe, Co, and Ni) for enhancing and modifying the unique characteristics of MgO nanoparticles and films [105 - 112]. Wu *et al.*. observed that the transition metals (Ni, Mn, Co and Fe)-doped MgO have smaller band gap than pure MgO with ferromagnetic nature [113]. Prada *et al.*. reported the substitutional doping (Ni, Li and Al) in MgO ultrathin films on metals, and confirmed the possibility of enhancing the metal oxide properties [114]. The structural and optical properties of Mn-doped MgO NPs were investigated by Singh *et al..*, the band gap energy was decreased from 5.37eV upto 5.06 eV with increasing concentration of Mn ions [115]. Kumar *et al..* confirmed a close relation between ferromagnetism, oxygen vacancies, and adsorbed hydrogen, where oxygen vacancy could induce RTFM spin-order in MgO nanostructure [116]. Yang *et al..* reported that increasing magnetization of MgO nanocrystals is attributed to the increase in Mg^{2+} vacancies and oxygen vacancies by dopant ions (Al^{3+}) into MgO crystal [117]. Varshney *et al.*, confirmed that the M–H curve of MgO, and MgO doped with Zn- were diamagnetic in nature at room temperature, which is attributed to surface defects, and the saturation magnetization is found to increase with decreased particle size [118]. Hanish *et al..* investigated the influence of Ni and Cr on the magnetic behavior of MgO NPs, and found that the FM behavior in MgO-NPs was ascribed to the Mg vacancy at room temperature, as well as they observed the paramagnetic ordering upon doping ions [119].

ANTIMICROBIAL ACTIVITIES OF METAL OXIDE NANOPARTICLES

Nanomaterials may exhibit potential antibacterial activity due to their intrinsic properties. Antibacterial agent plays a significant role in different fields such as medicine, food packaging, textile industry and water disinfection. Nanomaterials are considered emerging tools for preventing and controlling infection without the development of any resistance. Antimicrobial agents have the possibility to be utilized in numerous applications in the biomedical field. Many types of nanomaterials were used as strong antimicrobial properties towards human pathogens responsible for nosocomial infections including microorganisms, bacteria, virus infections, and fungus. Nanomaterials can reduce microbial growth or kill microorganisms without affecting surrounding normal tissues, nanomaterials do not necessarily kill all microorganisms, but might be applied to living tissue or skin to reduce the probability of infections. Generally, agents can be categorized as (a) bactericidal, which is having the capability to kill bacteria, and (b) bacteriostatic, which is stopping or inhibiting bacterial growth and multiplication [120]. The Gram-positive and Gram-negative organisms have sensitivity differences in various forms of nanoparticles [121]. Several metal and metal oxide-based nanomaterials are known for their highly potent antibacterial effect, which controls effectively the infection against pathogens. MgO, ZnO, and TiO_2 metal oxides NPs were considered antimicrobial agents, which are declining bacterial growth by generating reactive oxygen species [122 - 124]. Nanoparticles have proved attention in the context of combating bacterial and microbial resistance, in which nanoparticles display improvement in the properties according to their high ratio of surface area/volume, which leads to the appearance of a new chemical, mechanical, magnetic, optical, magneto-optical, electrical, and electrical-optical properties of nanoparticles that are different from bulk material properties. There is a growing interest in the use of NPs as an alternate to antibiotics for treating the infection of bacteria. Subhan *et al..* investigated the antibacterial action of $Ag_2O_3.SnO_2.Cr_2O_3$ NPs, and found that standard bacteriological tests have shown excellent action against *Staphylococcus aureus* (positive-gram bacteria) and *Escherichia coli* (negative-gram bacteria) [125]. Sangwan *et al..* pointed out the antibacterial activity of Cr_2O_3 nanoparticles on *Enterococcus faecalis* bacteria, and the effect of concentration on the bacterial growth and antibacterial activity of Cr_2O_3 metal oxides nanoparticles [126]. Al-Kalifawi *et al..* confirmed that Cr_2O_3 crystallite nanoparticles could be used as an antibacterial agent for human pathogenic bacteria [127]. Gupta *et al..* studied Cr_2O_3 nanoparticles as antibacterial effect on *E. coli* bacteria, which is degrading the activity of *E. coli* bacteria [128]. Sangwan *et al..*, studied the antimicrobial activity of Cr_2O_3 NPs against *Klebsiella pneumoniae* bacteria by determining the zone of inhibition (ZOI), colony forming units (CFU), and optical density (OD) on solid agar media and liquid medium [129]. Kanakalakshmi *et al..* reported that Cr_2O_3 nanoparticles inhibited the growth of pathogenic bacteria *S. aureus* and *E. coli* [130]. Ramesh *et al..* reported that Cr_2O_3 nanoparticles have an efficient antibacterial agent against *E. coli* [131]. Chang *et al..* reported that In_2O_3 and $CaIn_2O_4$ displayed high antibacterial performance, whereas In_2O_3 and $CaIn_2O_4$ possess excellent bactericidal characteristics against infectious bacteria such as *S. aureus* (gram-positive) and *A. baumannii* (gram-negative bacteria) [132]. Ramesh *et al..* discussed the antibacterial activity of In_2O_3/ZnO NPs against a bioluminescent bacterium *via* minimum inhibitory concentration (MIC) [133]. Skorb *et al..* examined the antibacterial efficiency of $TiO_2:In_2O_3$ nanocomposite against *Pseudomonas fluorescens* (Gram-negative bacterium) and *Lactococcus lactis* (Gram-positive bacterium) [134]. Bindhu *et al..* pointed out that MgO nanoparticles showed a remarkable antimicrobial activity against *Pseudomonas aeruginosa* and *Staphylococcus aureus*, which usually existed in water. This influence occurred due to the binding of surface oxide ion to the bacteria, whereas the surface area of MgO particles increased with increasing surface oxide ion concentration, leading to damage to the cytoplasmic membrane [135]. Consequently, the cell wall of the bacteria is affected by the particle size, when the particle size of MgO decreases the antibacterial activity increases. Nguyen *et al..* confirmed that the particle size of MgO had affected its antimicrobial activities, and proved that the minimum inhibitory concentration (MIC) of MgO NPs was higher for *E. coli* than for *S. aureus*, referring to increasing concentrations of MgO nanoparticles, which might inhibit the growth of Gram-positive bacteria and put to death the Gram-negative bacteria [136]. Stoimenov *et al..* confirmed the excellent action of MgO NPs against *E. coli* (negative-gram bacteria) [137]. Jin *et al..* reported a strong activity of MgO NPs against *E. coli*

[138]. Pugazhendhi *et al..* pointed out that MgO NPs have strong antimicrobial activities against human pathogenic bacteria [139]. Almontasser *et al..* performed a comparative study on the structural, optical, and antibacterial properties of pure Cr_2O_3 and Cr_2O_3 doped with Ni, Co, and Fe nanoparticles, and confirmed that this type of metal oxide nanoparticles has the ability to inhibit the bacterial growth of *E. coli* more than *S. aureus* [140].

Role of the Bacterial Cell Wall

Nanoparticles have various antibacterial mechanisms against bacteria. The effectiveness of nanoparticles and their defense mechanisms for treating drug-resistant bacteria and bacteria change according to the bacterial cell structure, functions, components and so on. The bacterial cell wall is divided into two major classes: (a) Gram-negative bacteria such as *E. coli* which is associated with the contagious diseases such as urinary tract, bacteriemia, and gastrointestinal infection, and (b) Gram-positive bacteria such as *S. aureus* which is associated with the contagious diseases such as wound infection, skin infection, nosocomial pneumonia, abscess, bacteremia, endocarditis and toxic shock syndrome. The microbial pathogenic properties (mechanisms of resistance) of *S. aureus* are complex, in which *S. aureus* can resist different types of antibiotics, thus becoming a prime hurdle in the treatment [141]. *S. aureus* strains which are very difficult to treat can resist the new antibiotic drugs easily, as a result of the abuse of antibiotics and the evolution of *S. aureus* bacteria. Thus, the rate of infection increases with a gradual increase in the resistance of *S. aureus* [142]. The differences in Gram-positive cell wall and Gram-negative cell wall might be the reason for this phenomenon to occur.

Bacterial cell of Gram-negative is surrounded by a layer of lipopolysaccharides (LPS) about (1–3 μm in thickness) and a thin layer peptidoglycan (outer lipid membrane) of around 8 nm thickness. This configuration may facilitate the entry of releasing ions from nanoparticles into the cell. The bacterial cell walls of Gram-positive possess a peptidoglycan which is a thick layer that can act to protect these bacteria, whereas in the cell of Gram-positive bacteria there is no outer lipid membrane. The diagram below illustrates the differences in the structure of the cell of Gram-positive and Gram-negative bacteria (see Fig. **12**). Another possible reason for the sensitivity of Gram-positive bacteria to NPs less than Gram-negative bacteria is that the lipopolysaccharide (endotoxin) molecules cover the Gram-negative bacteria, in which these molecules are negative charge. So, these molecules possess a high attractiveness to the positive ions that are released from NPs, causing accumulation and increased absorption of ions, which then leads to intracellular destruction [143, 144].

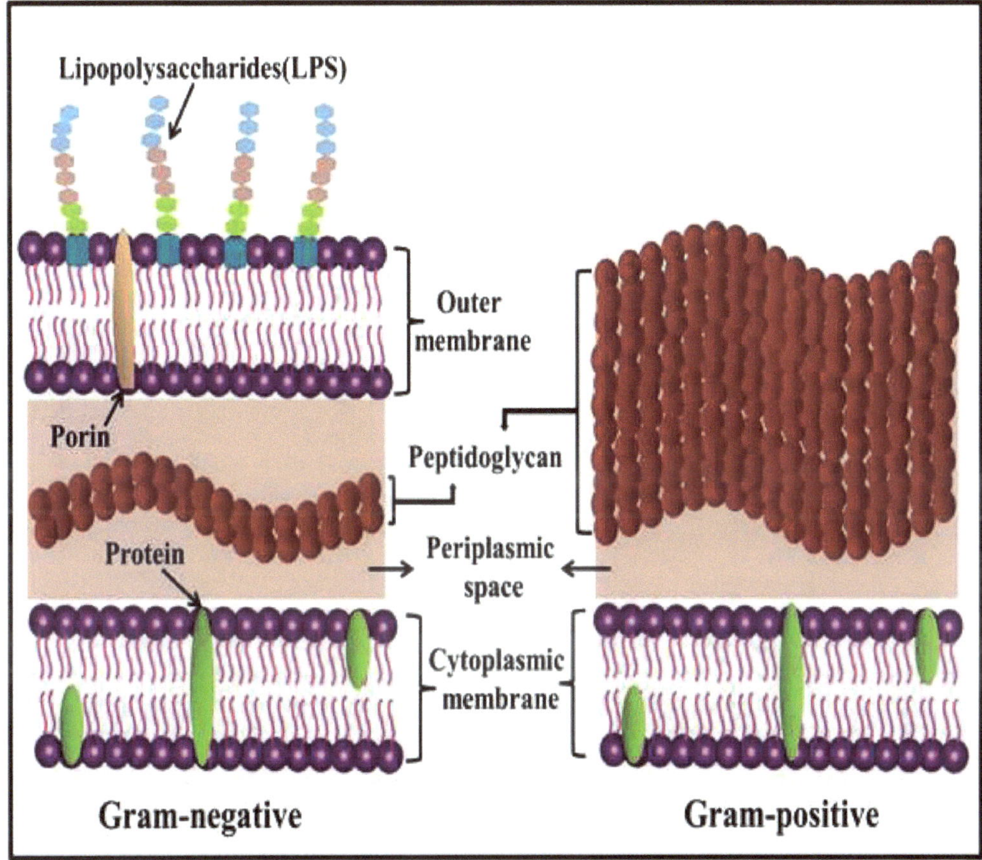

Fig. (12). Diagram illustrated the bacterial cell wall structure of Gram- negative and Gram-positive bacteria.

Antibacterial Activity Mechanism

Many researches have proved that Gram-positive bacteria could resist the nanoparticles [145, 146]. The exact mechanism of antibacterial activity is still under discussion, some intrinsic mechanisms have been suggested such as metal-ion release [147], reactive oxygen species (ROS) formation [148], internalization of metal oxide nanoparticles within the bacterial cell [149], and destruction of membrane and bacterial cell wall [150]. Some of the more important antibacterial activity mechanisms of nanoparticles are explained below. A schematic diagram representing antibacterial activity mechanisms (Fig. **13**).

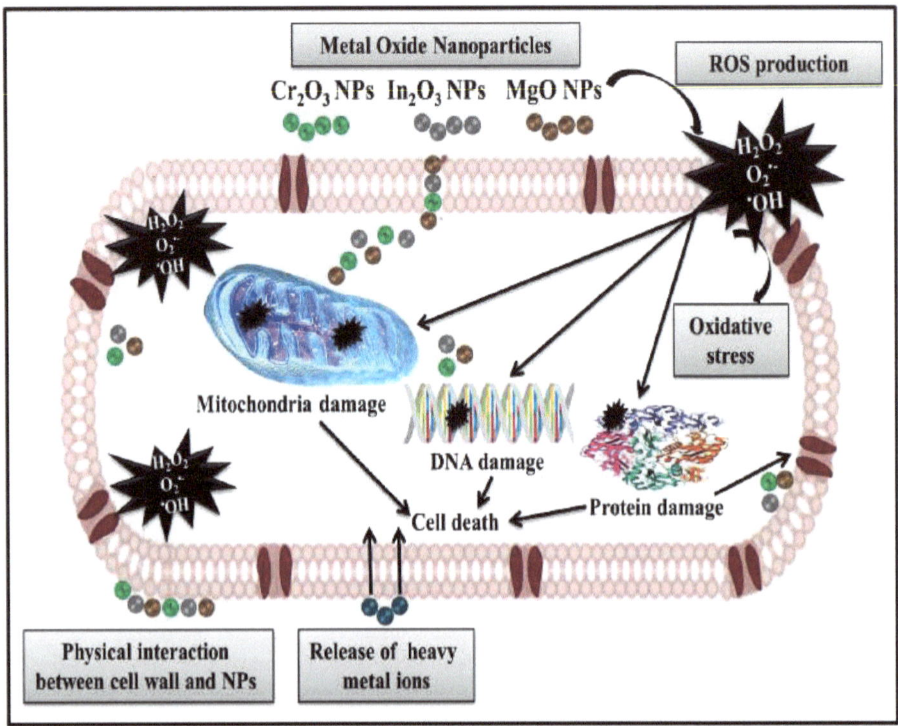

Fig. (13). Schematic representation antibacterial activity mechanisms of metal oxide nanoparticles.

Ion Release From Nanoparticles

Nanoparticles are continually subjected to dissolution as a result of the electrochemical potential in the solution, which shows that the antibacterial activity of nanoparticles is dependent on the ion release, although there are other mechanisms that could be involved like-wise. Metallic nanoparticle cytotoxicity hastwo influences: (a) the nanoparticle-derived influence due to its high mass-specific surface and (b) the influence of the metal ions solutions from nanoparticles in a biological environment or suspending medium [151]. Nanoparticle concentrations have an efficient effect on toxicity [152, 153], in which more ions are released with a concomitant increase over time when the amount of nanoparticles is large [154]. When the bacterial cell exposes to the metal ions that are released from nanoparticles, these ions distribute uniformly surrounding the bacterial cell without specific position. NPs that interact with bacterial cells release ions continuously, leading to toxicity of bacterial cells, so with the creation of more ion concentration, the ions can penetrate the bacterial cell.

Bacterial Cell Wall Interactions and Cell Penetration

This mechanism clears the damaged membrane of the bacterial cell which is treated with nanoparticles, resulting from the adsorption which sometimes penetrates nanoparticles into the cell [155, 156]. Previous studies indicated that the essential reason for cell toxicity is adsorption of NPs on the surface of the cell wall following its deconstruction [157, 158]. Adsorption of nanoparticles causes depolarization of the bacterial cell wall, leading to changes in the negative charge of the cell wall to be most penetrable. Leung *et al.*. studied the toxicity of MgO NPs and their antibacterial activity, they examined the toxicity of various samples of MgO, and the results have shown the great toxicity for two samples of MgO nanoparticle towards *Escherichia coli* bacterial cells without the involvement reactive oxygen species (ROS). This referred to that the essential mechanism of cell death is associated with damage to the cell membrane [159]. The extracellular NPs aggregation can cause cell envelope damage and alterations in the thickness and smoothness of the cell [160].

Reactive Oxygen Species (ROS)

The formation of ROS is the primary antimicrobial mechanism, ROS production (such as the singlet oxygen ($^1O_2$), hydroxyl radical ($^{\cdot}OH$), superoxide ($O_2^{\cdot-}$), and hydrogen peroxide (H_2O_2)) in the intracellular and extracellular possess the ability to damage cell membranes [161]. ROS are generated by sequential reduction of oxygen, which transforms the oxygen into a number of free radicals (peroxides, superoxides, hydroxyl) [162]. ROS lead to changes in the cell membrane through lipid oxidation. In the case of *S. aureus* (gram positive), because of the thicker cell wall structure, the lipid is not influenced as expected [163]. The thick layer of peptidoglycan plays a major role in protecting Gram- positive bacteria from nanoparticles, so the antibacterial property of metal oxide nanoparticles might affect Gram-negative bacteria more than Gram-positive bacteria. In ROS mechanisms there are free radicals such as hydroxyl (OH) radicals that have negative charge, and are unable to penetrate the negative charged cell membrane easily, but H_2O_2 radicals have the ability to penetrate the cell membrane and cause bacterial membrane damage resulting in the bacterial cell death. Metal oxide NPs are considered a source for induced ROS production [164], which is produced by the combination of oxides with metals. Consequently, ROS is a prime mechanism for the antibacterial activity of metal oxide NPs [165]. These metal oxide NPs can increase the inducing ROS dramatically, resulting in to damage bacterial cell, this is called oxidative stress [166]. Páez *et al.*. studied the oxidative stress and antibacterial activity of chromium (III) metal NPs on *S. aureus* and *E. coli*, the results found that *E. coli* (Gram-negative bacteria) was less resistant than *S. aureus* (Gram-positive bacteria) for production of ROS by antibacterial, resulting in less formation of ROS in Gram- positive bacteria [167]. He *et al..,* studied the various mechanisms for the action of MgO NPs on bacteria *Campylobacter jejuni, Escherichia coli, and Salmonella* bacteria, and they found that MgO nanoparticles continuously create hydrogen peroxide (H_2O_2) leading to the production oxidative stress in the bacterial cells. Moreover, the interaction between NPs and cell wall deactivates bacterial membrane and leads to leakage of the membrane. Furthermore, a higher concentration of MgO metal oxide nanoparticles cause cell membrane damage, and biomolecules oxidation (lipids, proteins, and nucleic acids (DNA)) resulting in the death of bacterial cells [168].

CONCLUSION AND FUTURE DIRECTION

Metal oxide nanoparticles play a significant role in biomedical applications. In this chapter, some of the developments in metal oxide nanoparticles for biomedical applications have been reported. The antibacterial mechanisms of these materials were described in detail. Most metal oxide NPs are safe and biocompatible to be used in humans, but still, some of the metal oxide NPs have limited use in biomedical applications because of their toxicity at higher concentrations. This obstacle is considered the ongoing challenge facing the researchers in removing the toxicity of metal oxide NPs. On account of the intrinsic properties of metal oxide NPs have an important impact, especially in antibacterial activity, hyperthermia therapy, drug delivery, and others. An overview of the structural, optical, and magnetic properties along with the antibacterial activity of Cr_2O_3, In_2O_3, and MgO nanoparticles, was presented. The concept of using the magnetic NPs in drug delivery originated from the progress of nanotechnology technique, Nanotechnology has been developed to produce magnetic NPs in a tiny size with specified magnetic properties, which are extensively exploited in medicine as drug delivery agents to target the desired area with using the external magnetic field. Metal oxide nanoparticles have good activity against bacteria causing diseases. Antibacterial activity of Cr_2O_3, In_2O_3, and MgO metal oxide nanoparticles was observed as the best inhibitory activity against infectious bacteria. The recent research clearly demonstrates that nanotechnology has opened up the door for new future uses of metal oxide NPs in biomedical applications.

CONSENT FOR PUBLICATION

Not applicable.

CONFLICT OF INTEREST

The author declares no conflict of interest, financial or otherwise.

ACKNOWLEDGEMENTS

Declared none.

REFERENCES

[1]   Mansoori, G.A. An Introduction to Nanoscience & Nanotechnology. In: *Nanoscience & Plant Soil System*; Springer, **1961**; 42, pp. 196-198.https://doi.org/DOI
        [http://dx.doi.org/10.1007/978-3-319-46835-8]

[2]   Rao, C.N.R.; Cheetham, A.K. Science and Technology of Nanomaterials: Current Status and Future Prospects. *J. Mater. Chem.,* **2001**, *11*(12), 2887-2894.
        [http://dx.doi.org/10.1039/b105058n]

[3]   Daniel, M.C.; Gold Nanoparticles, D.A. Assembly, Supramolecularchemistry, Quantum-Size-Related Properties, and Applications. *Chem. Rev.,* **2004**, *104*, 293-346.
        [http://dx.doi.org/10.1021/cr030698+] [PMID: 14719978]

[4]   Nasrollahzadeh, M.; Sajadi, S.M.; Sajjadi, M.; Issaabadi, Z. An Introduction to Nanotechnology. In: *An Introduction to Green Nanotechnology*; Interface Science and Technology. Elsevier Ltd, **2019**; 28, pp. 1-27.
        [http://dx.doi.org/10.1016/B978-0-12-813586-0.00001-8]

[5]   Li, Y.; Lee, N.H.; Hwang, D.S.; Song, J.S.; Lee, E.G.; Kim, S.J. Synthesis and characterization of nano titania powder with high photoactivity for gas-phase photo-oxidation of benzene from $TiOCl_2$ aqueous solution at low temperatures. *Langmuir,* **2004**, *20*(25), 10838-10844.
        [http://dx.doi.org/10.1021/la0489716] [PMID: 15568831]

[6]   Dabbs, D.M.; Aksay, I.A. Self-assembled ceramics produced by complex-fluid templation. *Annu. Rev. Phys. Chem.,* **2000**, *51*(1), 601-622.
        [http://dx.doi.org/10.1146/annurev.physchem.51.1.601] [PMID: 11031294]

[7]   Iijima, S.; Brabec, C.; Maiti, A.; Bernholc, J. Structural Flexibility of Carbon Nanotubes. *J. Chem. Phys.,* **1996**, *104*(5), 2089-2092.
        [http://dx.doi.org/10.1063/1.470966]

[8]   Hobson, D.W. Nanotechnology. In: *Comprehensive Biotechnology, Second Edition*; 683-697., **2011**; 3, pp.
        [http://dx.doi.org/10.1016/B978-0-08-088504-9.00228-2]

[9]   Xue, Y.; Mansoori, G.A. Self-assembly of diamondoid molecules and derivatives (MD simulations and DFT calculations). *Int. J. Mol. Sci.,* **2010**, *11*(1), 288-303.
        [http://dx.doi.org/10.3390/ijms11010288] [PMID: 20162016]

[10]   Agrawal, D.C. *Introduction to Nanoscience and Nanomaterials*; World Scientific, **2013**.
        [http://dx.doi.org/10.1142/8433]

[11]   Bréchignac, C.; Houdy, P.; Lahmani, M. *Nanomaterials and Nanochemistry*; Springer-Verlag Berlin Heidelberg, **2007**.
        [http://dx.doi.org/10.1007/978-3-540-72993-8]

[12]   Malhotra, B.D.; Ali, M.A. Nanomaterials in Biosensors: Fundamentals and Applications. In: *Nanomaterials For Biosensors*; William Andrew, **2018**; pp. 1-74.
        [http://dx.doi.org/10.1016/B978-0-323-44923-6.00001-7]

[13]   Ashoori, R.C. Electrons in Artificial Atoms. *Nature,* **1996**, *104*(5)

[http://dx.doi.org/10.1038/35016189]

[14]   Cockins, L.; Miyahara, Y.; Bennett, S.D.; Clerk, A.A.; Studenikin, S.; Poole, P.; Sachrajda, A.; Grutter, P. Energy levels of few-electron quantum dots imaged and characterized by atomic force microscopy. *Proc. Natl. Acad. Sci. USA,* **2010**, *107*(21), 9496-9501.
[http://dx.doi.org/10.1073/pnas.0912716107] [PMID: 20457938]

[15]   Alivisatos, A.P. Semiconductor Clusters, Nanocrystals, and Quantum Dots. *Science,* **1996**, *271*(5251), 933-936.http://science.sciencemag.org/content/271/5251/933
[http://dx.doi.org/10.1126/science.271.5251.933]

[16]   Shehzad, K.; Xu, Y.; Gao, C.; Duan, X. Three-dimensional macro-structures of two-dimensional nanomaterials. *Chem. Soc. Rev.,* **2016**, *45*(20), 5541-5588.
[http://dx.doi.org/10.1039/C6CS00218H] [PMID: 27459895]

[17]   Biju, V.; Itoh, T.; Anas, A.; Sujith, A.; Ishikawa, M. Semiconductor quantum dots and metal nanoparticles: syntheses, optical properties, and biological applications. *Anal. Bioanal. Chem.,* **2008**, *391*(7), 2469-2495.
[http://dx.doi.org/10.1007/s00216-008-2185-7] [PMID: 18548237]

[18]   Sukhanova, A.; Devy, J.; Venteo, L.; Kaplan, H.; Artemyev, M.; Oleinikov, V.; Klinov, D.; Pluot, M.; Cohen, J.H.M.; Nabiev, I. Biocompatible fluorescent nanocrystals for immunolabeling of membrane proteins and cells. *Anal. Biochem.,* **2004**, *324*(1), 60-67.
[http://dx.doi.org/10.1016/j.ab.2003.09.031] [PMID: 14654046]

[19]   Åkerman, M.E.; Chan, W.C.W.; Laakkonen, P.; Bhatia, S.N.; Ruoslahti, E. Nanocrystal targeting *in vivo. Proc. Natl. Acad. Sci. USA,* **2002**, *99*(20), 12617-12621.
[http://dx.doi.org/10.1073/pnas.152463399] [PMID: 12235356]

[20]   Chan, W.C.W.; Nie, S. Quantum dot bioconjugates for ultrasensitive nonisotopic detection. *Science,* **1998**, *281*(5385), 2016-2018.
[http://dx.doi.org/10.1126/science.281.5385.2016] [PMID: 9748158]

[21]   Daniel, M.C.M.; Astruc, D. Gold nanoparticles: assembly, supramolecular chemistry, quantum-siz--related properties, and applications toward biology, catalysis, and nanotechnology. *Chem. Rev.,* **2004**, *104*(1), 293-346.
[http://dx.doi.org/10.1021/cr030698+] [PMID: 14719978]

[22]   Vasudevan, D.; Gaddam, R.R.; Trinchi, A.; Cole, I. Core-Shell Quantum Dots: Properties and Applications. *J. Alloys Compd.,* **2015**, *636*(February), 395-404.
[http://dx.doi.org/10.1016/j.jallcom.2015.02.102]

[23]   Leslie-Pelecky, D.L.; Rieke, R.D. Magnetic Properties of Nanostructured Materials. *Chem. Mater.,* **1996**, *8*(8), 1770-1783.
[http://dx.doi.org/10.1021/cm960077f]

[24]   Sengupta, A.; Kumar, C. *Engineering Materials Introduction to Nano,* **2015**, https://doi.org/DOI 10.1007/978-3-662-47314-6

[25]   Leeg, T.; Leeg, T. Advanced Magnetic Nanostructures. *Springer Science,* **2019**, *53*
[http://dx.doi.org/10.5771/9783845289892-15]

[26]   Maitra, U.; Das, B.; Kumar, N.; Sundaresan, A.; Rao, C.N.R. Ferromagnetism exhibited by nanoparticles of noble metals. *ChemPhysChem,* **2011**, *12*(12), 2322-2327.
[http://dx.doi.org/10.1002/cphc.201100121] [PMID: 21744458]

[27]   Roduner, E. Size matters: why nanomaterials are different. *Chem. Soc. Rev.,* **2006**, *35*(7), 583-592.
[http://dx.doi.org/10.1039/b502142c] [PMID: 16791330]

[28]   Hernandez, S.C. Measuring the Size Dependence of the Magnetic Properties of Alkanethiol-Coated Gold Nanocrystals. *Materials (Basel),* **2006**, 66-67.

[29]   Cullity, B.D.; Graham, C.D. *Ntroduction to Magnetic Materials*; Wiley-IEEE Press, **2008**.

[http://dx.doi.org/10.1002/9780470386323]

[30] Li, H.; Huang, M.; Cao, G. Magnetic Properties of Atomic 3d Transition-Metal Chains on S-Vacanc--Line Templates of Monolayer MoS_2: Effects of Substrate and Strain. *J. Mater. Chem. C Mater. Opt. Electron. Devices,* **2017**, *5*(18), 4557-4564.
[http://dx.doi.org/10.1039/C6TC04672J]

[31] Gubin, S.P. *Magnetic Nanoparticles*; WILEY-VCH Verlag GmbH & Co: KGaA, Germany, **2016**.

[32] Jiang, F.; Peng, L.; Liu, T. Ni-Doped SnO_2 Dilute Magnetic Semiconductors: Morphological Characteristics and Optical and Magnetic Properties. *J. Supercond. Nov. Magn.,* **2020**, *33*(10), 3051-3058.
[http://dx.doi.org/10.1007/s10948-020-05533-y]

[33] Kasuya, T.; Yanase, A. Anomalous Transport Phenomena in Eu-Chalcogenide Alloys. *Rev. Mod. Phys.,* **1968**, *40*(4), 684-696.
[http://dx.doi.org/10.1103/RevModPhys.40.684]

[34] Baltzer, P.K.; Lehmann, H.W.; Robbins, M. Insulating Ferromagnetic Spinels. *Phys. Rev. Lett.,* **1965**, *15*(13), 572-572.
[http://dx.doi.org/10.1103/PhysRevLett.15.572.3]

[35] Tanaka, M.; Ohya, S.; Hai, P.N. Recent Progress in III-V Based Ferromagnetic Semiconductors. *Appl. Phys. Rev.,* **2014**, 1-42.https://doi.org/https://doi.org/10.1063/1.4840136

[36] Furdyna, J.K. Diluted Magnetic Semiconductors. *J. Appl. Phys.,* **1988**, *2012*(29), 29-64.
[http://dx.doi.org/10.1063/1.341700]

[37] Bououdina, M.; Song, Y.; Azzaza, S. *Nano-Structured Diluted Magnetic Semiconductors*; Elsevier Ltd., **2016**.
[http://dx.doi.org/10.1016/B978-0-12-803581-8.09223-7]

[38] Twardowski, A. Magnetic Properties of Fe-Based Diluted Magnetic Semiconductors (Invited). *J. Appl. Phys.,* **1990**, *67*(9), 5108-5113.
[http://dx.doi.org/10.1063/1.344685]

[39] Mycielski, A. Fe-Based Semimagnetic Semiconductors (Invited). *J. Appl. Phys.,* **1988**, *63*(8), 3279-3284.
[http://dx.doi.org/10.1063/1.340813]

[40] Hamdani, F.; Lascaray, J.P.; Coquillat, D.; Bhattacharjee, A.K.; Nawrocki, M.; Golacki, Z. Magnetoreflectance and magnetization of the Co-based wurtzite-structure diluted magnetic semiconductor Cd1-xCoxSe. *Phys. Rev. B Condens. Matter,* **1992**, *45*(23), 13298-13306.
[http://dx.doi.org/10.1103/PhysRevB.45.13298] [PMID: 10001411]

[41] Lewicki, A.; Schindler, A.I.; Furdyna, J.K.; Giriat, W. Magnetic susceptibility of $Zn_{1-x}CoxS$ and $Zn_{1-x}Co_xSe$ alloys. *Phys. Rev. B Condens. Matter,* **1989**, *40*(4), 2379-2382.
[http://dx.doi.org/10.1103/PhysRevB.40.2379] [PMID: 9992122]

[42] Pekarek, T.M. J. E. L.; I. Miotkowski, and B. C. C. Magnetization and Heat-Capacity Measurements on Zn, „Cr„Te. *Phys. Rev. B,* **1994**, *23*, 915-920.https://doi.org/https://doi.org/10.1103/PhysRevB.50.16914

[43] Mac, W.; Khoi, N.T.; Twardowski, A.; Gaj, J.A.; Demianiuk, M. Ferromagnetic p-d exchange in Zn1-xCrxSe diluted magnetic semiconductor. *Phys. Rev. Lett.,* **1993**, *71*(14), 2327-2330.
[http://dx.doi.org/10.1103/PhysRevLett.71.2327] [PMID: 10054645]

[44] Ohno, H.; Munekata, H.; Penney, T.; Chang, L.L.; Chang, L.L. Magnetotransport properties of p-type (In,Mn)As diluted magnetic III-V semiconductors. *Phys. Rev. Lett.,* **1992**, *68*(17), 2664-2667.
[http://dx.doi.org/10.1103/PhysRevLett.68.2664] [PMID: 10045456]

[45] Ohno, Y.; Arata, I.; Matsukura, F.; Ohno, H.; Young, D.K.; Beschoten, B.; Awschalom, D.D. Electrical Spin Injection in Ferromagnetic/Nonmagnetic Semiconductor Heterostructures. *Physica E,*

2001, *10*(1–3), 489-492.
[http://dx.doi.org/10.1016/S1386-9477(01)00143-6]

[46]   Dietl, T.; Ohno, H.; Matsukura, F.; Cibert, J.; Ferrand, D. Zener Model Description of Ferromagnetism in Zinc-Blende Magnetic Semiconductors. *Science,* **2000**, *287*, 1019-1022.
[http://dx.doi.org/10.1126/science.287.5455.1019]

[47]   Khamlich, S.; Srinivasu, V.V.; Konkin, A.; Cingo, N. Photoinduced Electron Spin Resonance Phenomenon   in α-Cr_2O_3 Nanospheres.   *Nanomater,*   **2015**, 1-8.https://doi.org/https://doi.org/10.1155/2015/831065

[48]   Abdullah, M. M.; Fahd, M. Structural and Optical Characterization of Cr2O3 Nanostructures : Evaluation of Its Dielectric Properties. *AIP Advances,* **2014**.
[http://dx.doi.org/10.1063/1.4867012]

[49]   Palermo, V.; Gardey, M.C.; Merino, J.A.A.; Romanelli, G.; Rodriguez, M. S. L. Studies on Structural, Morphological, and Optical Properties of Cr_2O_3 Nanoparticles: Synthesized *via* One Step Combustion Process by Different Fuels. *Materia (Rio J.),* **2020**, *24*(5500)

[50]   Gibot, P.; Vidal, L. Original Synthesis of Chromium (III) Oxide Nanoparticles. *J. Eur. Ceram. Soc.,* **2010**, *30*(4), 911-915.
[http://dx.doi.org/10.1016/j.jeurceramsoc.2009.09.019]

[51]   Ma, H.; Xu, Y.; Rong, Z.; Cheng, X.; Gao, S.; Zhang, X.; Zhao, H.; Huo, L. Highly Toluene Sensing Performance Based on Monodispersed Cr_2O_3 Porous Microspheres. *Sens. Actuators B Chem.,* **2012**, *174*, 325-331.
[http://dx.doi.org/10.1016/j.snb.2012.08.073]

[52]   Gupta, R.K.; Mitchell, E.; Candler, J.; Kahol, P.K.; Ghosh, K.; Dong, L. Facile Synthesis and Characterization of Nanostructured Chromium Oxide. *Powder Technol.,* **2014**, *254*, 78-81.
[http://dx.doi.org/10.1016/j.powtec.2014.01.014]

[53]   Inturi, S.N.R.; Suidan, M.; Smirniotis, P.G. Influence of Synthesis Method on Leaching of the Cr-TiO_2 Catalyst for Visible Light Liquid Phase Photocatalysis and Their Stability. *Appl. Catal. B,* **2016**, *180*, 351-361.
[http://dx.doi.org/10.1016/j.apcatb.2015.05.046]

[54]   Gibot, P.; Schnell, F.; Spitzer, D. $Ca_3(PO_4)_2$ Biomaterial: A Non Toxic Template to Prepare Highly Porous Cr_2O_3. *Mater. Lett.,* **2015**, *161*, 172-174.
[http://dx.doi.org/10.1016/j.matlet.2015.07.015]

[55]   Almontasser, A.; Parveen, A. Preparation and Characterization of Chromium Oxide Nanoparticles. *3rd International Conference on Condensed Matter and Applied Physics (Icc-2019),* **2020**.
[http://dx.doi.org/10.1063/5.0001685]

[56]   Anandan, K.; Rajendran, V. Studies on Structural, Morphological, Magnetic and Optical Properties of Chromium Sesquioxide (Cr_2O_3) Nanoparticles: Synthesized *via* Facile Solvothermal Process by Different Solvents. *Mater. Sci. Semicond. Process.,* **2014**, *19*(1), 136-144.
[http://dx.doi.org/10.1016/j.mssp.2013.12.004]

[57]   Mohanapandian, K.; Krishnan, A. Synthesis, Structural, Morphological and Optical Properties of Cu [2+] Doped Cr_2O_3 Nanoparticles. *Int. J. Adv. Eng. Technol.,* **2016**, *7*(2), 273-279.

[58]   Zekaik, A.; Benhebal, H.; Benrabah, B.; Chibout, A.; Tayebi, N.; Kharroubi, A.; Ammari, A.; Dalache, C. Sol–Gel Synthesis of Nickel-Doped Cr_2O_3 Thin Films. *J. Mol. Eng. Mater.,* **2018**, *05*(04), 1750012.
[http://dx.doi.org/10.1142/S2251237317500125]

[59]   Tian, S.; Ye, X.; Dong, Y.; Li, W.; Zhang, B.; Li, B.; Feng, H. Production and Characterization of Chromium Oxide (Cr_2O_3) *via* a Facile Combination of Electrooxidation and Calcination. *Int. J. Electrochem. Sci.,* **2019**, *14*(9), 8805-8818.
[http://dx.doi.org/10.20964/2019.09.21]

[60]   Asma., Almontasser; Azra., Parveen Optical Properties of Chromium Oxide (III) Nanoparticles and Assessment of Their Antibacterial Activity. *AIP Conference Proceedings,* **2020**.https://doi.org/https://doi.org/10.1063/5.0025751

[61]   Chen, J.; Zou, H.; Yao, Q.; Luo, M.; Li, X.; Lu, Z.H. Cr_2O_3 -Modified NiFe Nanoparticles as a Noble-Metal-Free Catalyst for Complete Dehydrogenation of Hydrazine in Aqueous Solution. *Appl. Surf. Sci.,* **2019**, *2020*(501), 144247.
[http://dx.doi.org/10.1016/j.apsusc.2019.144247]

[62]   Yang, S.; Feng, C.; Spence, D.; Al Hindawi, A.M.A.A.; Latimer, E.; Ellis, A.M.; Binns, C.; Peddis, D.; Dhesi, S.S.; Zhang, L.; Zhang, Y.; Trohidou, K.N.; Vasilakaki, M.; Ntallis, N.; MacLaren, I.; de Groot, F.M.F. Robust Ferromagnetism of Chromium Nanoparticles Formed in Superfluid Helium. *Adv. Mater.,* **2017**, *29*(1), 1-6.
[http://dx.doi.org/10.1002/adma.201604277] [PMID: 27787938]

[63]   Bañobre-López, M.; Vázquez-Vázquez, C.; Rivas, J.; López-Quintela, M.A. Magnetic Properties of Chromium (III) Oxide Nanoparticles. *Nanotechnology,* **2003**, *14*(2), 318-322.
[http://dx.doi.org/10.1088/0957-4484/14/2/342]

[64]   Grempel, D.R. Surface Magnetic Ordering in Chromium. *Phys. Rev. B Condens. Matter,* **1981**, *24*(7), 3928-3938.
[http://dx.doi.org/10.1103/PhysRevB.24.3928]

[65]   Klebanoff, L.E.; Robey, S.W.; Liu, G.; Shirley, D.A. Observation of a Surface Magnetic Phase Transition on Cr(100). *Phys. Rev. B Condens. Matter,* **1984**, *30*(2), 1048-1051.
[http://dx.doi.org/10.1103/PhysRevB.30.1048]

[66]   Sugawara, E.; Nikaido, H. Properties of AdeABC and AdeIJK efflux systems of Acinetobacter baumannii compared with those of the AcrAB-TolC system of Escherichia coli. *Antimicrob. Agents Chemother.,* **2014**, *58*(12), 7250-7257.
[http://dx.doi.org/10.1128/AAC.03728-14] [PMID: 25246403]

[67]   Meier, F.; Pescia, D.; Schriber, T. Oxygen-Induced Magnetism of the Nonreconstructed Chromium (100) Surface. *Phys. Rev. Lett.,* **1982**, *48*(9), 645-648.
[http://dx.doi.org/10.1103/PhysRevLett.48.645]

[68]   Si, P.Z.; Wang, X.L.; Xiao, X.F.; Chen, H.J.; Liu, X.Y.; Jiang, L.; Liu, J.J.; Jiao, Z.W.; Ge, H.L. Structure and Magnetic Properties of Cr_2O_3/CrO_2 Nanoparticles Prepared by Reactive Laser Ablation and Oxidation under High Pressure of Oxygen. *J. Magn.,* **2015**, *20*(3), 211-214.
[http://dx.doi.org/10.4283/JMAG.2015.20.3.211]

[69]   Jin, C.H.; Si, P.Z.; Xiao, X.F.; Feng, H.; Wu, Q.; Ge, H.L.; Zhong, M. Structure and Magnetic Properties of $Cr/Cr_2O_3/$ CrO_2 Microspheres Prepared by Spark Erosion and Oxidation under High Pressure of Oxygen. *Mater. Lett.,* **2013**, *92*, 213-215.
[http://dx.doi.org/10.1016/j.matlet.2012.10.126]

[70]   Lei, S.; Peng, X.; Liang, Z.; Li, X.; Wang, C.; Cheng, B.; Xiao, Y.; Zhou, L. Self-Template Formation and Properties Study of Cr_2O_3 Nanoparticle Tubes. *J. Mater. Chem.,* **2012**, *22*(4), 1643-1651.
[http://dx.doi.org/10.1039/C1JM13127C]

[71]   Vázquez-Vázquez, C.; Bañobre-López, M.; López-Quintela, M.A.; Hueso, L.E.; Rivas, J. Evidence of Weak Ferromagnetism in Chromium(III) Oxide Particles. *J. Magn. Magn. Mater.,* **2004**, *272–276*, 1547-1548.
[http://dx.doi.org/10.1016/j.jmmm.2003.12.260]

[72]   Yang, Z. a Jing Zhang, a Daqiang Gao, a Zhonghua Zhu, b Guijin Yang, c and D. X. Unexpected Surface Superparamagnetism in Antiferromagnetic Cr_2O_3 Nanoparticles. *RSC Advances,* **2015**, *3*, 10715-10722.
[http://dx.doi.org/10.1039/b000000x]

[73]   Nagata, T. Indium Oxide: In_2O_3. Single Crystals of Electronic Materials: Growth and Properties ,

2018; pp. 523-546.
[http://dx.doi.org/10.1016/B978-0-08-102096-8.00015-X]

[74]   Klein, A.; Körber, C.; Wachau, A.; Säuberlich, F.; Gassenbauer, Y.; Harvey, S.P.; Proffit, D.E.; Mason, T.O. Transparent Conducting Oxides for Photovoltaics: Manipulation of Fermi Level, Work Function and Energy Band Alignment. *Materials (Basel),* **2010**, *3*(11), 4892-4914.
[http://dx.doi.org/10.3390/ma3114892] [PMID: 28883359]

[75]   Gopchandran, K.G.; Joseph, B.; Abraham, J.T. *Preparation of Transparent Electrically Conducting Indium Oxide Films by Reactive Vacuum Evaporation.,* **1997**, (97), 547-550.

[76]   Hamberg, I.; Granqvist, C.G. Evaporated Sn-Doped In_2O_3 Films: Basic Optical Properties and Applications to Energy-Efficient Windows. *J. Appl. Phys.,* **1986**, *60*(11), R123-R160.
[http://dx.doi.org/10.1063/1.337534]

[77]   Ohhata, Y.; Shinoki, F.; Yoshida, S. Optical Properties of r.f. Reactive Sputtered Tin-Doped In_2O_3 Films. *Thin Solid Films,* **1979**, *59*(2), 255-261.
[http://dx.doi.org/10.1016/0040-6090(79)90298-0]

[78]   Xiangying, Chen a; Zhongjie, Zhang a; Xingfa, Zhang a Single-Source Approach to the Synthesis of In_2S_3 and In_2O_3 Crystallites and Their Optical Properties. *Chemical Physics Letters,* **2005**, *407*, 482-486.
[http://dx.doi.org/10.1016/j.cplett.2005.03.141]

[79]   Phokha, S.; Seraphin, S. Indium Oxide (In_2O_3). *Nanoparticles Using Aloe Vera Plant Extract : Synthesis and Optical Properties,* **2015**, *2*(April), 161-165.

[80]   Ye, E.; Zhang, S.Y.; Lim, S.H.; Liu, S.; Han, M.Y. Morphological tuning, self-assembly and optical properties of indium oxide nanocrystals. *Phys. Chem. Chem. Phys.,* **2010**, *12*(38), 11923-11929.
[http://dx.doi.org/10.1039/c0cp00138d] [PMID: 20820551]

[81]   Farvid, S.S.; Dave, N.; Wang, T.; Radovanovic, P.V. Dopant-Induced Manipulation of the Growth and Structural Metastability of Colloidal Indium Oxide Nanocrystals. *J. Phys. Chem. C,* **2009**, *113*(36), 15928-15933.
[http://dx.doi.org/10.1021/jp905281k]

[82]   Forsh, E.A.; Marikutsa, A.V.; Martyshov, M.N.; Forsh, P.A.; Rumyantseva, M.N.; Gaskov, A.M.; Kashkarov, P.K. Charge Carrier Transport Mechanisms in Nanocrystalline Indium Oxide. *Thin Solid Films,* **2014**, *558*, 320-325.
[http://dx.doi.org/10.1016/j.tsf.2014.02.064]

[83]   Ivanovskaya, M.; Gurlo, A.; Bogdanov, P. Mechanism of O_3 and NO_2 Detection and Selectivity of In_2O_3 Sensors. *Sens. Actuators B Chem.,* **2001**, *77*(1–2), 264-267.
[http://dx.doi.org/10.1016/S0925-4005(01)00708-0]

[84]   Shen, X.; Liu, H.; Fan, X.; Jiang, Y.; Hong, J.; Xu, Z. Construction and Photoluminescence of In_2O_3 Nanotube Array by CVD-Template Method. *J. Cryst. Growth,* **2005**, *276*(3-4), 471-477.
[http://dx.doi.org/10.1016/j.jcrysgro.2004.11.394]

[85]   Hosamani, G.; Pattar, J.; Manjanna, J.; Jagadale, B.N. Influence of Oxidant to Fuel Ratio on the Structural and Optical Properties of In2O3 Nanoparticles Synthesized by Auto-Combustion Method. *Mater. Today Proc.,* **2017**, *4*(11), 12075-12079.
[http://dx.doi.org/10.1016/j.matpr.2017.09.133]

[86]   Almontasser, A.; Parveen, A. Synthesis and Characterization of Indium Oxide Nanoparticles. *3Rd International Conference on Condensed Matter and Applied Physics (Icc-2019),* **2020**.
[http://dx.doi.org/10.1063/5.0001683]

[87]   Prakash, R.; Kumar, S.; Ahmed, F.; Lee, C.G.; Song, J. Il. Room Temperature Ferromagnetism in Ni Doped In_2O_3 Nanoparticles. *Thin Solid Films,* **2011**, *519*(23), 8243-8246.
[http://dx.doi.org/10.1016/j.tsf.2011.03.105]

[88]   Wongsaprom, K.; Sonsupap, S.; Maensiri, S.; Kidkhunthod, P. Room-Temperature Ferromagnetism in

Fe-Doped In2O3 Nanoparticles. *Appl. Phys., A Mater. Sci. Process.,* **2015**, *121*(1), 239-244.
[http://dx.doi.org/10.1007/s00339-015-9416-5]

[89] Singhal, A.; Achary, S.N.; Manjanna, J.; Jayakumar, O.D.; Kadam, R.M.; Tyagi, A.K. Colloidal Fe-Doped Indium Oxide Nanoparticles: Facile Synthesis, Structural, and Magnetic Properties. *J. Phys. Chem. C,* **2009**, *113*(9), 3600-3606.
[http://dx.doi.org/10.1021/jp8097846]

[90] Meng, X.; Tang, L.; Li, J. Room-Temperature Ferromagnetism in Co-Doped In_2O_3 Nanocrystals. *J. Phys. Chem. C,* **2010**, *114*(41), 17569-17573.
[http://dx.doi.org/10.1021/jp106767n]

[91] Yoo, Y.K.; Xue, Q.; Lee, H.C.; Cheng, S.; Xiang, X.D.; Dionne, G.F.; Xu, S.; He, J.; Chu, Y.S.; Preite, S.D.; Lofland, S.E.; Takeuchi, I. Bulk Synthesis and High-Temperature Ferromagnetism of (In $_{1-x}$Fe $_x$) $_2O_{3-\sigma}$ with Cu Co-Doping. *Appl. Phys. Lett.,* **2005**, *86*(4), 8-10.
[http://dx.doi.org/10.1063/1.1854720]

[92] Jiang, F.X.; Feng, Q.; Quan, Z.Y.; Ma, R.R.; Heald, S.M.; Gehring, G.A.; Xu, X.H. The Role of Cu Codoping on the Fe Metal Clustering and Ferromagnetism in Fe-Doped In2O3 Films. *Mater. Res. Bull.,* **2013**, *48*(9), 3178-3182.
[http://dx.doi.org/10.1016/j.materresbull.2013.04.066]

[93] Sun, Q.; Zeng, Y.P.; Jiang, D. High Magnetic Field Inducing Magnetic Transitions of Fe $^{3+}$ and Ni $^{2+}$ Doped In_2O_3 Nanocubes. *Solid State Commun.,* **2011**, *151*(18), 1220-1223.
[http://dx.doi.org/10.1016/j.ssc.2011.06.004]

[94] Prakash, R.; Song, J. Il; Kumar, S.; Lee, C. G. Study of Structural and Magnetic Properties of Co-Doped In_2O_3 Nanoparticles. *Int. J. Nanosci.,* **2011**, *10*(4–5), 961-965.
[http://dx.doi.org/10.1142/S0219581X11008721]

[95] Li, X.; Xia, C.; Pei, G.; He, X. Synthesis and Characterization of Room-Temperature Ferromagnetism in Fe- and Ni-Co-Doped In_2O_3. *J. Phys. Chem. Solids,* **2007**, *68*(10), 1836-1840.
[http://dx.doi.org/10.1016/j.jpcs.2007.05.019]

[96] Naik, M.Z.; Salker, A.V. Nanoparticles and Their Applications A Systematic Study of Cobalt Doped In_2O_3 Nanoparticles and Their Applications. *Materials Research Innovations I,* **2016**, *8917*, 1-7.
[http://dx.doi.org/10.1080/14328917.2016.1207044]

[97] Badar, N.; Chayed, N.F.; Rusdi, R.; Kamarudin, N.; Kamarulzaman, N. Band Gap Energies of Magnesium Oxide Nanomaterials Synthesized by the Sol-Gel Method. *Adv. Mat. Res.,* **2012**, *545*, 157-160.
[http://dx.doi.org/10.4028/www.scientific.net/AMR.545.157]

[98] Roessler, D.M.; Walker, W.C. Electronic Spectrum and Ultraviolet Optical Properties of Crystalline MgO. *Phys. Rev.,* **1967**, *159*(3), 733-738.
[http://dx.doi.org/10.1103/PhysRev.159.733]

[99] Stankic, S.; Müller, M.; Diwald, O.; Sterrer, M.; Knözinger, E.; Bernardi, J. Size-dependent optical properties of MgO nanocubes. *Angew. Chem. Int. Ed.,* **2005**, *44*(31), 4917-4920.
[http://dx.doi.org/10.1002/anie.200500663] [PMID: 15999373]

[100] Wobbe, M.C.; Kerridge, A.; Zwijnenburg, M.A. Optical excitation of MgO nanoparticles; a computational perspective. *Phys. Chem. Chem. Phys.,* **2014**, *16*(40), 22052-22061.
[http://dx.doi.org/10.1039/C4CP03442B] [PMID: 25207882]

[101] Hadia, N.M.A.; Mohamed, H.A.H. Characteristics and Optical Properties of MgO Nanowires Synthesized by Solvothermal Method. *Mater. Sci. Semicond. Process.,* **2015**, *29*, 238-244.
[http://dx.doi.org/10.1016/j.mssp.2014.03.049]

[102] Niedermeier, C.A.; Råsander, M.; Rhode, S.; Kachkanov, V.; Zou, B.; Alford, N.; Moram, M.A. Band gap bowing in NixMg1-xO. *Sci. Rep.,* **2016**, *6*(1), 31230.
[http://dx.doi.org/10.1038/srep31230] [PMID: 27503808]

[103]   Asakura, K.; Iwasawa, Y. A Structure Model as the Origin of Catalytic Properties of Metal-Doped MgO Systems. *Mater. Chem. Phys.,* **1988**, *18*(5–6), 499-512.
[http://dx.doi.org/10.1016/0254-0584(88)90019-3]

[104]   Almontasser, A.; Parveen, A.; Azam, A. Synthesis, Characterization and Antibacterial Activity of Magnesium Oxide (MgO) Nanoparticles. *IOP Conf. Series Mater. Sci. Eng.,* **2019**, *577*(1), 012051.
[http://dx.doi.org/10.1088/1757-899X/577/1/012051]

[105]   Sarmadian, N.; Saniz, R.; Lamoen, D.; Partoens, B. Influence of Al Concentration on the Optoelectronic Properties of Al-Doped MgO. *Phys. Rev. B Condens. Matter Mater. Phys.,* **2012**, *86*(20), 1-5.
[http://dx.doi.org/10.1103/PhysRevB.86.205129]

[106]   Sadjadi, S.; Farzaneh, V.; Shirvani, S.; Ghashghaee, M. Preparation of Cu-MgO Catalysts with Different Copper Precursors and Precipitating Agents for the Vapor-Phase Hydrogenation of Furfural. *Korean J. Chem. Eng.,* **2017**, *34*(3), 692-700.
[http://dx.doi.org/10.1007/s11814-016-0344-7]

[107]   Benedetti, S.; Nilius, N.; Valeri, S. Chromium-Doped MgO Thin Films: Morphology, Electronic Structure, and Segregation Effects. *J. Phys. Chem. C,* **2015**, *119*(45), 25469-25475.
[http://dx.doi.org/10.1021/acs.jpcc.5b09033]

[108]   Suzuki, T.; Ohishi, Y.; Tani, T. Structural and Fluorescence Properties of Ni:MgO-SiO2 Particles Synthesized by Flame Spray Pyrolysis. *Materials Science and Engineering B. Mater. Sci. Eng. B,* **2006**, *128*(1–3), 151-155.
[http://dx.doi.org/10.1016/j.mseb.2005.11.035]

[109]   Tardío, M.M.; Ramírez, R.; González, R.; Chen, Y. P-Type Semiconducting Properties in Lithium-Doped MgO Single Crystals. *Phys. Rev. B Condens. Matter,* **2002**, *66*(13), 1-8.
[http://dx.doi.org/10.1103/PhysRevB.66.134202]

[110]   Benedetti, S.; Nilius, N.; Valeri, S.; Tosoni, S.; Albanese, E.; Pacchioni, G. Dopant-Induced Diffusion Processes at Metal-Oxide Interfaces Studied for Iron- and Chromium-Doped MgO/Mo(001) Model Systems. *J. Phys. Chem. C,* **2016**, *120*(25), 13604-13609.
[http://dx.doi.org/10.1021/acs.jpcc.6b04182]

[111]   Mishra, D.; Mandal, B.P.; Mukherjee, R.; Naik, R.; Lawes, G.; Nadgorny, B. Oxygen Vacancy Enhanced Room Temperature Magnetism in Al-Doped MgO Nanoparticles. *Appl. Phys. Lett.,* **2013**, *102*(18), 1-5.
[http://dx.doi.org/10.1063/1.4804425]

[112]   Pesci, M.; Gallino, F.; Di Valentin, C.; Pacchioni, G. Nature of Defect States in Nitrogen-Doped MgO. *J. Phys. Chem. C,* **2010**, *114*(2), 1350-1356.
[http://dx.doi.org/10.1021/jp9097556]

[113]   Wu, P.; Cao, G.; Tang, F.; Huang, M. Electronic and Magnetic Properties of Transition Metal Doped MgO Sheet: A Density-Functional Study. *Comput. Mater. Sci.,* **2014**, *86*, 180-185.
[http://dx.doi.org/10.1016/j.commatsci.2014.01.052]

[114]   Prada, S.; Giordano, L.; Pacchioni, G. Li, Al, and Ni Substitutional Doping in MgO Ultrathin Films on Metals: Work Function Tuning *via* Charge Compensation. *J. Phys. Chem. C,* **2012**, *116*(9), 5781-5786.
[http://dx.doi.org/10.1021/jp211363q]

[115]   Singh, J.; Paul, S.; Bhardwaj, P.; Kumar, R. Investigation of structural and optical properties of MnxMg$_{1-x}$ O (x=0.00, 0.03, 0.07) Nanoparticles Prepared by Wet Chemical Technique. *AIP Conference Proceedings,* **2020**, 1-5.https://doi.org/https://doi.org/10.1063/5.0017156

[116]   Kumar, A.; Kumar, J.; Priya, S. Defect and Adsorbate Induced Ferromagnetic Spin-Order in Magnesium Oxide Nanocrystallites. *Appl. Phys. Lett.,* **2012**, *100*(19), 1-5.
[http://dx.doi.org/10.1063/1.4712058]

[117]   Yang, S.W.; Lim, D.H.; Yoo, D.J.; Kang, Y.; Lee, C.G.; Kang, G.M. Opto-Magnetic Properties of

Nano-Structured MgO:Al Powders Prepared in a Micro Drop Fluidized Reactor. *Adv. Powder Technol.,* **2018**, *29*(3), 499-505.
[http://dx.doi.org/10.1016/j.apt.2018.02.022]

[118]   Varshney, D.; Dwivedi, S. On the Synthesis, Structural, Optical and Magnetic Properties of Nano-Size Zn-MgO. *Superlattices Microstruct.,* **2015**, *85*, 886-893.
[http://dx.doi.org/10.1016/j.spmi.2015.07.016]

[119]   Hanish, H.H.; Edrees, S.J.; Shukur, M.M. The Effect of Transition Metals Incorporation on the Structural and Magnetic Properties of Magnesium Oxide Nanoparticles. *International Journal of Engineering, Transactions A: Basics,* **2020**, *33*(4), pp. 647-656.
[http://dx.doi.org/10.5829/ije.2020.33.04a.16]

[120]   Saidin, S.; Jumat, M.A.; Mohd Amin, N.A.A.; Saleh Al-Hammadi, A.S. Organic and Inorganic Antibacterial Approaches in Combating Bacterial Infection for Biomedical Application. *Mater. Sci. Eng. C,* **2020**, *2021*(118), 1-19.
[http://dx.doi.org/10.1016/j.msec.2020.111382] [PMID: 33254989]

[121]   Marambio-Jones, C.; Hoek, E.M.V. A Review of the Antibacterial Effects of Silver Nanomaterials and Potential Implications for Human Health and the Environment. *J. Nanopart. Res.,* **2010**, *12*(5), 1531-1551.
[http://dx.doi.org/10.1007/s11051-010-9900-y]

[122]   Xu, C.; Zheng, J.; Wu, A. Antibacterial Applications of TiO_2 Nanoparticles. In: *TiO2 Nanoparticles: Applications in Nanobiotechnology and Nanomedicine*; , **2020**; pp. 105-132.

[123]   Zhang, L.; Jiang, Y.; Ding, Y.; Povey, M.; York, D. *Investigation into the Antibacterial Behaviour of Suspensions of ZnO Nanoparticles (ZnO Nanofluids)*; Nanoparticle Research, **2007**, pp. 479-489.
[http://dx.doi.org/10.1007/s11051-006-9150-1]

[124]   Sawai, J.; Kojima, H.; Igarashi, H.; Hashimoto, A.; Shoji, S.; Sawaki, T.; Hakoda, A.; Kawada, E.; Kokugan, T.; Shimizu, M. Antibacterial Characteristics of Magnesium Oxide Powder. *World J. Microbiol. Biotechnol.,* **2000**, *16*(2), 187-194.https://doi.org/https://doi.org/10.1023/A:1008916209784
[http://dx.doi.org/10.1023/A:1008916209784]

[125]   Subhan, M.A.; Jhuma, S.S.; Chandra Saha, P.; Alam, M.M.; Asiri, A.M.; Al-Mamun, M.; Attia, S.A.; Emon, T.H.; Azad, A.K.; Rahman, M.M. Efficient Selective 4-Aminophenol Sensing and Antibacterial Activity of Ternary $Ag_2O \cdot SnO_2 \cdot Cr_2O_3$ Nanoparticles. *New J. Chem.,* **2019**, *43*(26), 10352-10365.
[http://dx.doi.org/10.1039/C9NJ01760G]

[126]   Sangwan, P.; Kumar, H.; Purewal, S.S. Antibacterial activity of chemically synthesized chromium oxide nanoparticles against enterococcus faecalis. *International journal of engineering & science,* **2016**, *4*(8), 492-499.

[127]   Esam, J. Characterization of Cr2O3 Nanoparticles Prepared by Using Different Plant Extracts. *Academia Journal of Biotechnology,* **2018**, *8*(3), 052-058.

[128]   Gupta, N.; Sp, R. Synthesis and Characterization of Chromium Oxide Nanoparticles by Chemical Route and Its Antibacterial Activity against E. Coli Bacteria. *International Research Journal of Engineering and Technology,* **2019**, *6*(12), 2695-2700.

[129]   Sangwan, P.; Kumar, H. Synthesis, Characterization, and Antibacterial Activities of Chromium Oxide Nanoparticles against Klebsiella Pneumoniae. *Asian J. Pharm. Clin. Res.,* **2017**, *10*(2), 206-209.
[http://dx.doi.org/10.22159/ajpcr.2017.v10i2.15189]

[130]   Kanakalakshmi, A.; Janaki, V.; Shanthi, K.; Kamala-Kannan, S. Biosynthesis of Cr(III) nanoparticles from electroplating wastewater using chromium-resistant Bacillus subtilis and its cytotoxicity and antibacterial activity. *Artif. Cells Nanomed. Biotechnol.,* **2017**, *45*(7), 1304-1309.
[http://dx.doi.org/10.1080/21691401.2016.1228660] [PMID: 27608920]

[131]   Ramesh, C.; Mohan Kumar, K.T.; Latha, N.; Ragunathan, V. Green Synthesis of Cr_2O_3 Nanoparticles Using Tridax Procumbens Leaf Extract and Its Antibacterial Activity on Escherichia Coli. *Curr.*

Nanosci., **2012**, *8*(4), 603-607.
[http://dx.doi.org/10.2174/157341312801784366]

[132] Chang, W.K.; Sun, D.S.; Chan, H.; Huang, P.T.; Wu, W.S.; Lin, C.H.; Tseng, Y.H.; Cheng, Y.H.; Tseng, C.C.; Chang, H.H. Visible light-responsive core-shell structured In_2O_3@$CaIn_2O_4$ photocatalyst with superior bactericidal properties and biocompatibility. *Nanomedicine,* **2012**, *8*(5), 609-617.
[http://dx.doi.org/10.1016/j.nano.2011.09.016] [PMID: 22033083]

[133] Ramesh, R.K.; Chottanahalli, S.P.K.; Madegowda, N.M.; Rai, V.R.; Ananda, S. Electrochemical Synthesis of Hierarchal Flower-like Hierarchical In_2O_3/ZnO Nanocatalyst for Textile Industry Effluent Treatment, Photo-Voltaic, [Rad]OH Scavenging and Anti-Bacterial Studies. *Catal. Commun.,* **2017**, *89*, 25-28.
[http://dx.doi.org/10.1016/j.catcom.2016.10.006]

[134] Skorb, E.V.; Antonouskaya, L.I.; Belyasova, N.A.; Shchukin, D.G.; Möhwald, H.; Sviridov, D.V. Antibacterial Activity of Thin-Film Photocatalysts Based on Metal-Modified TiO_2 and TiO_2:In_2O_3 Nanocomposite. *Appl. Catal. B,* **2008**, *84*(1–2), 94-99.
[http://dx.doi.org/10.1016/j.apcatb.2008.03.007]

[135] Bindhu, M.R.; Umadevi, M.; Kavin Micheal, M.; Arasu, M.V.; Abdullah Al-Dhabi, N. Structural, Morphological and Optical Properties of MgO Nanoparticles for Antibacterial Applications. *Mater. Lett.,* **2016**, *166*, 19-22.
[http://dx.doi.org/10.1016/j.matlet.2015.12.020]

[136] Nguyen, N. T.; Grelling, N.; Wetteland, C. L.; Rosario, R. Antimicrobial Activities and Mechanisms of Magnesium Oxide Nanoparticles (NMgO) against Pathogenic Bacteria, Yeasts, and Biofilms. *Sci,* **2018**, 1-23.
[http://dx.doi.org/10.1038/s41598-018-34567-5]

[137] Stoimenov, P.K.; Klinger, R.L.; Marchin, G.L.; Klabunde, K.J. Metal Oxide Nanoparticles as Bactericidal Agents. *Langmuir,* **2002**, *18*(17), 6679-6686.
[http://dx.doi.org/10.1021/la0202374]

[138] Jin, T.; He, Y. Antibacterial Activities of Magnesium Oxide (MgO) Nanoparticles against Foodborne Pathogens. *J. Nanopart. Res.,* **2011**, *13*(12), 6877-6885.
[http://dx.doi.org/10.1007/s11051-011-0595-5]

[139] Pugazhendhi, A.; Prabhu, R.; Muruganantham, K.; Shanmuganathan, R.; Natarajan, S. Anticancer, antimicrobial and photocatalytic activities of green synthesized magnesium oxide nanoparticles (MgONPs) using aqueous extract of Sargassum wightii. *J. Photochem. Photobiol. B,* **2019**, *190*, 86-97.
[http://dx.doi.org/10.1016/j.jphotobiol.2018.11.014] [PMID: 30504053]

[140] Almontasser, A.; Parveen, A.; Hashim, M.; Ul-Hamid, A.; Azam, A. Structural, Optical, and Antibacterial Properties of Pure and Doped (Ni, Co, and Fe) Cr_2O_3 Nanoparticles: A Comparative Study. *Appl. Nanosci.,* **2021**, *11*(2), 583-604.
[http://dx.doi.org/10.1007/s13204-020-01590-w]

[141] Torgersen, H.; Lassen, J.; Jelsoe, E.; Rusanen, T.; Nielsen, T.H. Antimicrobial Resistance: The Example of SA. *J. Biolaw Bus.,* **2000**, *3*(3), 53-59.
[http://dx.doi.org/10.1172/JCI200318535.In]

[142] Guo, Y.; Song, G.; Sun, M.; Wang, J.; Wang, Y. Prevalence and Therapies of Antibiotic-Resistance in *Staphylococcus aureus. Front. Cell. Infect. Microbiol.,* **2020**, *10*(March), 107.
[http://dx.doi.org/10.3389/fcimb.2020.00107] [PMID: 32257966]

[143] Wang, L.; Hu, C.; Shao, L. The-Antimicrobial-Activity-of-Nanoparticles--Present-Situati. *Int. J. Nanomedicine,* **2017**, *12*, 1227-1249.
[http://dx.doi.org/10.2147/IJN.S121956] [PMID: 28243086]

[144] Sirelkhatim, A.; Mahmud, S.; Seeni, A.; Kaus, N.H.M.; Ann, L.C.; Bakhori, S.K.M.; Hasan, H.; Mohamad, D. Review on Zinc Oxide Nanoparticles: Antibacterial Activity and Toxicity Mechanism. *Nano-Micro Lett.,* **2015**, *7*(3), 219-242.

[http://dx.doi.org/10.1007/s40820-015-0040-x] [PMID: 30464967]

[145] Feng, Q.L.; Wu, J.; Chen, G.Q.; Cui, F.Z.; Kim, T.N.; Kim, J.O. A mechanistic study of the antibacterial effect of silver ions on Escherichia coli and Staphylococcus aureus. *J. Biomed. Mater. Res.,* **2000**, *52*(4), 662-668.
[http://dx.doi.org/10.1002/1097-4636(20001215)52:4<662::AID-JBM10>3.0.CO;2-3]   [PMID: 11033548]

[146] Slavin, Y.N.; Asnis, J.; Häfeli, U.O.; Bach, H. Metal nanoparticles: understanding the mechanisms behind antibacterial activity. *J. Nanobiotechnology,* **2017**, *15*(1), 65.
[http://dx.doi.org/10.1186/s12951-017-0308-z] [PMID: 28974225]

[147] Dizaj, S.M.; Lotfipour, F.; Barzegar-Jalali, M.; Zarrintan, M.H.; Adibkia, K. Antimicrobial activity of the metals and metal oxide nanoparticles. *Mater. Sci. Eng. C,* **2014**, *44*, 278-284.
[http://dx.doi.org/10.1016/j.msec.2014.08.031] [PMID: 25280707]

[148] Kaweeteerawat, C.; Ivask, A.; Liu, R.; Zhang, H.; Chang, C.H.; Low-Kam, C.; Fischer, H.; Ji, Z.; Pokhrel, S.; Cohen, Y.; Telesca, D.; Zink, J.; Mädler, L.; Holden, P.A.; Nel, A.; Godwin, H. Toxicity of metal oxide nanoparticles in Escherichia coli correlates with conduction band and hydration energies. *Environ. Sci. Technol.,* **2015**, *49*(2), 1105-1112.
[http://dx.doi.org/10.1021/es504259s] [PMID: 25563693]

[149] Li, Y.; Yang, D.; Wang, S.; Li, C.; Xue, B.; Yang, L.; Shen, Z.; Jin, M.; Wang, J.; Qiu, Z. The Detailed Bactericidal Process of Ferric Oxide Nanoparticles on E. coli. *Molecules,* **2018**, *23*(3), E606.
[http://dx.doi.org/10.3390/molecules23030606] [PMID: 29518002]

[150] Stankic, S.; Suman, S.; Haque, F.; Vidic, J. Pure and multi metal oxide nanoparticles: synthesis, antibacterial and cytotoxic properties. *J. Nanobiotechnology,* **2016**, *14*(1), 73.
[http://dx.doi.org/10.1186/s12951-016-0225-6] [PMID: 27776555]

[151] Xia, T.; Kovochich, M.; Liong, M.; Mädler, L.; Gilbert, B.; Shi, H.; Yeh, J.I.; Zink, J.I.; Nel, A.E. Comparison of the mechanism of toxicity of zinc oxide and cerium oxide nanoparticles based on dissolution and oxidative stress properties. *ACS Nano,* **2008**, *2*(10), 2121-2134.
[http://dx.doi.org/10.1021/nn800511k] [PMID: 19206459]

[152] Soltani Nezhad, S.; Rabbani Khorasgani, M.; Emtiazi, G.; Yaghoobi, M.M.; Shakeri, S. Isolation of copper oxide (CuO) nanoparticles resistant Pseudomonas strains from soil and investigation on possible mechanism for resistance. *World J. Microbiol. Biotechnol.,* **2014**, *30*(3), 809-817.
[http://dx.doi.org/10.1007/s11274-013-1481-3] [PMID: 24146307]

[153] Tamayo, L.A.; Zapata, P.A.; Vejar, N.D.; Azócar, M.I.; Gulppi, M.A.; Zhou, X.; Thompson, G.E.; Rabagliati, F.M.; Páez, M.A. Release of silver and copper nanoparticles from polyethylene nanocomposites and their penetration into Listeria monocytogenes. *Mater. Sci. Eng. C,* **2014**, *40*, 24-31.
[http://dx.doi.org/10.1016/j.msec.2014.03.037] [PMID: 24857461]

[154] Rousk, J.; Ackermann, K.; Curling, S.F.; Jones, D.L. Comparative toxicity of nanoparticulate CuO and ZnO to soil bacterial communities. *PLoS One,* **2012**, *7*(3), e34197.
[http://dx.doi.org/10.1371/journal.pone.0034197] [PMID: 22479561]

[155] Pelletier, D.A.; Suresh, A.K.; Holton, G.A.; McKeown, C.K.; Wang, W.; Gu, B.; Mortensen, N.P.; Allison, D.P.; Joy, D.C.; Allison, M.R.; Brown, S.D.; Phelps, T.J.; Doktycz, M.J. Effects of engineered cerium oxide nanoparticles on bacterial growth and viability. *Appl. Environ. Microbiol.,* **2010**, *76*(24), 7981-7989.
[http://dx.doi.org/10.1128/AEM.00650-10] [PMID: 20952651]

[156] Stoimenov, P.K.; Klinger, R.L.; Marchin, G.L.; Klabunde, K.J. *Metal Oxide Nanoparticles as Bactericidal Agents.,* **2002**, (13), 6679-6686.

[157] Simon-deckers, L.; Herlin-boime, N.; Marie, P. Shape-Dependent Toxicological Impact of Metal Oxide Nanoparticles and Carbon Nanotubes toward Bacteria. *American Chemical Society,* **2009**, *18*(17), 8423-8429.

[158] McQuillan, J.S.; Infante, H.G.; Stokes, E.; Shaw, A.M. Silver nanoparticle enhanced silver ion stress response in Escherichia coli K12. *Nanotoxicology,* **2012**, *6*(8), 857-866.
[http://dx.doi.org/10.3109/17435390.2011.626532] [PMID: 22007647]

[159] Leung, Y.H.; Ng, A.M.C.; Xu, X.; Shen, Z.; Gethings, L.A.; Wong, M.T.; Chan, C.M.N.; Guo, M.Y.; Ng, Y.H.; Djurišić, A.B.; Lee, P.K.H.; Chan, W.K.; Yu, L.H.; Phillips, D.L.; Ma, A.P.Y.; Leung, F.C.C. Mechanisms of antibacterial activity of MgO: non-ROS mediated toxicity of MgO nanoparticles towards Escherichia coli. *Small,* **2014**, *10*(6), 1171-1183.
[http://dx.doi.org/10.1002/smll.201302434] [PMID: 24344000]

[160] Yu, J.; Zhang, W.; Ghasaban, S.; Atai, M.; Imani, M.; Matharu, R. K.; Ciric, L. Enhanced Bioactivity of ZnO Nanoparticles — an Antimicrobial Study. *Science and Technology of Advanced Materials,* *9*(4)
[http://dx.doi.org/10.1088/1468-6996/9/3/035004]

[161] Shkodenko, L.; Kassirov, I.; Koshel, E. Metal Oxide Nanoparticles Against Bacterial Biofilms: Perspectives and Limitations. *Microorganisms,* **2020**, *8*(10), 1-21.
[http://dx.doi.org/10.3390/microorganisms8101545] [PMID: 33036373]

[162] Memar, M.Y.; Ghotaslou, R.; Samiei, M.; Adibkia, K. Antimicrobial use of reactive oxygen therapy: current insights. *Infect. Drug Resist.,* **2018**, *11*, 567-576.
[http://dx.doi.org/10.2147/IDR.S142397] [PMID: 29731645]

[163] Franci, G.; Falanga, A.; Galdiero, S.; Palomba, L.; Rai, M.; Morelli, G.; Galdiero, M. Silver nanoparticles as potential antibacterial agents. *Molecules,* **2015**, *20*(5), 8856-8874.
[http://dx.doi.org/10.3390/molecules20058856] [PMID: 25993417]

[164] Behera, N.; Arakha, M.; Priyadarshinee, M.; Pattanayak, B.S.; Soren, S.; Jha, S.; Mallick, B.C. Oxidative Stress Generated at Nickel Oxide Nanoparticle Interface Results in Bacterial Membrane Damage Leading to Cell Death. *RSC Advances,* **2019**, *9*(43), 24888-24894.
[http://dx.doi.org/10.1039/C9RA02082A]

[165] Raghunath, A.; Perumal, E. Metal oxide nanoparticles as antimicrobial agents: a promise for the future. *Int. J. Antimicrob. Agents,* **2017**, *49*(2), 137-152.
[http://dx.doi.org/10.1016/j.ijantimicag.2016.11.011] [PMID: 28089172]

[166] Kumar, A.; Pandey, A.K.; Singh, S.S.; Shanker, R.; Dhawan, A. Engineered ZnO and TiO(2) nanoparticles induce oxidative stress and DNA damage leading to reduced *via*bility of Escherichia coli. *Free Radic. Biol. Med.,* **2011**, *51*(10), 1872-1881.
[http://dx.doi.org/10.1016/j.freeradbiomed.2011.08.025] [PMID: 21920432]

[167] Páez, P.L.; Bazán, C.M.; Bongiovanni, M.E.; Toneatto, J.; Albesa, I.; Becerra, M.C.; Argüello, G.A. Oxidative stress and antimicrobial activity of chromium(III) and ruthenium(II) complexes on Staphylococcus aureus and Escherichia coli. *BioMed Res. Int.,* **2013**, *2013*, 906912.
[http://dx.doi.org/10.1155/2013/906912] [PMID: 24093107]

[168] He, Y.; Ingudam, S.; Reed, S.; Gehring, A.; Strobaugh, T.P., Jr; Irwin, P. Study on the mechanism of antibacterial action of magnesium oxide nanoparticles against foodborne pathogens. *J. Nanobiotechnology,* **2016**, *14*(1), 54.
[http://dx.doi.org/10.1186/s12951-016-0202-0] [PMID: 27349516]

CHAPTER 8

Analysis of the Effect of Load Direction on the Stress Distribution in Orthopaedic Implants

Nihar R. Mahapatra[1,*], **Manit Gosalia**[1] and **Ishika Tulsian**[1]

[1] *Department of Physics, Aditya Birla World Academy, Mumbai, India*

Abstract: Orthopaedic implant materials have an important role in the field of medical science. Characteristics of implant materials such as rigidity, corrosion, biocompatibility, surface morphology, tissue receptivity, and stability are the key factors that influence the choice of the implant material. The mechanical properties of the implants are one of the significant factors for bone substitution. To understand the mechanical properties of these solid substitutes, the use of solid mechanics, which is intended for general structural analysis of 2-dimensional and 3-dimensional bodies is vital. In this study, 3-dimensional modelling of implant and simulation using the finite element analysis software were incorporated to investigate the effect of load direction on the stress distribution in different orthopaedic implant materials.

Keywords: Finite element analysis, Load-bearing capacity, Load direction, Mechanical properties, Metal alloys, Orthopedic implant, Simulation, Solid mechanics, Stress distribution, Volume displacement, Von Mises stress distribution.

INTRODUCTION

Orthopaedic implants have developed critically through the years. Its sole purpose remains the same – to replace bone or articulating surfaces of a joint [1] so that patients can regain full range of motion. In 1965, Brånemark in Sweden conceptualised orthopaedic implants while supervising and leading a research project at the University of Gothenburg [2]. The original research of Brånemark was based on the microcirculation of blood flow in rabbit tibias. For this investigation, a small titanium optical chamber was inserted in a bone to investigate the rabbit's blood supply. Later, Brånemark found it impossibly difficult to remove the blood chamber as it was now integrated into the bone [2], leading to Brånemark evidently realising that the integration between these and

---

* **Corresponding author Nihar R. Mahapatra:** Department of Physics, Aditya Birla World Academy, Mumbai, India; Tel:+919899098966; E-maill: niharmahapatra@gmail.com

particular metals and the bone is possible, eventually leading to the concept of orthopaedic implants. Today, dental implants are extremely beneficial; for example, the osseointegration of implants provides great benefits and advantages to the patient when compared to removable prosthesis dentures; it gives better support to the denture [3], and improves aesthetics. However, the decision to use an orthopaedic implant must be made while considering an analysis of the biological conditions of the patient, in order to avoid potentially harmful complications that may occur to worsen the condition.

This paper focuses on the effectiveness of various possible implant materials by analysing to find the one that adheres most to an ideal bone substitution material. Several implant material characteristics (rigidity, corrosion, biocompatibility, surface morphology, tissue receptivity, and stability) are vital to ensure it is an ideal candidate for bone substitution. The mechanical properties of the material are critical to confirm it has the intended load-bearing capacity to structural support the surrounding bones, without causing much volume displacement of the pre-existing bone and tissue around the implant and relatively uniform stress distribution. It is imperative that the material used for the orthopaedic implant has the required mechanical and structural properties to support the load so that the patient has no difficulty moving.

The implant material will be the independent variable in this investigation, differentiating between various alloys. The five possible orthopaedic materials are Ti-6Al-4V, Nitinol, CoCrMo, CoCr, and 316L. There are multiple deciding factors when it comes to the properties of these alloys, some of which are used to simulate, for example, Young's modulus, Poisson ratio and Density. The 3-dimensional modelling of the implant consists of two cylinders at the top and bottom, with the orthopaedic implant being the smaller cylinder in the middle. The upper and lower bone is kept to be the same throughout the simulations and structural analysis, only changing the implant material with a different alloy each time. A load was applied tangentially to a specific side of the upper bone, Fig. (**4**), with the constraint of keeping the base, Fig. (**3**), fixed (immobile) the model was analysed to see the stress distribution across it and its volume displacement. Force distributions, bending moment, torque and flexural rigidity are determined by structural analysis of the osseointegrated implant system using finite element analysis (FEA) [4].

The finite element analysis (FEA) is largely known to be used for damage evaluation [5], material characterisation [6], and biomechanical applications [7] for many years. It is also used to predict biomechanical occurrences on the bone-implant interface [8].

In this study, we analyse the effects of external loading applied tangentially and its stress distribution on different implant materials using 3D FEA. This study can further be used to evaluate and analyse the consequences of forces applied on osseointegrated implants and also understand how the properties of implant materials affect its behaviour and its biocompatibility with surrounding tissue and bone.

Geometrical Model

The three-dimensional geometrical model of the bone-implant structure, Fig. **(3)**, is analysed using the finite element methodology. The upper and lower big cylindrical structures, Fig. **(1)**, are modelled to be bone. The upper cylinder was given the dimensions: 1cm radius, 3 cm height and the position coordinates (0,0,4). The lower cylinder was given the dimensions: 1cm radius, 3 cm height and the position coordinates (0,0,0). The implant is presented as a smaller cylinder in between the bone, Fig. **(2)**, with the dimensions: 0.5cm radius, 3cm height and the position coordinates (0,0,2).

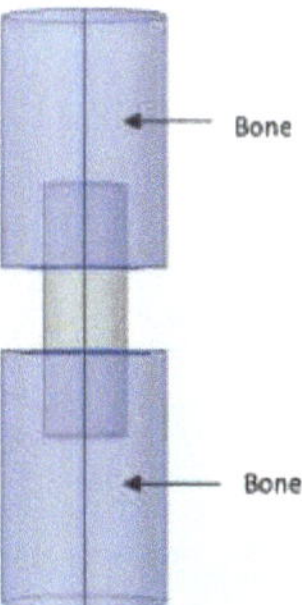

Fig. (1). Bone structures indicated by purple.

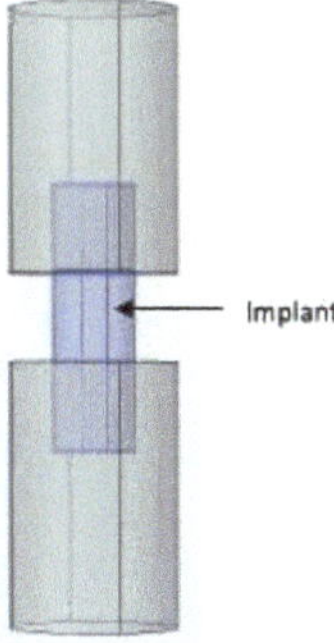

Fig. (2). Orthopaedic implant indicated by purple.

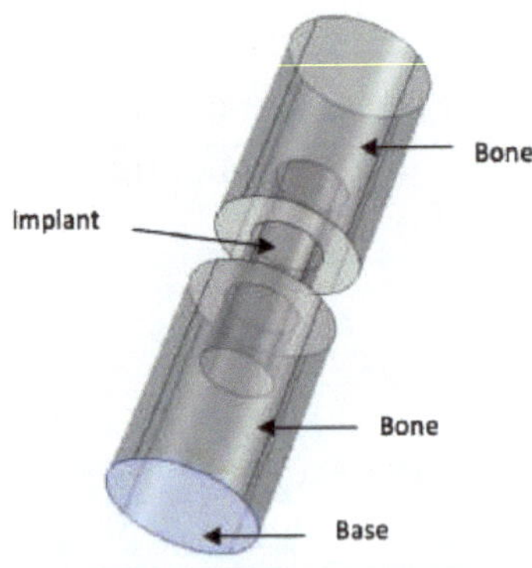

Fig. (3). 3D model of the structure.

Boundary Conditions

The implant load boundary is applied orthogonally to the side of the upper bone. The grey area in Fig. (**4**) indicates the free boundaries. A constraint is applied to the simulation, requiring the base to stay fixed in its position, as shown in Fig. (**3**). Fig. (**5**) gives a vector-based representation of the boundary load being applied to the upper bone.

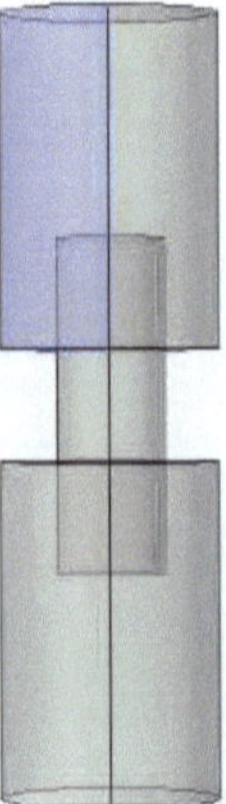

Fig. (4). Load applied tangentially.

RESULTS

For the above 5 materials (Table **1**), we need to find the effect of load distribution on the stress distribution of different orthopaedic implant materials.

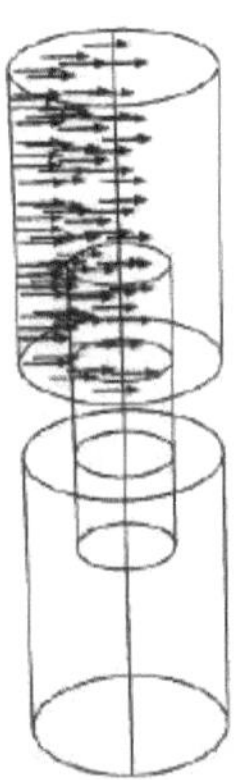

Fig. (5). 3D directional representation of the boundary load.

Table 1. Elastic Properties of the materials being used in this study.

Material	Young's Modulus (*GPa*)	Density (*gm/cm³*)	Poisson Ratio
Bone	10	3.05	0.3
Ti 6Al 4V	110	4.43	0.31
316L	193	8	0.27
Ni Ti (Nitinol)	75	6.45	0.23
Co Cr	210	10	0.29
Co Cr Mo	230	8.4	0.29

The stress and strain fields caused by the applied force were obtained using a simulation that used finite element analysis. For each of the 5 materials, the simulation was carried out and results relating to Von Mises stress on the bone were obtained. We evaluate these stress and strain fields for axial implants.

316l Implant

316l is an iron-based alloy containing chromium, nickel, molybdenum, and small quantities of carbon. Chromium creates an oxidation layer on the implant which resists corrosion. The first set of results corresponds to the orthopaedic implants for 316L stainless steel, which is also referred to as A4 stainless steel. As Fig. (6) shows, the stress distribution and volume displacement of the pair of bones and axial implant, for the 316L implant for a leg joint. As the force is applied orthogonally to the joint the Von Mises stress can be seen in Fig. (6). Clearly seen, there is a small equivalent stress concentration at the lower region of the bottom bone and the higher region of the bone further along the z-axis. The stress concentration is higher at the instance where the implant is connecting both bones and has maximum stress of 246 N/m^2 near the sides of the bottom end of the exposed region of the implant. Also, it was noticed that there is a wider range of stress to this exposed region of the implant and varies from approximately 60 to 246 N/m^2. The maximum value of the volume displacement is recorded as 9.41 x $10^{-16}cm$ which is low in magnitude and hence qualifies 316l implant as a good substitution for the bone.

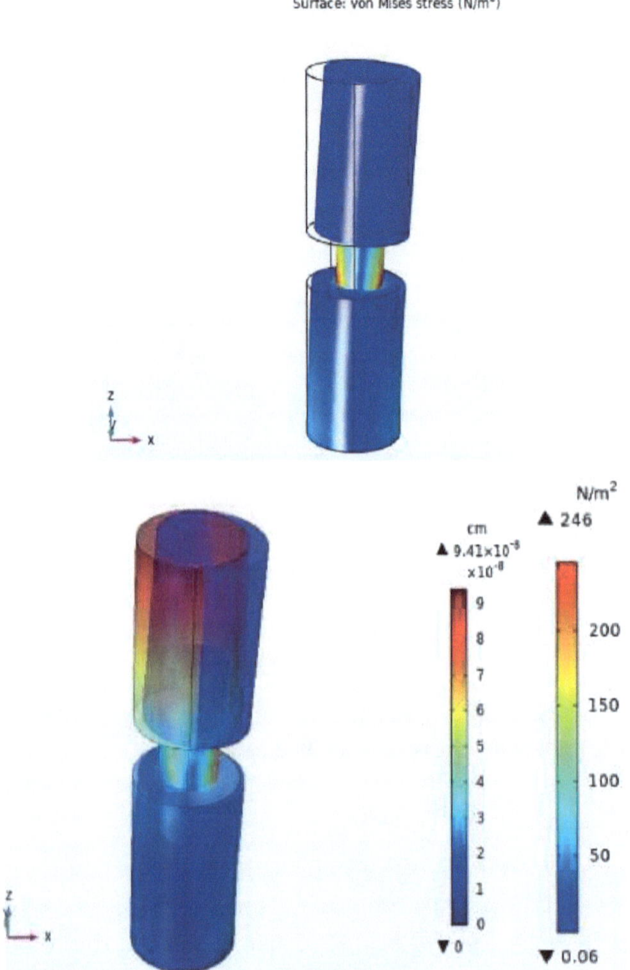

Fig. (6). Von Mises stress distribution on a bone with 316L implant.

CoCr Implant

These next set of results correspond to the orthopaedic implants for CoCr – Cobalt-chrome – a metal alloy with high corrosion resistance making it a recognized biomaterial used mainly in dental implants [9]. As seen in Fig. (7), the Finite element analysis simulation showed similar stress distribution as that of the 316L implant. There is a high concentration of Von Mises stress in the exposed region of the implant where the bones themselves do not exert a stress of large magnitude. In fact, it's also observed that the volume displacement along the x-axis is of low magnitude making it a suitable biomaterial as there is low stress and force then applied to surrounding tissue and either the lateral condyle at the bottom of the femur or the greater trochanter at the top. Also seen in Fig. (7) is the maximum stress is again found at the osseointegration – the functional and structural connection between the actual living bone and the surface of the load-bearing artificial implant. The maximum stress again for this implant is 246 N/m^2. However, unlike the previous implant discussed here in the exposed region, the stress distribution has a lower minimum being that of approximately 40 N/m^2.

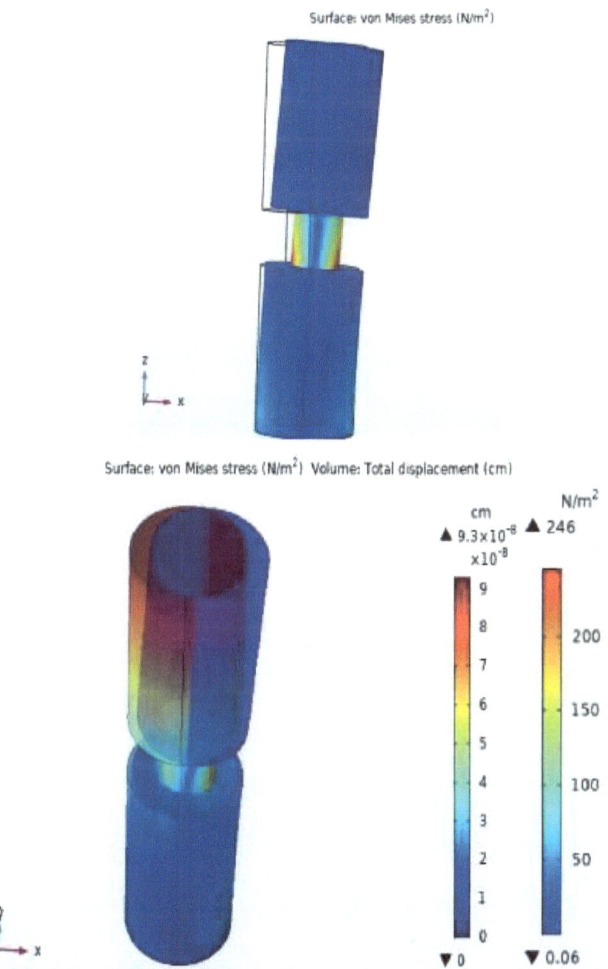

Fig. (7). Von Mises stress distribution on a bone with CoCr implant.

CoCrMo Implant

These next set of results correspond to the orthopaedic implants of CoCrMo – Cobalt-Chromium-Molybdenum – an essential metal alloy that is mainly used in total hip replacements and implants [10]. This alloy has a much higher tolerance in the human body than cobalt or nickel and therefore less toxicity making it a suitable biomaterial for hip implants. Below, Fig. (**8**) shows the volume displacement and stress concentration distribution for the CoCrMo alloy as an implant. As seen in the case of CoCr and 316L, the maximum magnitude of stress concentration occurs at the osseointegration. The exposed implant has a higher Von Mises stress magnitude around and the stress concentration increases as you go further along the y axis and at the lower end of the implant or lower end of the z-axis. The volume displacement is lowest compared to the other implants validating that this makes the implant most suitable for bones near the hip. The small volume displacement reduces this stress on surrounding tissues and the Acetabular Rim. However, the CoCrMo alloy undergoes electrochemical corrosion and wear mechanism in biological corrosion environment.

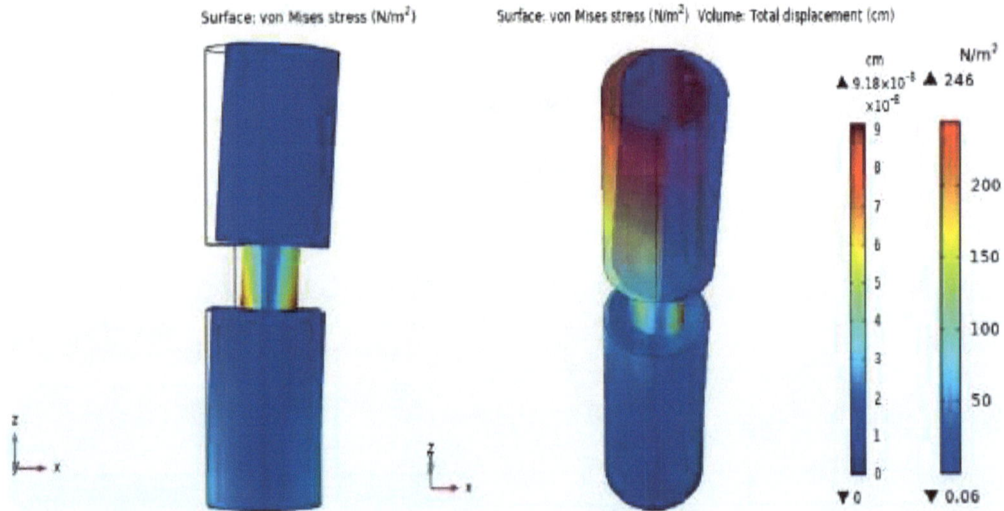

Fig. (8). Von Mises stress distribution on a bone with CoCrMo implant.

Nitinol Implant

These set of results correspond to the orthopaedic implants of Nitinol – Nickel and Titanium alloy – which is an alloy with the ability to change shape and is superelastic or pseudoelastic [11]. It also is an alloy that shows great elasticity when under stress. As seen in Fig. (**9**), the volume displacement for the Nitinol implant is very low showing why this superelastic alloy is used in foot and ankle implants [12] where the amount of space, the bone-implant can displace, is less and the placement is in a tight concise area. The stress distribution is the same as in previous simulations with maximum stress near the osseointegration. The magnitude of maximum stress is also higher for the Nitinol implant peaking at 252 N/m^2. The maximum volume displacement observed from the simulation is $1.21 \times 10^{-15} cm$.

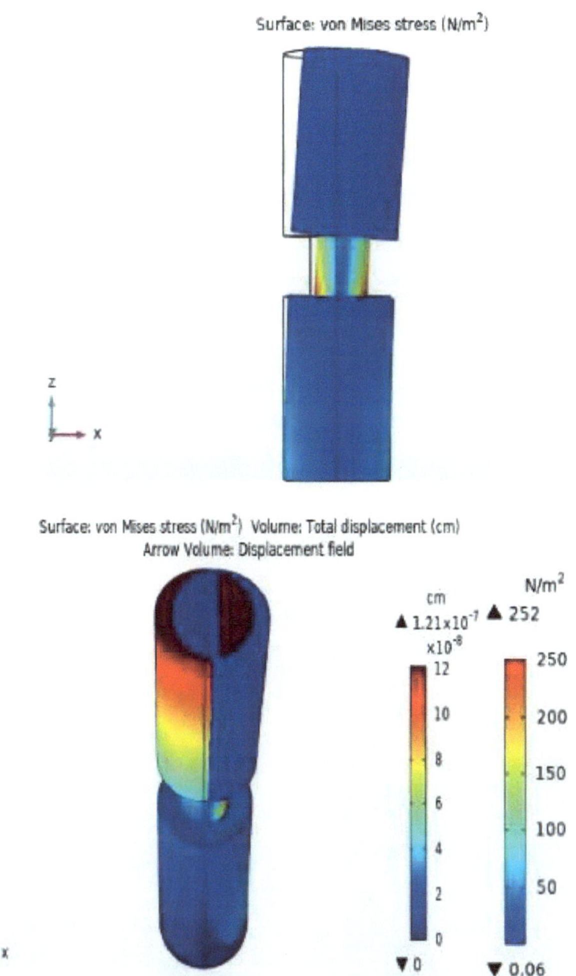

Fig. (9). Von mises stress distribution on a bone with Nitinol implant.

Ti-6Al-4V Implant

These final set of results correspond to the orthopaedic implants of Ti-6Al-4V – a Titanium alloy also known as T64. This alpha-beta titanium alloy has a low density and its biocompatibility makes it an assuring prosthetic device in orthopaedic and dental implants. These are most often used in joint implants [13] and dental implants [14]. As shown in Fig. (**10**), the stress distribution is similar to those discussed earlier where the maximum stress of 247 *N/m²* is present at the osseointegration at the same time this magnitude is slightly larger than the ones for CoCr and 316L. The maximum stress at this osseointegration is also smaller in magnitude than the maximum stress values of a Maxilla proving Ti-6Al-4V to be an effective alloy for dental implants. The volume displacement is also the least for this Titanium alloy suggesting the force and stress applied to the cancellous and cortical bone. Fig. (**8**) also shows the volume displacement field to show the volume deformation. The volume displacement for T64 is recorded as 1.06 x $10^{-15}cm$ which is comparatively larger than other alloys considered for the simulation. However, it is considered a suitable implant material as the surface of Ti-6Al-4V can be modified to mitigate the release of metal ions into the bloodstream as well as its biocompatibility can be improved.

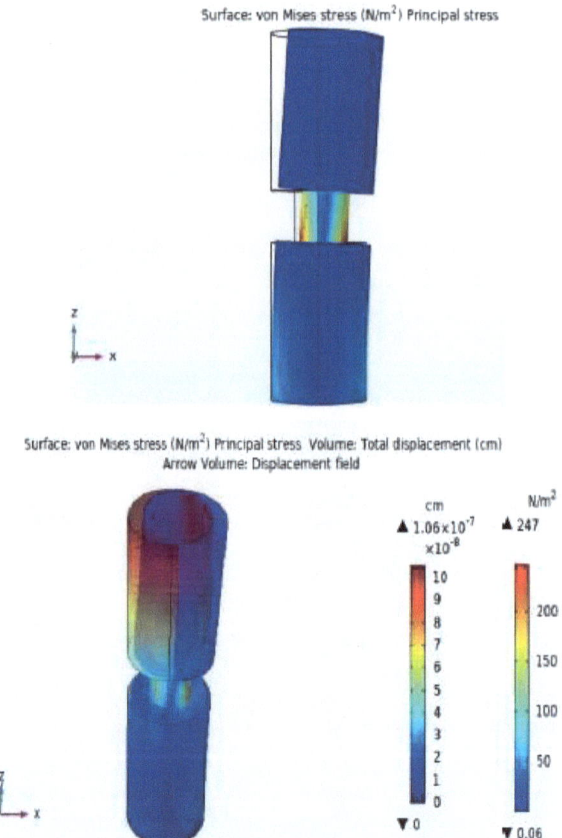

Fig. (10). Von Mises stress distribution on a bone with Ti-6Al-4V implant.

CONCLUSION

This study was carried out to analyze and evaluate the effect of load direction on the stress distribution in orthopaedic implants. Evidently, the study showed that at the osseointegration of a bone and an artificial implant, there is maximum stress that needs to be studied for different processes of fixation of implants and the shape of implants. However, in all the above alloys of the same dimensions, the volume displaced for all the implants is very small, making them suitable for a certain implant at a certain joint or location in the skeletal system. For example, a Ti-6Al-4V is most suitable for a dental implant due to its low volume displacement and biocompatibility. Similarly, Nitinol implants are suitable for foot and ankle implants due to their pseudoelasticity and biocompatibility. However, these alloys should be studied in the biological environment to find the behaviour of these materials in long term use as a prosthesis. In future works, the authors aim to analyze the stress distribution for different shapes, designs and processes of fixation of implant materials used at different positions of typical human bones.

CONSENT FOR PUBLICATION

Not applicable.

CONFLICT OF INTEREST

The author declares no conflict of interest, financial or otherwise.

ACKNOWLEDGEMENTS

Declared none.

REFERENCES

[1]   Rouhi, G.; Amani, M. A Brief Introduction Into Orthopaedic Implants: Screws, Plates, and Nails. **2012**.

[2]   Elias, C. N. Factors Affecting the Success of Dental Implants; Implant Dentistry - A Rapidly Evolving Practice. **2011**.

[3]   Jacobs, R.; Van-Steenberghe, D.; Nys, M.; Naert, I. Maxillary bone resorption in patients with mandibular implant-supported overdenture or fixed prostheses. *J. Prosthet. Dent.*, **1993**, *70*(2), 135-140.
[http://dx.doi.org/10.1016/0022-3913(93)90008-C]

[4]   Pogosian, A. *Finite Element Analysis of Osseointegrated Transfemoral Implant; Degree Project, Dept. Med. Eng*; KTH Royal Institute of Technology: Stockholm, **2018**.

[5]   Asdrúbal, A.; Graciano, C.; González-Estrada, O.A. *Resistencia de vigas esbeltas de acero inoxidable bajo cargas concentradas mediante análisis por elementos finitos*; Revista UIS Ingenierías, **2017**.

[6]   Nadal, E.; Rupérez, M.J.; Martínez-Sanchis, S.; Monserrat, C.; Tur, M.; Fuenmayor, F.J. *Evaluación basada en el método del gradiente de las propiedades elásticas de tejidos humanos in vivo*; Revista

UIS Ingenierías, **2017**.

[7]   Valencia-Aguirre, F.; Mejía-Echeverria, C.; Erazo-Arteaga, V. *Desarrollo de una prótesis de rodilla para amputaciones transfemorales usando herramientas computacionales*; Revista UIS Ingenierías, **2017**.

[8]   Geng, J.P.; Tan, K.B.C.; Liu, G.R. Application of finite element analysis in implant dentistry: a review of the literature. *J. Prosthet. Dent.,* **2001**, *85*(6), 585-598.
[http://dx.doi.org/10.1067/mpr.2001.115251] [PMID: 11404759]

[9]   Rubo, H.; Jose, S. Edson Finite-Element Analysis of Stress on Dental Implant Prosthesis. **2009**.

[10]   Panigrahi, P.; Liao, Y.; Mathew, M.T.; Fischer, A.; Wimmer, M.A.; Jacobs, J.J.; Marks, L.D. Intergranular pit- ting corrosion of CoCrMo biomedical implant alloy. *J. Biomed. Mater. Res. B Appl. Biomater.,* **2014**, *102B*(4), 850-859.
[http://dx.doi.org/10.1002/jbm.b.33067]

[11]   McNaney, J.M. *An Experimental Study of the Superelastic Effect in a Shape-Memory Nitinol Alloy under Biaxial Loading,* **2003**.
[http://dx.doi.org/10.1016/S0167-6636(02)00310-1]

[12]   Schillinger, M.; Sabeti, S.; Loewe, C.; Dick, P.; Amighi, J.; Mlekusch, W.; Schlager, O.; Cejna, M.; Lammer, J.; Minar, E. Balloon angioplasty versus implantation of nitinol stents in the superficial femoral artery. *N. Engl. J. Med.,* **2006**, *354*(18), 1879-1888.
[http://dx.doi.org/10.1056/NEJMoa051303] [PMID: 16672699]

[13]   Murr, L.E. *Microstructure and Mechanical Behavior of Ti–6Al–4V Produced by Rapid-Layer Manufacturing, for Biomedical Applications,* **2008**.

[14]   Elias, C.N.; Lima, J.H.C.; Valiev, R.; Meyers, M.A. Biomedical applications of titanium and its alloys. *J. Miner. Met. Mater. Soc.,* **2008**, *60*(3), 46-49.
[http://dx.doi.org/10.1007/s11837-008-0031-1]

CHAPTER 9

Advanced Materials and Nanosystems for Catalysis, Sensing and Wastewater Treatment

Shikha Rana[1,*] and **Mahavir Singh**[1]

[1] *Department of Physics, Himachal Pradesh University, Shimla, India*

Abstract: This chapter accentuates the latest breakthrough in the advanced and intelligent materials' catalysis, sensing and wastewater treatment applications. The various engineered materials are securing the interest of researchers for optimized technical utilizations. This chapter discusses the number of catalytic and sensing operations of advanced and intelligent materials in detail. Catalysis and sensing phenomena involve the conversion of obtained signals into a readable format, and advanced materials with their superior optical, semiconducting or physical properties are studied widely. However, wastewater treatment needs adsorption and advanced oxidation of the different types of contaminants by the advanced materials. The list of advanced materials includes many organic/inorganic and natural/synthetic platforms with desired properties. These advanced materials have high biocompatibility and easy biodegradable characteristics. With the latest synthesis and functionalization methods, these advanced materials are becoming nanohybrid systems. This chapter covers implementing these nanohybrid systems for catalysis, wastewater treatment and sensing. The first half of the chapter focuses on introducing the basic catalytic, sensing and wastewater processes based on the application of advanced materials. However, the second half includes introducing various advanced materials in the techniques mentioned above.

Keywords: Advanced Oxidation, Carbon Dot, Carbon Nanotube, Ferrite, Gold Nanoparticles, Metal and Metal Oxides, Nanomaterial, Photocatalysts, Semiconductor, Wastewater Treatment, Water Splitting.

INTRODUCTION

Material science and nanotechnology advancements are currently resulting in novel materials for improving in situ remediations of the most ubiquitous and persistent contaminants, as well as providing accessible sensing techniques. The earth faces two significant threats: rapid consumption of non-sustainable energy resources and an extended increase in pollution concentrations in the ecosystem.

* **Corresponding author Shikha Rana**: Department of Physics, Himachal Pradesh University, Shimla, India; Tel: 8263829966; E-mail: shikharana2807@gmail.com

Recently, numerous studies on alternative energy conservation and generation methods have been accepted and studied in laboratories to fulfill the need for energy [1, 2]. Nowadays, with the increase in population and pollution every moment, the obligation of researchers has also been raised to develop new materials with extra unequivocal features like quick recovery, excellent reusability, facile degradation, immediate resolution, easy processing, and less processing expensive, desired dimensions along with others. For the last few decades, nanotechnology has been catering to all these needs of researchers and levering the expansion in material science domains. Nano ranged dimensions of these particles have resulted in compelling transition due to their peculiarity in analogy to similar bulk components. Generally, nano dimensional particles can be formed by following two methods: Top-Bottom or Bottom-Up, where the first one includes breakdown and grinding like methods, and the second one incorporates proportionate mixing of precursors [3]. Some of the commonly used nanoparticle synthesis techniques are listed in Fig. (1). Nanoparticles have high surface reactivity compared to bulk materials, and their size and shape depend on the preparation technique. Nowadays, plant-based green synthesis methods provide cost-effective and chemical-free techniques for nanoparticle synthesis.

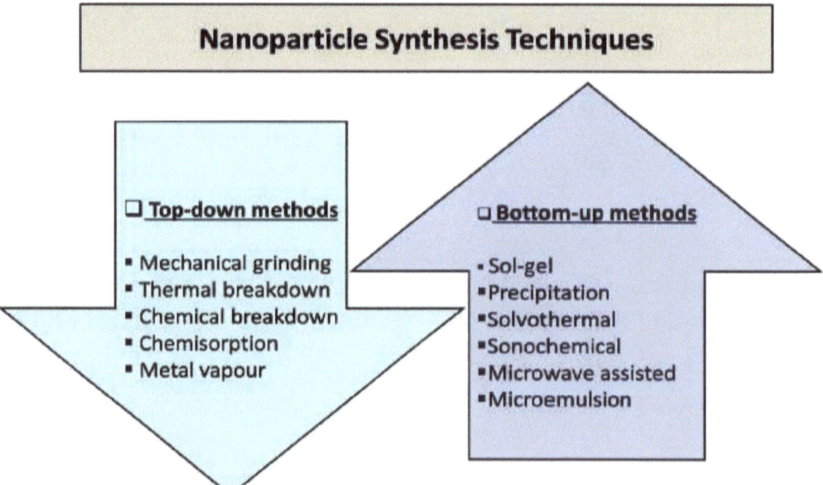

Fig. (1). Nanoparticle synthesis techniques.

Catalytic reactions necessitate the involvement of a fascinating material named a catalyst to expedite a chemical process without leaving any perpetual transition in the material. For the last few decades, catalytic reactions and advanced engineered catalysts have been investigated in detail and prepared by chemical and bio-inspired synthesis techniques. Green synthesis techniques have elaborated ingenious strategies to establish solar light-driven systems for afferent energy production [4]. Catalytic processes effectively perform their role in supportable ecological systems and energy sources for forthcoming centuries, as depicted in fuel cells and catalytic converters [5]. Many nano ranged functionalized substances are used as a substrate for catalytic studies and have also been found to increase the efficacy of the reaction immensely. Various types of homogeneous and heterogeneous catalytic materials, including photocatalysts, Gold nanocatalysts, metal and metal oxide-based nanocatalysts, nanoclusters, carbon-based catalysts, etc. are engineered by combining catalysis studies with material and surface science [6, 7]. Many nano ranged functionalized substances are used as a substrate for catalytic studies and have also been found to increase the efficacy of the reaction immensely. Various types of homogeneous and heterogeneous catalytic materials, including photocatalysts, gold nanocatalysts, metal and metal oxide-based nanocatalysts, nanoclusters, carbon-based catalysts, etc. are engineered by combining catalysis studies with material and surface science [8].

This chapter includes a brief discussion about advanced innovational materials such as nano catalytic systems, sensing probes and wastewater treatment. Nanoparticles have a common characteristic of heterogeneity as different sizes and shapes of these nanoparticles impart different catalytic activity rates to each particle [9, 10]. Catalytic reactions are also found to play an essential role in sensing and wastewater treatments. The sensing mechanism involves stimuli-responsive systems attached to a signal converter, whereas wastewater treatment includes the degradation of toxic, harmful, organic, or inorganic materials by certain substances using various mechanisms. An efficient sensor should have high selectivity and sensitivity with an easy manufacturing process. Nanoparticles have facile characteristics, due to which they can be easily employed in the sensing field. They have nano ranged dimensions which come in the close range of visible light wavelength and can show electron collective motion known as SPR. SPR phenomena lead to the development of the real-time and label-free sensing of different targets, including harmful gases, pesticides and bio-organisms. SPR sensors include the collective electron excitation by incident photons on the surface. The resonance frequency of SPR oscillations depends on the refractive index of the surface, type and morphology of the nanoparticles [11]. As we are quite aware, the availability of potable water is a primary concern of human beings nowadays. Due to the mass industrialization required for human lives, surfaces and groundwater resources are getting polluted daily. Various remediation techniques based on different mechanisms are used to remove contaminants. Naturally, minerals provide a low-cost way to remove toxic and harmful contaminants from the wastewater, but with advancements in scientific processes, new low-cost methods are developed using advanced nanomaterials [12]. All wastewater treatment methods are categorized into conventional and non-conventional types. The conventional methods involve separation, filtration, adsorption and biological treatment of the wastewater, and the non-conventional methods involve Fenton oxidation, photocatalytic reduction and ozonation of wastewater. The conventional methods have many advantages as they are simple and most cost-effective than non-conventional methods. Besides, they are more time consuming and work over small volumes of wastewater. Highly efficient advanced oxidation processes overthrow these limitations of conventional methods.

Generally, wastewater treatment is done by the selective membrane separation technique. This technique includes microfiltration and osmosis treatment of wastewater by using permeable membranes. Traditional wastewater treatment methods have many limitations like difficult removal of subsurface contamination and poor mixing of remediation substances. The various remediation technologies are used for wastewater treatments based on extraction, congelation, immobilization, Fenton reactions and advanced oxidation process as given in Fig. (**2**). The advanced oxidation process is one of the promising technologies used nowadays for remediation purposes. The advanced oxidation process involves the degradation of organic water pollutants by Fenton reactions assisted by metal radicals [13].

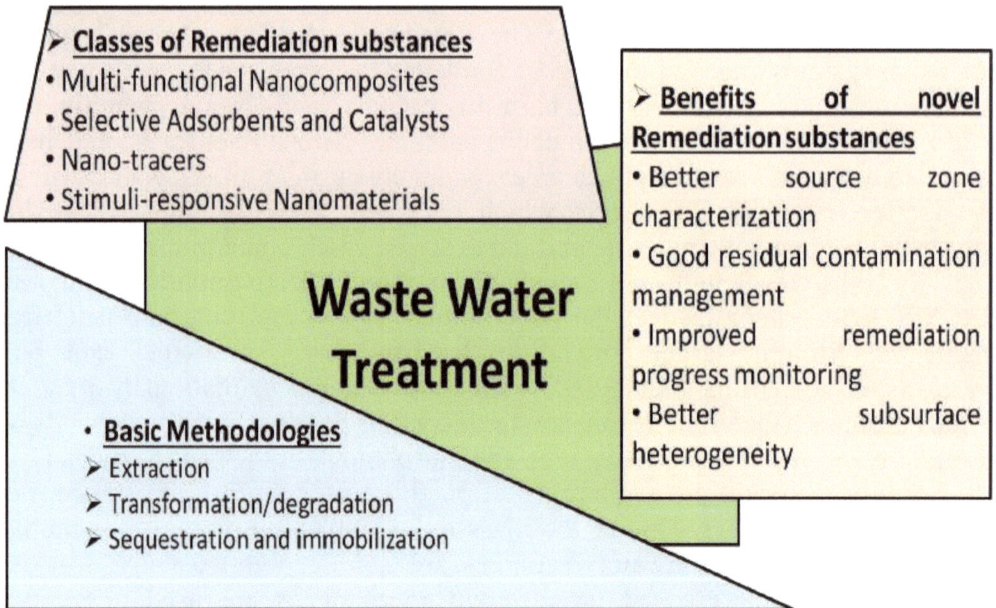

Fig. (2). Recently used water remediation substances, methodologies and their benefits.

TYPES OF ADVANCED NANOMATERIALS

Metal and Metal Oxide Nanoparticles

Magnetic nanoparticles of metal or metal oxides are a widely studied group of ceramic materials with easy tracking and targeting characteristics under externally applied magnetic fields. On removing an external field, they lose magnetization and show random thermal spin orientation. For various in vivo and in vitro biomedical applications, magnetic nanoparticles are functionalized with biological molecules to maintain their pH stability, biocompatibility and non-toxicity in nature. Whereas sensing and catalytic application of magnetic nanoparticles is attributed to the high surface area, narrow bandgap, SPR effect and electrical properties [14].

Magnetic or non-magnetic metal oxide-based catalysts are also utilized for wastewater treatment [15, 16]. Various polymer composites with metal oxide materials enhance their environmental remediation process [17]. Carbon allotropic form named Graphene is also associated with these metal oxides for catalytic treatment of wastewater [18]. Many different types of metal oxide structures are used for the same purpose, which also helps maintain underwater life [19 - 21]. Metal oxide nanostructures have propitious band gap and electronic carrier transference needed for photocatalysis in the visible-UV wavelength regions. These nano metal oxide substances help in the photo-degradation of water pollutants by redox reaction [16]. Nanoparticles of magnetic metal and metal oxides of desired dimensions are synthesized and characterized by various latest techniques. Below a particular size known as critical size, these magnetic nanoparticles possess a single domain structure [22].

Most commonly, iron oxide and doped iron oxide is, known as ferrites, are recently investigated intensively for sensing and catalytic studies. Ferrites are face centred cubic (fcc) structured magnetic substances which can have normal, mixed and inverse spinel-type geometrical arrangements [23]. Magnetic characteristics are attributed to the unique structure of these substances and facilitate multiple applicative fields for them like biosensing and imaging, magnetic fluids, in vivo medical studies, magnetic storage, enzyme immobilization, catalysis, microwave and electromagnetic devices [24 - 26]. Their fcc structure includes tetrahedral and octahedral sites. The occupation of these sites by different transition or divalent doped ions leads to unique electrical and magnetic behaviour below a specific size they can show superparamagnetic characteristics, which is needed for their biomedical applications [27]. Fig. (**3**) shows the spinel structure of ferrite. Magnetic ferrites are of two types: first soft type and another hard type. Soft materials have less magnetic hysteresis loss than hard magnetic materials. Hard magnetic ferrites have more retention of the magnetic field than soft materials. The magnetic response time of ferrites was found to depend on size. Commonly magnetic materials are Fe, Ni and Co like magnetic transition elements or their oxides [28]. The type and concentration of dopant with optimized synthesis technique lead to the preparation of functionalized ferrites for variant applications [29]. Metal, Metal oxide and ferrite composites with quantum dots are studied for multiple exciton generation and dye-sensitized solar cells. Recently, Carbon quantum dots doped with Nitrogen, Zinc and Titanium oxides have been found to possess good photocatalytic characteristics than undoped samples. Their photocatalytic behaviour depends on their stoichiometric ratios in the prepared composites [30 - 33].

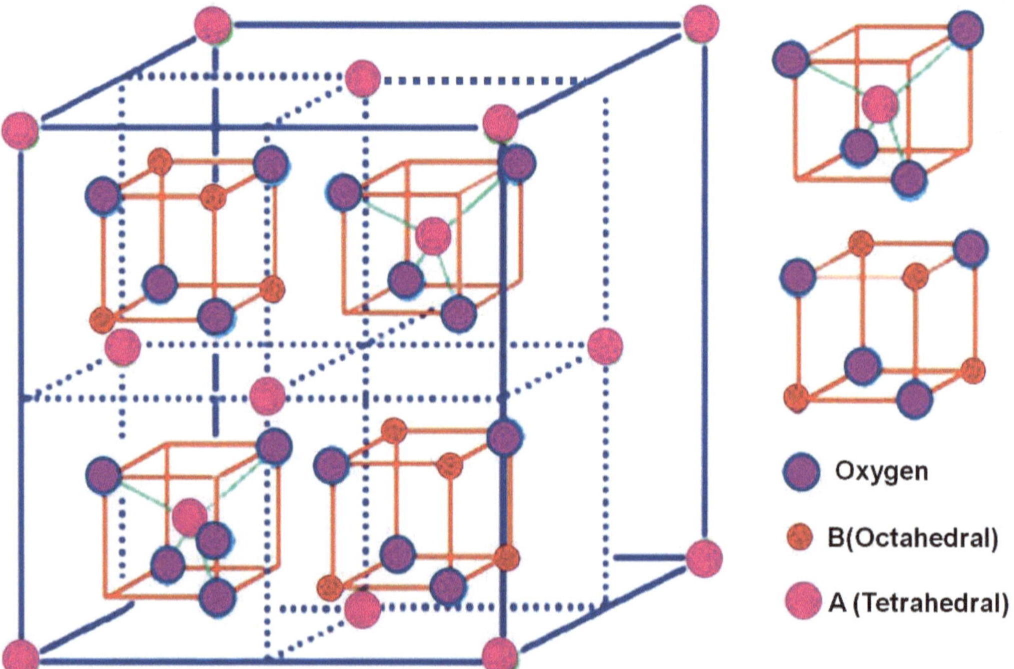

Fig. (3). Spinel structure of ferrite.

SOLAR DRIVEN CATALYSTS OR PHOTOCATALYSTS

The solar energy based catalytic process is one of the effective advanced oxidation processes used to remove non-biodegradable substances from wastewater. Photocatalytic degradation of organic materials in drain water could also be recycled as an energy source in solar-driven fuel cells [34]. Generally, photocatalysts are semiconducting, as semiconductors are light receptive with a filled outermost shell, leading to easy conduction [35, 36]. In this reaction, when photons fall on the surface of the anode (semiconductor photocatalyst) leads to the generation of photo-electrons and hole pairs. The generated holes react with water to produce highly reactive hydroxyl ions, and these hydroxyl ions further react with organic compounds to decompose them into carbon dioxide and water. The free electrons present in the conduction band reduce oxygen to form superoxide ions. Fig. (**4**) depicts the semiconductor photocatalytic process.

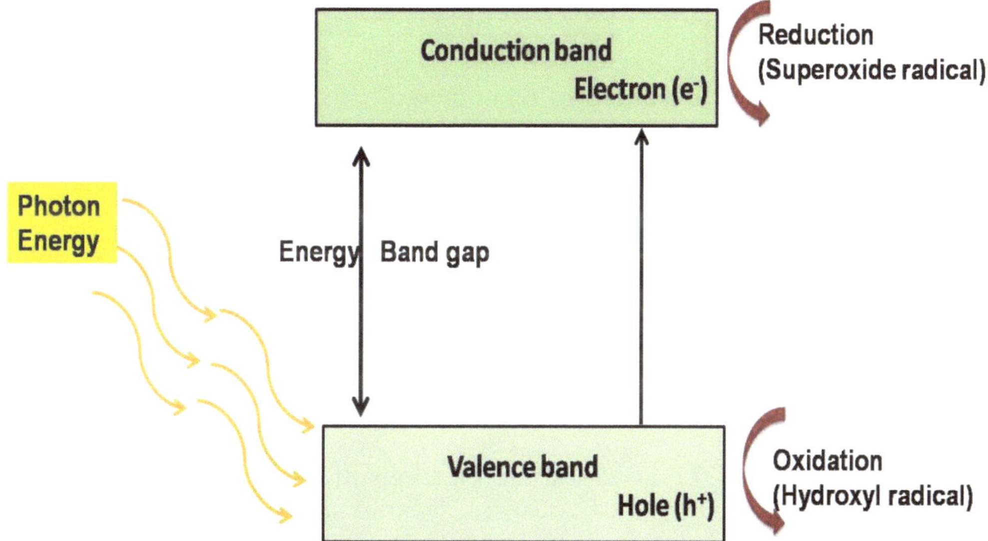

Fig. (4). Semiconducting photocatalysis mechanism.

Photocatalysis based degradation of organic compounds present in wastewater has many benefits over other fuel cells as they convert waste substances into an energy source, and they can easily be regulated by handling the falling photon energy [37]. The physiochemical properties of the electrodes are compelling operative criteria. The photochemical nature of the photocatalyst is determined by the physicochemical properties, nanosurface defects and type of the light source used. Two possible types of photocatalytic fuel cells are used one electrode and dual electrodes systems [5]. The anode, a photocatalyst, determines the energy generation rate and degradation of substrate materials. TiO_2 is the firstly used and highly investigated photoanode employed for water pollutant remediation [38]. Different types of noble ceramics of metal and non-metal type elements are deposited on the surface of TiO_2 to develop Schottky potential at the interface of the semiconductor, which improves the electron flow inside the cell. Also, they show the SPR phenomenon, which helps extend the energy band gap to the visible region. TiO_2 composites are used as photo-anode in the fuel cell and are also found to show better photocurrent generation. Titanium oxide samples doped with silver, Iodine, Boron and Nitrogen have better photo-electric characteristics due to more photocurrent generation after doping [39, 40]. TiO_2 has high oxidation power and good environmental stability, and its photocatalytic mechanism is a light-induced surface charge separation. TiO_2 is readily available and has good solvent solubility with high photo-corrosion degradation stability, increasing its technical uses in medicine and the food industry [41, 42]. To increase its efficiency in the visible band gap and to increase its recovery rate, various metallic and non-metallic materials are added to the TiO_2 structure to form hybrid TiO_2. The mesoporous type of Titanium oxide has a macroporous structure that provides transfer paths to photon energy and enhances the efficiency of photocells [43]. The photocatalytic behaviour of titanium oxide has remarkably increased by the doping of noble metals. This change is attributed to decreased electron-hole pair recombination and prominent charge separation [44]. Transition-metal ions doped samples show visible-light absorption and ozone absorption [45]. Iron doping in titanium oxide shifts its active region to visible light [46]. Three-dimensional porous aerogel substances are recently used to support titanium oxide-based photocells for effective air pollutant removal [47].

Copper dioxide and Manganese dioxide are the other two types of generally used heterogeneous photocatalysts [20]. Manganese dioxide has a narrow energy bandgap (1 to 2 eV), low cost, better redox activities and outstanding electrical behaviour, making it useful in various applications in biosensing, energy storage and ion exchange areas [48]. It has different layered and tunnel structures. These nanoparticles' energy bandgap and degradation efficiency depend on their geometrical structure and surface area. The geometrical structure could easily be modified by controlling temperature, dopant and precursors concentration.

Another efficient use of photocatalysis is nowadays studied as an energy source by hydrogen production by water splitting. Commonly metal oxide-based semiconducting type photocatalysts are used for this application. The electron-hole pair generated by falling photons performs oxidation and reduction reactions. The photogenerated electrons perform reduction, and holes perform oxidation. To enhance the hydrogen evolution rate, the conduction band and valence band need to behave with more negative and positive charges, respectively, compared to the production and oxidation levels of hydrogen and water. Fig. (**5**) shows the factors responsible for the efficient photocatalytic splitting of water. Small bandgap, low recombination rate, significant temperature, large light intensity and small size like factors enhance the photocatalytic action and hydrogen evolution rate.

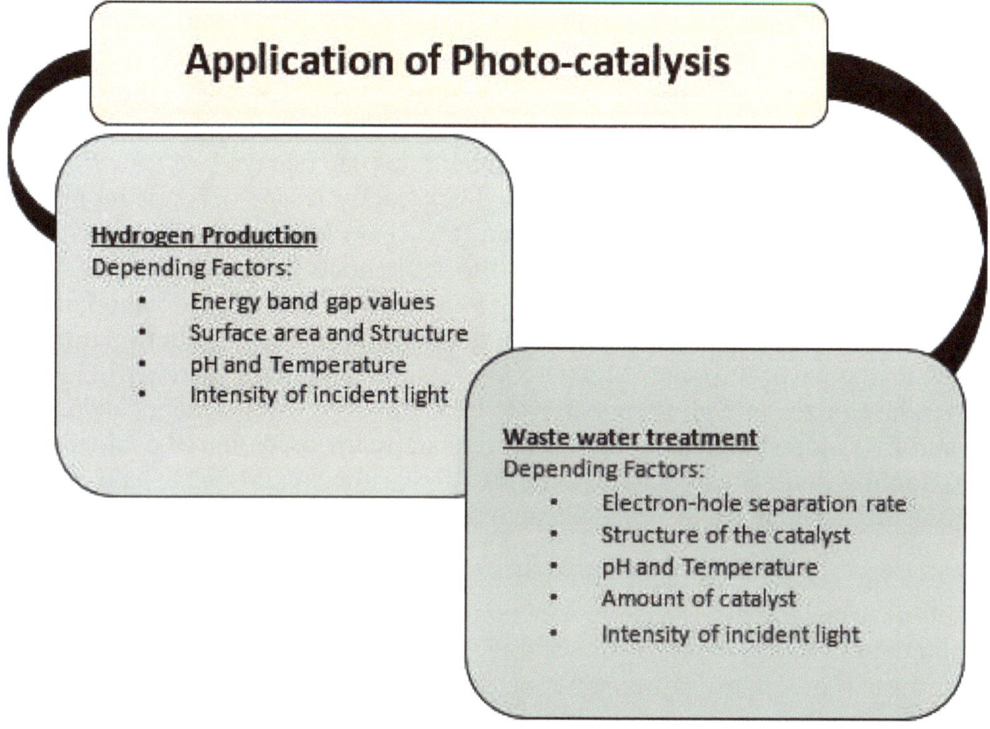

Fig. (5). The governing factors of the photocatalysts in wastewater treatment and hydrogen production [34].

GOLD NANOPARTICLES

Gold nanoparticles have shown significant and distinctive catalytic behaviour for eclectic hydrogenation processes for unsaturated carbonyl substances, alkynes, amides and the like [49, 50]. Also, their unique optical and chemical characteristics made their use more possible in sensing applications. Gold metal has shown exclusive chemoselectivity, which provides uncommon behaviour at the nano level as it does not show any catalytic nature at the bulk level [51]. Four decades earlier, the catalytic behaviour of gold nanoparticles was first investigated for hydrogenation [52]. Later on, gold nanoparticles were investigated for hydrogenation of unsaturated aldehydes, too [53]. Recently, Gold nanoparticles were functionalized and employed for various gas and liquid medium hydrogenation processes [54]. Metal ions like palladium and gold nanoparticles are incorporated on organic or inorganic supports to form hydrogenation catalysts with enhanced enantio-selectivity and reusability in acetylene and dienes hydrogenation [55 - 57].

Gold nanocatalysts have size dependence toxicity; hence they are supported by metal, polymer, biomolecules, nanorods, silica, etc [58 - 60]. Gold nanocatalysts and colloidal solutions have excellent sensitivity and selectivity for various biochemical processes. Functionalized gold nanoparticles have shown good catalytic characteristics for organic processes like CO oxidation, Hydrosilylation, Alkylation, and others [61 - 64]. Gold nanoparticles can behave like Lewis acid to promote various C-C coupling processes. Their catalytic rate depends on precise morphology, size, and synthesis technique [65]. They can replace the traditional oxidation process by Hal-oxo acids and d-block elements by performing oxidation of alcohols, alkanes and olefins [66 - 68]. The catalytic rate of gold nanoparticles in these organic reactions is highly dependent on particle size, functionalization method and type of support. Metal oxides doped with gold nanoparticles are found to have size depending on catalytic behaviour for 4-nitrophenol reduction. TiO_2 and ZrO_2 doped gold samples are used to study the oxidation of cyclohexene. The oxidation state of gold nanoparticles affects their catalytic behaviour and increases their selectivity towards alkene oxides [69, 70].

They exhibit excellent optical characteristics as they can strongly show light adsorption/scattering in the visible wavelength range, due to which they could be implemented in optical probes for sensing. Gold-based colourimetric sensors are of four basic types: aggregation, etching, growth and nanoenzyme based sensors. Gold nanoparticles evince surface plasmon resonance characteristics that depend on the size and show colour change according to size. The aggregation based sensors are categorized in two parts: label and label-free. In the first type, gold nanoparticles directly get linked with ligands, whereas in the second type, the electrostatic interactions among the gold nanoparticles and ligands result in colloidal stabilization. The different shapes of gold nanoparticles get attached by etching based sensors, and both the transverse and longitudinal SPR types are found to depend on the shapes and aspects of gold nanorods [71]. Also, they can show SPR phenomena that could be shifted to the near Infrared region and highly depends on their microstructure, which makes them a potential candidate for in vivo sensing applications. Colourimetric sensors can detect the presence of heavy metal ions by responding to a variation in chemical or physical properties. They have shown an increase in their sensitivity rate when gold nanoparticles are used as adsorbents [72]. Gold nanoparticles' change in SPR phenomena is generally observed in two ways: either adsorption or scattering after their interaction with the targeted group. The core-shell type structure of Gold nanoparticles is generally employed for sensing application and their surface chemistry plays a vital role in setting the detection limit of the sensors [73]. Gold nanoparticles incorporating colourimetric assay are cheap and have a fast and easy detection rate which could impressively recognize, amplify and inspect real-time sensing information obtained from target-specific assay [74, 75]. Gold nanoparticles can be used as a substrate for catalytic reactions or show catalytic properties in biological processes when attached to proteins or nucleic acids. The film of gold nanoparticles helps sense cancer-related substances, too [76].

Attributed to their sensing and photocatalytic characteristics, Gold Nanoparticles are also helpful in removing water contamination [77]. They can sense the presence of pollutants like heavy metal ions, halocarbons, non-bio-degradable organic substances and toxic pesticides in water [78]. They can offer both their catalytic abilities and could also behave as a catalytic substrate for metal-catalyzed wastewater treatments [79]. In brief, an engineered hybrid combination of gold nanoparticles with metals and metal oxides of desired morphology and immense degradation ability can be synthesized by optimizing various synthesis processes and fabrication techniques [77, 80].

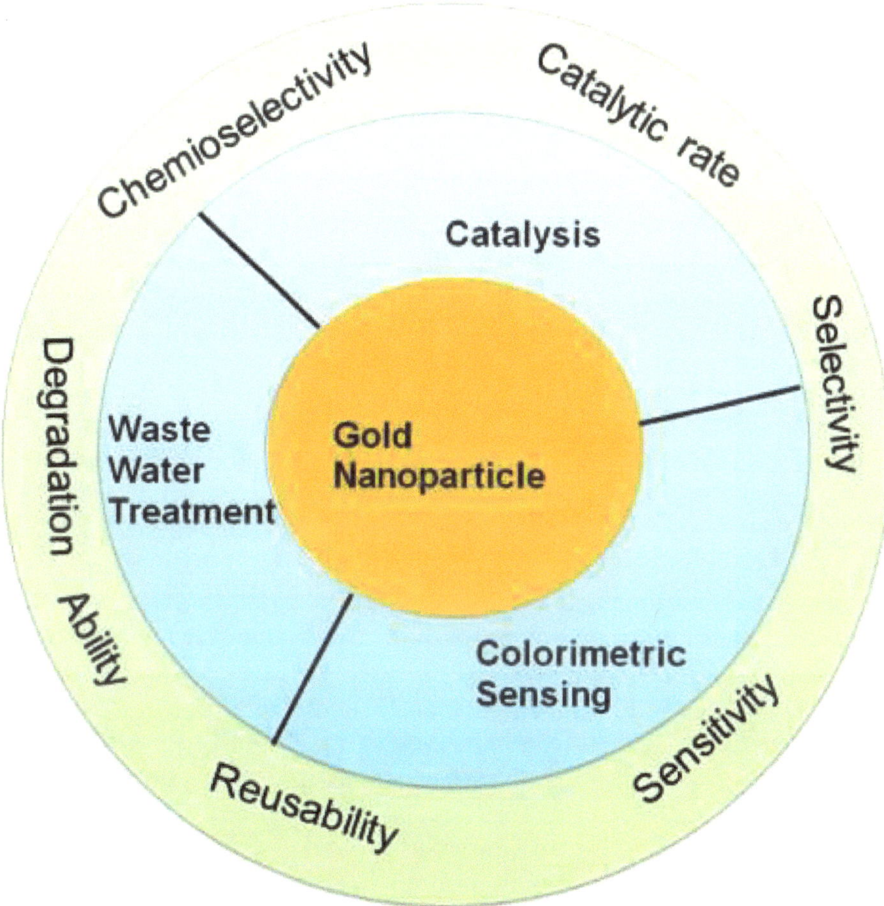

Fig. (6). Gold nanoparticles characteristics and various applications.

CARBON-BASED NANOPARTICLES

Carbon-based nanomaterials have intriguing morphological, electrical and mechanical characteristics through which they show unique heterogeneous catalytic behaviour in green and clean energy technologies. They have easy functionalization and synthesis procedures to produce nanotubes, nanowires and nanofibres of desirable shape, size and porosity with unique reaction stability. With increased research in this area, carbon nanocomposites with excellent electrical conductivity and mechanical stability are produced as catalysis support material [81, 82]. One of the commonly used carbon types is Graphene. Undoped or doped Graphene matrices have multiple applications in catalysis, sensing and wastewater treatment due to their definitive physiochemical and electrical nature. Graphitic carbon nitride has a layered structure with low toxic nature and distinctive electronic properties. It can be used in advanced oxidation wastewater treatment and photocatalysis [83]. The rolled form of Graphene sheet is named carbon nanotube, which generally has two types of structure: single or multi-walled. Carbon nanotubes contain sp^2 hybridized carbon atoms in a single dimension. Generally, they have excellent thermal, mechanical, electronic and physiochemical properties and are synthesized by chemical vapour deposition, arc-discharge and laser ablation methods. These nano-scaled tube structures are employed in various research areas of gas sensing, catalysis, environmental remediation and medicine. Depending on their size and chirality, these tubes can show metallic or semiconducting properties [84].

Carbon dots are noble fluorescent materials in different possible structures. They have quasi-spherical, flaky, amorphous type structures that offer excellent electron transfer characteristics with high chemical stability in sensing and catalysis. Carbon dots are used in electro-chemiluminescence sensors due to their exclusive optoelectronic characteristics [85]. Carbon dots have unique catalytic properties, and they can behave as enzyme mimics too. Their composites are water-soluble and biocompatible. Carbon dots are used in fluorescent probe sensing systems to determine hydrogen peroxide and o-phenylenediamine as they can behave like electron acceptors [86, 87]. Graphene Carbon nitrides are Carbon and nitrogen layered polymers used in many organic photocatalytic reactions. They are cheap and stable electrocatalysts employed generally for biosensing and catalysis. They have also employed carbon metal-free catalysts with tuneable electronic characteristics. To increase its photocatalytic behaviour, they are associated with metals like Boron, Phosphorus and Oxygen [88, 89]. Carbon nanocomposites formed by combining carbon with heteroatom and metal oxides have versatile applications in dehydrogenation and oxidation reactions [90]. Carbon and metal/metal oxide composites have dual benefits of high adsorption capacity due to carbon substrate area, and unique catalytic properties of nano ranged transition metal oxides. These nanocomposites have less agglomeration, high electron mobility, and an enhanced heat transfer rate. Carbon nanotubes in different metal matrices have good mechanical strength and less corrosion rate than bare samples [91]. Carbon composites with magnetic particles and metal oxides are used for dye degradation from wastewater and they have the potential application in bromophenol dye removal effectively up to 97% [92]. Besides this, Fe_2O_3 and carbon fibre composites are found to have good electrochemical behaviour. They have better stability and good discharge capacity in electrochemical cells [93, 94]. Transition element doped carbon nanotube have better corrosion resistance with fine fibre texture [95], and Manganese dioxide (MnO_2) with activated carbon composite is found to have better dye degradation characteristics. These composites have shown a better degradation rate up to 98.5% under both average and ultraviolet irradiation [96].

CONCLUSION

Nanomaterial has widespread effective utilization in diversified areas with inherited agreeable characteristics. The present energy consumption trends have corroborated the upcoming energy crisis and to outstrip such a situation, harvesting energy resources by employing recently developed cost-effective, ecologically viable and socially commensurable methods is necessary. Analytical estimation and quantifying inspection of the substances left undesignated so far and their degradation and transformation during a wastewater remediation process employ a nanoparticles-based sensing mechanism. This chapter discussed the latest classes of nano ranged materials and their application in catalysis, sensing and wastewater remediation techniques. Different functionalization, synthesis and characterization techniques are widely studied to investigate and appreciate the foremost action mechanism of these nanomaterials as redox couples in photocatalytic reactions. Some of the examples of these nanomaterials are discussed in this chapter.

CONSENT FOR PUBLICATION

Not applicable.

CONFLICT OF INTEREST

The author declares no conflict of interest, financial or otherwise.

ACKNOWLEDGEMENTS

Declared none.

REFERENCES

[1]   Feng, J.; Gao, C.; Yin, Y. Stabilization of noble metal nanostructures for catalysis and sensing. *Nanoscale,* **2018**, *10*(44), 20492-20504.
[http://dx.doi.org/10.1039/C8NR06757K] [PMID: 30398268]

[2]   Boukhvalov, D.W.; Paolucci, V.; D'Olimpio, G.; Cantalini, C.; Politano, A. Chemical reactions on surfaces for applications in catalysis, gas sensing, adsorption-assisted desalination and Li-ion batteries: opportunities and challenges for surface science. *Phys. Chem. Chem. Phys.,* **2021**, *23*(13), 7541-7552.
[http://dx.doi.org/10.1039/D0CP03317K] [PMID: 32926041]

[3]   Anu Mary Ealia, S.; Saravanakumar, M.P. A review on the classification, characterisation, synthesis of nanoparticles and their application *IOP Conference Series: Materials Science and Engineering, vol. 263*, p. 032019.**2017**,
[http://dx.doi.org/10.1088/1757-899X/263/3/032019]

[4]   Yan, X. Bio-inspired photosystem for green energy. *Green Energy & Environment,* **2017**, *2*(2), 66.
[http://dx.doi.org/10.1016/j.gee.2016.11.009]

[5]   Li, M.; Liu, Y.; Dong, L.; Shen, C.; Li, F.; Huang, M.; Ma, C.; Yang, B.; An, X.; Sand, W. Recent advances on photocatalytic fuel cell for environmental applications—The marriage of photocatalysis and fuel cells. *Sci. Total Environ.,* **2019**, *668*, 966-978.
[http://dx.doi.org/10.1016/j.scitotenv.2019.03.071] [PMID: 31018475]

[6]   Li, X.; Xie, J.; Jiang, C.; Yu, J.; Zhang, P. Review on design and evaluation of environmental photocatalysts. *Front. Environ. Sci. Eng.,* **2018**, *12*(5), 14.
[http://dx.doi.org/10.1007/s11783-018-1076-1]

[7]   Pan, J.H.; Dou, H.; Xiong, Z.; Xu, C.; Ma, J.; Zhao, X.S. Porous photocatalysts for advanced water purifications. *J. Mater. Chem.,* **2010**, *20*(22), 4512-4528.
[http://dx.doi.org/10.1039/b925523k]

[8]   Fogg, D.E.; dos Santos, E.N. Tandem catalysis: a taxonomy and illustrative review. *Coord. Chem. Rev.,* **2004**, *248*(21-24), 2365-2379.
[http://dx.doi.org/10.1016/j.ccr.2004.05.012]

[9]   Laoufi, I.; Saint-Lager, M-C.; Lazzari, R.; Jupille, J.; Robach, O.; Garaudée, S.; Cabailh, G.; Dolle, P.; Cruguel, H.; Bailly, A. Size and Catalytic Activity of Supported Gold Nanoparticles: An in Operando Study during CO Oxidation. *J. Phys. Chem. C,* **2011**, *115*(11), 4673-4679.
[http://dx.doi.org/10.1021/jp1110554]

[10]   Cuenya, B.R. Synthesis and catalytic properties of metal nanoparticles: Size, shape, support, composition, and oxidation state effects. *Thin Solid Films,* **2010**, *518*(12), 3127-3150.
[http://dx.doi.org/10.1016/j.tsf.2010.01.018]

[11]   Jana, J.; Ganguly, M.; Pal, T. Enlightening surface plasmon resonance effect of metal nanoparticles for practical spectroscopic application. *RSC Advances,* **2016**, *6*(89), 86174-86211.
[http://dx.doi.org/10.1039/C6RA14173K]

[12]   Amin, M.T.; Alazba, A.A.; Manzoor, U. A Review of Removal of Pollutants from Water/Wastewater Using Different Types of Nanomaterials. *Adv. Mater. Sci. Eng.,* **2014**, *2014*, 1-24.
[http://dx.doi.org/10.1155/2014/825910]

[13]   Yu, M.; Wang, J.; Tang, L.; Feng, C.; Liu, H.; Zhang, H.; Peng, B.; Chen, Z.; Xie, Q. Intimate coupling of photocatalysis and biodegradation for wastewater treatment: Mechanisms, recent advances and environmental applications. *Water Res.,* **2020**, *175*, 115673.
[http://dx.doi.org/10.1016/j.watres.2020.115673] [PMID: 32171097]

[14]   Guleria, A.; Priyatharchini, K.; Kumar, D. Biomedical Applications of Magnetic Nanomaterials. In: *Applications of Nanomaterials*; Mohan Bhagyaraj, S.; Oluwafemi, O.S.; Kalarikkal, N.; Thomas, S., Eds.; Woodhead Publishing, **2018**; pp. 345-389.
[http://dx.doi.org/10.1016/B978-0-08-101971-9.00013-2]

[15]   Sousa, V.S.; Ribau Teixeira, M. Metal-based engineered nanoparticles in the drinking water treatment systems: A critical review. *Sci. Total Environ.,* **2020**, *707*, 136077.
[http://dx.doi.org/10.1016/j.scitotenv.2019.136077] [PMID: 31863978]

[16]   Naseem, T.; Durrani, T. The role of some important metal oxide nanoparticles for wastewater and antibacterial applications: A review. *Environmental Chemistry and Ecotoxicology,* **2021**, *3*, 59-75.
[http://dx.doi.org/10.1016/j.enceco.2020.12.001]

[17]   Singh, N.B.; Susan, A.B.H. 21 - Polymer nanocomposites for water treatments. In: *Polymer-based Nanocomposites for Energy and Environmental Applications*; Jawaid, M.; Khan, M.M., Eds.; Woodhead Publishing, **2018**; pp. 569-595.
[http://dx.doi.org/10.1016/B978-0-08-102262-7.00021-0]

[18]   Upadhyay, R.K.; Soin, N.; Roy, S.S. Role of graphene/metal oxide composites as photocatalysts, adsorbents and disinfectants in water treatment: a review. *RSC Advances,* **2014**, *4*(8), 3823-3851.
[http://dx.doi.org/10.1039/C3RA45013A]

[19]   Tahir, N.; Zahid, M.; Bhatti, I.A.; Mansha, A.; Naqvi, S.A.R.; Hussain, T. Metal oxide-based ternary nanocomposites for wastewater treatment. In: *Aquananotechnology*; Abd-Elsalam, K.A.; Zahid, M., Eds.; Elsevier, **2021**; pp. 135-158.
[http://dx.doi.org/10.1016/B978-0-12-821141-0.00022-7]

[20]   Koiki, B.A.; Arotiba, O.A. Cu $_2$ O as an emerging semiconductor in photocatalytic and photoelectrocatalytic treatment of water contaminated with organic substances: a review. *RSC Advances,* **2020**, *10*(60), 36514-36525.
[http://dx.doi.org/10.1039/D0RA06858F] [PMID: 35517951]

[21]   Casbeer, E.; Sharma, V.K.; Li, X.Z. Synthesis and photocatalytic activity of ferrites under visible light: A review. *Separ. Purif. Tech.,* **2012**, *87*, 1-14.
[http://dx.doi.org/10.1016/j.seppur.2011.11.034]

[22]   Su, D. Advanced electron microscopy characterization of nanomaterials for catalysis. *Green Energy & Environment,* **2017**, *2*(2), 70-83.
[http://dx.doi.org/10.1016/j.gee.2017.02.001]

[23]   Dhiman, M.; Rana, S.; Singh, M.; Sharma, J.K. Magnetic studies of mixed Mg–Mn ferrite suitable for biomedical applications. *Integr. Ferroelectr.,* **2019**, *202*(1), 29-38.
[http://dx.doi.org/10.1080/10584587.2019.1674821]

[24]   Rana, S.; Sharma, A.; Kumar, A.; Kanwar, S.S.; Singh, M. Utility of Silane-Modified Magnesium-Based Magnetic Nanoparticles for Efficient Immobilization of Bacillus thermoamylovorans Lipase. *Appl. Biochem. Biotechnol.,* **2020**, *192*(3), 1029-1043.
[http://dx.doi.org/10.1007/s12010-020-03379-7] [PMID: 32638325]

[25]   Sharma, A.; Kumar, A.; Meena, K.R.; Rana, S.; Singh, M.; Kanwar, S.S. Fabrication and functionalization of magnesium nanoparticle for lipase immobilization in n -propyl gallate synthesis. *J. King Saud Univ. Sci.,* **2017**, *29*(4), 536-546.
[http://dx.doi.org/10.1016/j.jksus.2017.08.005]

[26]   Amiri, M.; Salavati-Niasari, M.; Akbari, A. Magnetic nanocarriers: Evolution of spinel ferrites for medical applications. *Adv. Colloid Interface Sci.,* **2019**, *265*, 29-44.
[http://dx.doi.org/10.1016/j.cis.2019.01.003] [PMID: 30711796]

[27]   Kefeni, K.K.; Msagati, T.A.M.; Nkambule, T.T.I.; Mamba, B.B. Spinel ferrite nanoparticles and nanocomposites for biomedical applications and their toxicity. *Mater. Sci. Eng. C,* **2020**, *107*, 110314.
[http://dx.doi.org/10.1016/j.msec.2019.110314] [PMID: 31761184]

[28]   Zhukov, A.; Inoue, M.; Phan, M.H.; Shavrov, V. Advanced Magnetic Materials. *Phys. Res. Int.,* **2012**, *2012*, 1-2.
[http://dx.doi.org/10.1155/2012/385396]

[29]   Kefeni, K.K.; Msagati, T.A.M.; Mamba, B.B. Ferrite nanoparticles: Synthesis, characterisation and

applications in electronic device. *Mater. Sci. Eng. B,* **2017**, *215*, 37-55.
[http://dx.doi.org/10.1016/j.mseb.2016.11.002]

[30] Zhang, Y.Q.; Ma, D.K.; Zhang, Y.G.; Chen, W.; Huang, S.M. N-doped carbon quantum dots for TiO2-based photocatalysts and dye-sensitized solar cells. *Nano Energy,* **2013**, *2*(5), 545-552.
[http://dx.doi.org/10.1016/j.nanoen.2013.07.010]

[31] Mahmoud, S.A.; Fouad, O.A. Synthesis and application of zinc/tin oxide nanostructures in photocatalysis and dye sensitized solar cells. *Sol. Energy Mater. Sol. Cells,* **2015**, *136*, 38-43.
[http://dx.doi.org/10.1016/j.solmat.2014.12.035]

[32] Han, Z.; Fei, J.; Li, J.; Deng, Y.; Lv, M.; Zhao, J.; Wang, C.; Zhao, X. Enhanced dye-sensitized photocatalysis for water purification by an alveoli-like bilayer Janus membrane. *Chem. Eng. J.,* **2021**, *407*, 127214.
[http://dx.doi.org/10.1016/j.cej.2020.127214]

[33] Habibi, M.H.; Habibi, A.H.; Zendehdel, M.; Habibi, M. Dye-sensitized solar cell characteristics of nanocomposite zinc ferrite working electrode: Effect of composite precursors and titania as a blocking layer on photovoltaic performance. *Spectrochim. Acta A Mol. Biomol. Spectrosc.,* **2013**, *110*, 226-232.
[http://dx.doi.org/10.1016/j.saa.2013.03.051] [PMID: 23571086]

[34] Tahir, M.B.; Sohaib, M.; Sagir, M.; Rafique, M. Role of Nanotechnology in Photocatalysis. In: *Reference Module in Materials Science and Materials Engineering*; Elsevier, **2020**.

[35] Astruc, D. Introduction: Nanoparticles in Catalysis. *Chem. Rev.,* **2020**, *120*(2), 461-463.
[http://dx.doi.org/10.1021/acs.chemrev.8b00696] [PMID: 31964144]

[36] Ahmed, S.N.; Haider, W. Heterogeneous photocatalysis and its potential applications in water and wastewater treatment: a review. *Nanotechnology,* **2018**, *29*(34), 342001.
[http://dx.doi.org/10.1088/1361-6528/aac6ea] [PMID: 29786601]

[37] Ge, J.; Zhang, Y.; Heo, Y.J.; Park, S.J. Advanced Design and Synthesis of Composite Photocatalysts for the Remediation of Wastewater: A Review. *Catalysts,* **2019**, *9*(2), 122.https://www.mdpi.com/2073-4344/9/2/122
[http://dx.doi.org/10.3390/catal9020122]

[38] Fujishima, A.; Honda, K. Electrochemical photolysis of water at a semiconductor electrode. *Nature,* **1972**, *238*(5358), 37-38.
[http://dx.doi.org/10.1038/238037a0] [PMID: 12635268]

[39] Deng, B.; Fu, S.; Zhang, Y.; Wang, Y.; Ma, D.; Dong, S. Simultaneous pollutant degradation and power generation in visible-light responsive photocatalytic fuel cell with an Ag-TiO2 loaded photoanode. *Nano-Structures & Nano-Objects,* **2018**, *15*, 167-172.
[http://dx.doi.org/10.1016/j.nanoso.2017.09.011]

[40] Szkoda, M.; Siuzdak, K.; Lisowska-Oleksiak, A. Non-metal doped TiO 2 nanotube arrays for high efficiency photocatalytic decomposition of organic species in water. *Physica E,* **2016**, *84*, 141-145.
[http://dx.doi.org/10.1016/j.physe.2016.06.004]

[41] Jafari, S.; Mahyad, B.; Hashemzadeh, H.; Janfaza, S.; Gholikhani, T.; Tayebi, L. Biomedical Applications of TiO_2 Nanostructures: Recent Advances. *Int. J. Nanomedicine,* **2020**, *15*, 3447-3470.
[http://dx.doi.org/10.2147/IJN.S249441] [PMID: 32523343]

[42] Musial, J.; Krakowiak, R.; Mlynarczyk, D.T.; Goslinski, T.; Stanisz, B.J. Titanium Dioxide Nanoparticles in Food and Personal Care Products—What Do We Know about Their Safety? *Nanomaterials (Basel),* **2020**, *10*(6), 1110.
[http://dx.doi.org/10.3390/nano10061110] [PMID: 32512703]

[43] Liu, Y.; Liu, C.; Rong, Q.; Zhang, Z. Characteristics of the silver-doped TiO2 nanoparticles. *Appl. Surf. Sci.,* **2003**, *220*(1-4), 7-11.
[http://dx.doi.org/10.1016/S0169-4332(03)00836-5]

[44] Sescu, A.M.; Favier, L.; Lutic, D.; Soto-Donoso, N.; Ciobanu, G.; Harja, M. TiO2 Doped with Noble

Metals as an Efficient Solution for the Photodegradation of Hazardous Organic Water Pollutants at Ambient Conditions. *Water,* **2020**, *13*(1), 19.https://www.mdpi.com/2073-4441/13/1/19
[http://dx.doi.org/10.3390/w13010019]

[45]   Suri, R.P.S.; Thornton, H.M.; Muruganandham, M. Disinfection of water using Pt- and Ag-doped TiO
       $_2$ photocatalysts. *Environ. Technol.,* **2012**, *33*(14), 1651-1659.
       [http://dx.doi.org/10.1080/09593330.2011.641590] [PMID: 22988625]

[46]   Riaz, N.; Mohamad Azmi, B-K.; Mohd Shariff, A. Iron Doped TiO2 Photocatalysts for Environmental
       Applications: Fundamentals and Progress. *Adv. Mat. Res.,* **2014**, *925*, 689-693.

[47]   Lee, S.Y.; Park, S.J. TiO2 photocatalyst for water treatment applications. *J. Ind. Eng. Chem.,* **2013**,
       19(6), 1761-1769.
       [http://dx.doi.org/10.1016/j.jiec.2013.07.012]

[48]   Chiam, S.-L.; Pung, S.-Y.; Yeoh, F.-Y. Recent developments in MnO2-based photocatalysts for
       organic dye removal: a review. *Environmental Science and Pollution Research,* **2020**, *27*, 5759-5778.
       [http://dx.doi.org/10.1007/s11356-019-07568-8]

[49]   Mitsudome, T.; Kaneda, K. Gold nanoparticle catalysts for selective hydrogenations. *Green Chem.,*
       2013, *15*(10), 2636-2654.
       [http://dx.doi.org/10.1039/c3gc41360h]

[50]   Haruta, M.; Yamada, N.; Kobayashi, T.; Iijima, S. Gold catalysts prepared by coprecipitation for low-
       temperature oxidation of hydrogen and of carbon monoxide. *J. Catal.,* **1989**, *115*(2), 301-309.
       [http://dx.doi.org/10.1016/0021-9517(89)90034-1]

[51]   Pan, Y.; Neuss, S.; Leifert, A.; Fischler, M.; Wen, F.; Simon, U.; Schmid, G.; Brandau, W.; Jahnen-
       Dechent, W. Size-dependent cytotoxicity of gold nanoparticles. *Small,* **2007**, *3*(11), 1941-1949.
       [http://dx.doi.org/10.1002/smll.200700378] [PMID: 17963284]

[52]   Yolles, R.; Wood, B.J.; Wise, H. Hydrogenation of alkenes on gold. *J. Catal.,* **1971**, *21*(1), 66-69.
       [http://dx.doi.org/10.1016/0021-9517(71)90121-7]

[53]   Pan, M.; Gong, J.; Dong, G.; Mullins, C.B. Model studies with gold: a versatile oxidation and
       hydrogenation catalyst. *Acc. Chem. Res.,* **2014**, *47*(3), 750-760.
       [http://dx.doi.org/10.1021/ar400172u] [PMID: 24635457]

[54]   Cárdenas-Lizana, F.; Keane, M.A. The development of gold catalysts for use in hydrogenation
       reactions. *J. Mater. Sci.,* **2013**, *48*(2), 543-564.
       [http://dx.doi.org/10.1007/s10853-012-6766-7]

[55]   Nikolaev, S.A.; Zanaveskin, L.N.; Smirnov, V.V.; Averyanov, V.A.; Zanaveskin, K.L. Catalytic
       hydrogenation of alkyne and alkadiene impurities from alkenes. Practical and theoretical aspects. *Russ.
       Chem. Rev.,* **2009**, *78*(3), 231-247.
       [http://dx.doi.org/10.1070/RC2009v078n03ABEH003893]

[56]   Campos, J. Dihydrogen and Acetylene Activation by a Gold(I)/Platinum(0) Transition Metal Only
       Frustrated Lewis Pair. *J. Am. Chem. Soc.,* **2017**, *139*(8), 2944-2947.
       [http://dx.doi.org/10.1021/jacs.7b00491] [PMID: 28186739]

[57]   Sárkány, A.; Horváth, A.; Beck, A. Hydrogenation of acetylene over low loaded Pd and Pd-Au/SiO2
       catalysts. *Appl. Catal. A Gen.,* **2002**, *229*(1-2), 117-125.
       [http://dx.doi.org/10.1016/S0926-860X(02)00020-0]

[58]   Gorbunova, M.; Apyari, V.; Dmitrienko, S.; Zolotov, Y. Gold nanorods and their nanocomposites:
       Synthesis and recent applications in analytical chemistry. *Trends Analyt. Chem.,* **2020**, *130*, 115974.
       [http://dx.doi.org/10.1016/j.trac.2020.115974]

[59]   Wang, K.; Weissmüller, J. Composites of nanoporous gold and polymer. *Adv. Mater.,* **2013**, *25*(9),
       1280-1284.
       [http://dx.doi.org/10.1002/adma.201203740] [PMID: 23288599]

[60] Liu, X.; Zhu, M.; Chen, S.; Yuan, M.; Guo, Y.; Song, Y.; Liu, H.; Li, Y. Organic-inorganic nanohybrids via directly grafting gold nanoparticles onto conjugated copolymers through the Diels-Alder reaction. *Langmuir,* **2008**, *24*(20), 11967-11974.
[http://dx.doi.org/10.1021/la8020639] [PMID: 18759505]

[61] Nikolaev, S.A.; Smirnov, V.V. Selective hydrogenation of phenylacetylene on gold nanoparticles. *Gold Bull.,* **2009**, *42*(3), 182-189.
[http://dx.doi.org/10.1007/BF03214932]

[62] Arrii, S.; Morfin, F.; Renouprez, A.J.; Rousset, J.L. Oxidation of CO on gold supported catalysts prepared by laser vaporization: direct evidence of support contribution. *J. Am. Chem. Soc.,* **2004**, *126*(4), 1199-1205.
[http://dx.doi.org/10.1021/ja036352y] [PMID: 14746491]

[63] Lantos, D.; Contel, M.; Sanz, S.; Bodor, A.; Horváth, I.T. Homogeneous gold-catalyzed hydrosilylation of aldehydes. *J. Organomet. Chem.,* **2007**, *692*(9), 1799-1805.
[http://dx.doi.org/10.1016/j.jorganchem.2006.10.025]

[64] Díez-González, S.; Nolan, S.P. Copper, silver, and gold complexes in hydrosilylation reactions. *Acc. Chem. Res.,* **2008**, *41*(2), 349-358.
[http://dx.doi.org/10.1021/ar7001655] [PMID: 18281951]

[65] Corma, A.; Garcia, H. Supported gold nanoparticles as catalysts for organic reactions. *Chem. Soc. Rev.,* **2008**, *37*(9), 2096-2126.
[http://dx.doi.org/10.1039/b707314n] [PMID: 18762848]

[66] Zhang, G.; Cui, L.; Wang, Y.; Zhang, L. Homogeneous gold-catalyzed oxidative carboheterofunctionalization of alkenes. *J. Am. Chem. Soc.,* **2010**, *132*(5), 1474-1475.
[http://dx.doi.org/10.1021/ja909555d] [PMID: 20050647]

[67] Segura, Y.; López, N.; Pérezramírez, J. Origin of the superior hydrogenation selectivity of gold nanoparticles in alkyne + alkene mixtures: Triple- versus double-bond activation. *J. Catal.,* **2007**, *247*(2), 383-386.
[http://dx.doi.org/10.1016/j.jcat.2007.02.019]

[68] Chiarucci, M.; Bandini, M. New developments in gold-catalyzed manipulation of inactivated alkenes. *Beilstein J. Org. Chem.,* **2013**, *9*, 2586-2614.
[http://dx.doi.org/10.3762/bjoc.9.294] [PMID: 24367423]

[69] Suchomel, P.; Kvitek, L.; Prucek, R.; Panacek, A.; Halder, A.; Vajda, S.; Zboril, R. Simple size-controlled synthesis of Au nanoparticles and their size-dependent catalytic activity. *Sci. Rep.,* **2018**, *8*(1), 4589.
[http://dx.doi.org/10.1038/s41598-018-22976-5] [PMID: 29545580]

[70] Ameur, N.; Bedrane, S.; Bachir, R.; Choukchou-Braham, A. Influence of nanoparticles oxidation state in gold based catalysts on the product selectivity in liquid phase oxidation of cyclohexene. *J. Mol. Catal. Chem.,* **2013**, *374-375*, 1-6.
[http://dx.doi.org/10.1016/j.molcata.2013.03.008]

[71] Chang, C-C.; Chen, C-P.; Wu, T-H.; Yang, C-H.; Lin, C-W.; Chen, C-Y. Gold Nanoparticle-Based Colorimetric Strategies for Chemical and Biological Sensing Applications. *Nanomaterials,* **2019**, *9*, 861.https://www.mdpi.com/2079-4991/9/6/861

[72] Sabela, M.; Balme, S.; Bechelany, M.; Janot, J.M.; Bisetty, K. A Review of Gold and Silver Nanoparticle-Based Colorimetric Sensing Assays. *Adv. Eng. Mater.,* **2017**, *19*(12), 1700270.
[http://dx.doi.org/10.1002/adem.201700270]

[73] Zeng, S.; Yong, K.T.; Roy, I.; Dinh, X.Q.; Yu, X.; Luan, F. A Review on Functionalized Gold Nanoparticles for Biosensing Applications. *Plasmonics,* **2011**, *6*(3), 491-506.
[http://dx.doi.org/10.1007/s11468-011-9228-1]

[74] Szunerits, S.; Spadavecchia, J.; Boukherroub, R. Surface plasmon resonance: signal amplification

using colloidal gold nanoparticles for enhanced sensitivity. *Rev. Anal. Chem.,* **2014**, *33*(3), 153-164.
[http://dx.doi.org/10.1515/revac-2014-0011]

[75]   Yu, L.; Song, Z.; Peng, J.; Yang, M.; Zhi, H.; He, H. Progress of gold nanomaterials for colorimetric sensing based on different strategies. *Trends Analyt. Chem.,* **2020**, *127*, 115880.
[http://dx.doi.org/10.1016/j.trac.2020.115880]

[76]   Yi, R.; Zhang, Z.; Liu, C.; Qi, Z. Gold-silver alloy film based surface plasmon resonance sensor for biomarker detection. *Mater. Sci. Eng. C,* **2020**, *116*, 111126.
[http://dx.doi.org/10.1016/j.msec.2020.111126] [PMID: 32806250]

[77]   Qian, H.; Pretzer, L.A.; Velazquez, J.C.; Zhao, Z.; Wong, M.S. Gold nanoparticles for cleaning contaminated water. *J. Chem. Technol. Biotechnol.,* **2013**, *88*(5), 735-741.
[http://dx.doi.org/10.1002/jctb.4030]

[78]   Das, M.; Shim, K.H.; An, S.S.A.; Yi, D.K. Review on gold nanoparticles and their applications. *Toxicol. Environ. Health Sci.,* **2011**, *3*(4), 193-205.
[http://dx.doi.org/10.1007/s13530-011-0109-y]

[79]   Lu, H.; Wang, J.; Stoller, M.; Wang, T.; Bao, Y.; Hao, H. An Overview of Nanomaterials for Water and Wastewater Treatment. *Adv. Mater. Sci. Eng.,* **2016**, *2016*, 1-10.
[http://dx.doi.org/10.1155/2016/4964828]

[80]   Bethi, B.; Sonawane, S.H.; Bhanvase, B.A.; Gumfekar, S.P. Nanomaterials-based advanced oxidation processes for wastewater treatment: A review. *Chem. Eng. Process.,* **2016**, *109*, 178-189.
[http://dx.doi.org/10.1016/j.cep.2016.08.016]

[81]   Liu, X.; Dai, L. Carbon-based metal-free catalysts. *Nat. Rev. Mater.,* **2016**, *1*(11), 16064.
[http://dx.doi.org/10.1038/natrevmats.2016.64]

[82]   Sadjadi, S.; Sadjadi, S. Electrospun carbon (nano) fibers for catalysis. In: *Emerging Carbon Materials for Catalysis*; Sadjadi, S., Ed.; Elsevier, **2021**; pp. 197-234.
[http://dx.doi.org/10.1016/B978-0-12-817561-3.00006-8]

[83]   Yang, Y.; Li, X.; Zhou, C.; Xiong, W.; Zeng, G.; Huang, D.; Zhang, C.; Wang, W.; Song, B.; Tang, X.; Li, X.; Guo, H. Recent advances in application of graphitic carbon nitride-based catalysts for degrading organic contaminants in water through advanced oxidation processes beyond photocatalysis: A critical review. *Water Res.,* **2020**, *184*, 116200.
[http://dx.doi.org/10.1016/j.watres.2020.116200] [PMID: 32712506]

[84]   Zhang, W.D.; Zhang, W.H. Carbon Nanotubes as Active Components for Gas Sensors. *J. Sens.,* **2009**, *2009*, 1-16.
[http://dx.doi.org/10.1155/2009/160698]

[85]   Chen, Y.; Cao, Y.; Ma, C.; Zhu, J-J. Carbon-based dots for electrochemiluminescence sensing. In: *Materials Chemistry Frontiers*; , **2020**; 4, .
[http://dx.doi.org/10.1039/C9QM00572B]

[86]   Chen, B.B.; Liu, M.L.; Huang, C.Z. Carbon dot-based composites for catalytic applications. *Green Chem.,* **2020**, *22*(13), 4034-4054.
[http://dx.doi.org/10.1039/D0GC01014F]

[87]   Mathivanan, D.; Tammina, S.K.; Wang, X.; Yang, Y. Dual emission carbon dots as enzyme mimics and fluorescent probes for the determination of o-phenylenediamine and hydrogen peroxide. *Mikrochim. Acta,* **2020**, *187*(5), 292.
[http://dx.doi.org/10.1007/s00604-020-04256-0] [PMID: 32347382]

[88]   Veerakumar, P.; Thanasekaran, P.; Subburaj, T.; Lin, K-C. A Metal-Free Carbon-Based Catalyst: An Overview and Directions for Future Research. **2018**, *4*, 54.https://www.mdpi.com/2311-5629/4/4/54

[89]   Patel, M. Carbon-Based, Metal-Free Catalysts for Chemical Catalysis. In: *Carbon☐Based Metal☐Free Catalysts*; , **2018**; pp. 597-657.

[90] Chandrasekaran, N.; Adarsh, K.S. Pristine, transition metal and heteroatom-doped carbon aerogels for catalytic and electrocatalytic applications. In: *Emerging Carbon Materials for Catalysis*; Sadjadi, S., Ed.; Elsevier, **2021**; pp. 235-253.
[http://dx.doi.org/10.1016/B978-0-12-817561-3.00007-X]

[91] Say, Y.; Guler, O.; Dikici, B. Carbon nanotube (CNT) reinforced magnesium matrix composites: The effect of CNT ratio on their mechanical properties and corrosion resistance. *Mater. Sci. Eng. A,* **2020**, *798*, 139636.
[http://dx.doi.org/10.1016/j.msea.2020.139636]

[92] Alorabi, A.Q.; Shamshi Hassan, M.; Azizi, M. Fe3O4-CuO-activated carbon composite as an efficient adsorbent for bromophenol blue dye removal from aqueous solutions. *Arab. J. Chem.,* **2020**, *13*(11), 8080-8091.
[http://dx.doi.org/10.1016/j.arabjc.2020.09.039]

[93] Li, Y.; Zhu, C.; Lu, T.; Guo, Z.; Zhang, D.; Ma, J.; Zhu, S. Simple fabrication of a Fe2O3/carbon composite for use in a high-performance lithium ion battery. *Carbon,* **2013**, *52*, 565-573.
[http://dx.doi.org/10.1016/j.carbon.2012.10.015]

[94] Sun, C.; Chen, S.; Li, Z. Controllable synthesis of Fe2O3-carbon fiber composites via a facile sol-gel route as anode materials for lithium ion batteries. *Appl. Surf. Sci.,* **2018**, *427*, 476-484.
[http://dx.doi.org/10.1016/j.apsusc.2017.08.070]

[95] Jyotheender, K.S.; Gupta, A.; Srivastava, C. Grain boundary engineering in Ni-carbon nanotube composite coatings and its effect on the corrosion behaviour of the coatings. *Materialia (Oxf.),* **2020**, *9*, 100617.
[http://dx.doi.org/10.1016/j.mtla.2020.100617]

[96] Khan, I.; Sadiq, M.; Khan, I.; Saeed, K. Manganese dioxide nanoparticles/activated carbon composite as efficient UV and visible-light photocatalyst. *Environ. Sci. Pollut. Res. Int.,* **2019**, *26*(5), 5140-5154.
[http://dx.doi.org/10.1007/s11356-018-4055-y] [PMID: 30607840]

CHAPTER 10

Advancement of Topological Nanostructures for Various Applications

Debarati Pal[1] **and Swapnil Patil**[1,*]

[1] *Department of Physics, Indian Institute of Technology (BHU), Varanasi-221005, India*

Abstract: Topological materials are characterized by a unique band topology that is prominently distinct from ordinary metals and insulators. This new type of quantum material exhibits insulating bulk and conducting surface states that are robust against time-reversal invariant perturbations. In 2009, Bi_2Se_3, Sb_2Te_3 and Bi_2Te_3 were predicted as 3D Topological insulators (TIs) with a single Dirac cone at the surface state. For application purposes, however, bulk conductivity due to Se vacancy in Bi_2Se_3 or anti-site defects in Bi_2Te_3 has been a challenging issue. In order to achieve an enhanced surface conductivity over the bulk, nanomaterials are irreplaceable. Nanostructures' high surface to volume ratio provides a good platform for investigating the topological existence of surface states. By tuning the position of Fermi level through field effect gating, it is also possible to terminate the bulk residual carriers. Moreover, the synthesis of nanomaterials allows for morphological, electronic, and chemical regulation, resulting in the ability to design structures with desired TI properties at the nanoscale. In this article, we review various technological applications of nanostructured topological insulators. We also survey the implementation of topological nanomaterials in the field of optoelectronic devices, p-n junction, superconducting materials, field effect transistor, memory device and spintronics, ultrafast photodetection, and quantum computations.

Keywords: Band topology, Dirac cone, Quantum computations, Quantum material, Topological nanostructure.

INTRODUCTION

Topological insulators, the recent discovery in condensed matter physics, contains insulating bulk and spin helical surface state with a unique band topology. The surface states are protected by time–reversal symmetry and caused by the inversion of the bulk bandgap, occurring by the strong spin–orbit coupling of the heavy (high Z) material. Fig. (**1**) manifests the 1D helical edge state and 2D helical surface states of 2D and 3D topological insulators. The exceptional electr-

* **Corresponding author Swapnil Patil**: Department of Physics, Indian Institute of Technology (BHU), Varanasi-221005, India; Tel:9421066673; E-mail:spatil.phy@itbhu.ac.in

Dibya Prakash Rai (Ed.)
All rights reserved-© 2022 Bentham Science Publishers

onic nature of the TI surface states assures spintronics devices and quantum computing, as well as guarantees applications in energy conversion such as thermoelectrics. Presently, the investigation of topological materials also includes Weyl semimetal(WSMs), Dirac semimetal(DSMs), and Nodal line semimetal (NDSMs) with different degeneracy and distribution of band crossings. Among them, the band crossing of Dirac semimetal possesses 4-fold degeneracy near the Fermi energy. When those Dirac points split into Weyl points with the breaking of either time reversal or inversion symmetry, Weyl semimetal comes into picture. The Weyl points always come into pairs with opposite chirality. Type 1 WSMs have point-like Fermi surfaces, whereas tilted Weyl cones that appear at the contact of electron pockets and hole pockets are Type 2 WSMs. For NLSMs, band crossings make closed loops instead of discrete points in the Brillouin zone.

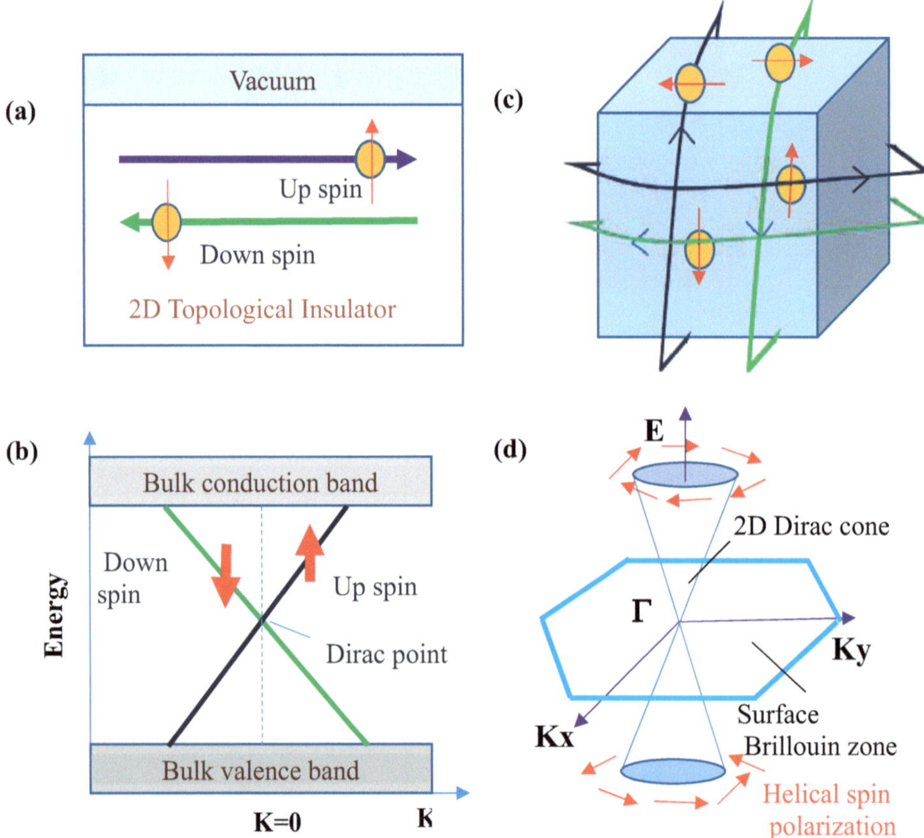

Fig. (1). (Modified from [9]. Edge and surface states of topological insulator with Dirac dispersions. **(a)** The 1D helical edge state of a 2D TI is shown schematically, **(b)** Energy dispersion of a 2D TI's spin non-degenerate edge state forming a 1D Dirac cone, **(c)** The 2D helical surface state of a 3D TI is shown schematically, **(d)** Energy dispersion of a 3D TI's spin non-degenerate surface state forming a 2D Dirac cone.

Fig. (2) represents the schematic diagram of normal metal and three types of topological semimetals. In this review, we will discuss the ongoing progress of nanostructured topological insulators and their implementations in future directions. The experimentally realized first predicted 2D TI was HgTe/CdTe quantum wells. Later $Bi_{1-x}Sb_x$ alloy was confirmed as 3D TI with a complex surface state; however, the A_2B_3 type of materials with a single Dirac cone signature was proved as 3D TI with simpler surface states.

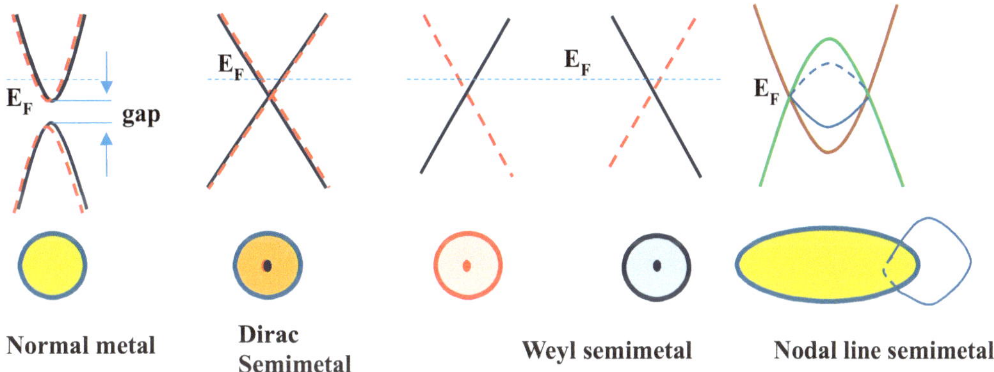

Fig. (2). Modified from [10]. Schematic diagram of Fermi surface and band structure for normal metal and three kinds of topological semimetals, The close path (blue thin circle) interlocked with the nodal ring (thick circle) is also shown.

Although the theoretical aspects of topological insulator materials are adequately recognized, there are numerous difficulties and ongoing efforts in solving the experimental challenges. One of the foremost experimental problems faced by the researchers associated with the synthesis of those materials in the bulk forms is unnecessary bulk residual carriers due to intrinsic defects. The benefit of nanostructured topological surface states is that they reduce the effect of unwanted bulk electronic states in transport measurements as much as possible, thus increasing the surface-to-volume ratio to facilitate the contribution of topological surface or edge states. Therefore, material dimensions are crucial in the surface-dominant transport regime with a view to take advantage of the surface electrons' invaluable properties. One more advantage of the nanostructured materials is the easy modulation of Fermi level *via* field effect gating (FEG) which is difficult in the case of bulk crystal because of the large volume. Together with the sizeable surface to volume ratio in nano TI, FEG ensures surface dominated electron transport. Additionally, the extraordinary morphology of TI nanomaterials provides control over the topological surface states in reduced dimensions such as nanoribbons [1, 2], nanoplates [3], nanorods [4], nanowires [5], nanotubes [6], nonorings [1] and quantum dots [7]. Bi_2Te_3

nanoribbons on $SrTiO_3$ substrate serve substantial surface state conduction. Moreover, a superconducting transition has also been noticed in TI nanostructures by combining a non-superconducting material with a superconductor (proximity-induced superconductivity).

Current materials problems that obstruct control and easy access of topological surface states are also established in this article, and novel solutions that TI nanostructures can offer to resolve these [8] challenges are proposed. We also survey the recent progress of TI in various applications.

APPLICATIONS

Proximity Induced Superconductivity

The superconducting proximity effect (SPE) has proved extremely potent for extensive research in electronic transport. The realization of Majorana Fermions (MFs) at the interface between a topological insulator and superconducting material, has proved significant interests in fundamental science as well as in quantum computing. Topological superconductor (SCs) can also be created by forming a Josephson junction where an TI is sandwiched between two superconductors. Cooper pairs can be transmitted dissipationlessly from one superconductor to the other due to the coupling between both superconducting condensates. As a result, with a doublet of topologically protected gapless Andreev bound states, an unconventional induced p-wave superconductivity emerges. These bound states are phase-coherent superpositions of electron and hole excitations, and with the superconducting phase difference across the junction, their energies differ 4π-periodically. By incorporating superconducting metallic electrodes into nanodevices, proximity-induced superconductivity can be easily established in TI nanoparticles. The first step towards the search for this quasiparticle was the discovery of supercurrents through topological surface state. However, the presence of bulk, intrinsic defects and impurities were experimental obstructions in the search for such supercurrents. It was observed that sample dimension is crucial for the observed feature and minimizing the size of the flake to the nanoscale facilitates the observation of the Majorana fermions in a single channel TI Josephson junction [11, 12]. Superconducting proximity effect was found in Bi_2Se_3/Bi_2Te_3 nanoflakes [13 - 16] and Sn-Bi_2Se_3 junction [17]. Fig. (**3a**) represents a schematic of a topological Josephson junction that uses two superconducting leads on Bi_2Se_3 TI. However, residual bulk conductance dominating electronic transport in these Bi-based materials prevents the analysis of topological superconductivity. The bulk of strained HgTe is effectively insulating, unlike Bi-based topological insulators. A HgTe barrier used (Fig. **3b**) in between a TI and superconducting material to facilitate the purpose of Josephson junction [18, 19]. A recent work on BSTS topological insulator nanoribbons reported the proximity effect using Josephson junction with Nb as a superconducting contact [8]. Many approaches were dedicated towards the SPE using topological heterostructure [20 - 23] but the difficulty arises in accessing the Majorana bound states (MBE) at the interface of those materials, the problem was resolved by researchers with an alternative idea of using Pb thin film grown on TI TI BiSe$_2$ [24]. Theoretical work based on Bi$_2$Se$_3$ TI and SC PdTe heterostructures presented a new inquiry into the chemical potential and thickness dependent phenomena of Majorana spectra [25].

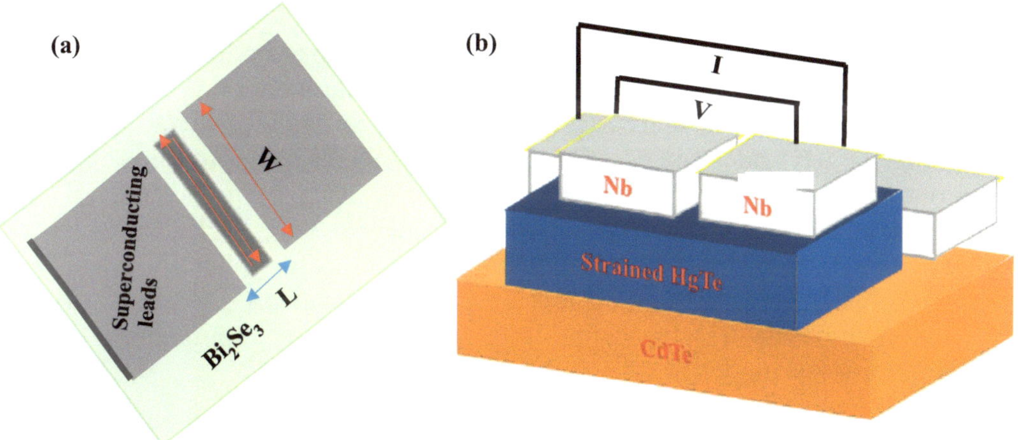

Fig. (3). (a) Schematic of a Josephson junction with TI is shown. Two superconducting leads are put on top of Bi$_2$Se$_3$ creating a junction with length L and width W, a one-dimensional wire of Majorana fermions in between superconducting leads is formed (modified from [14]), **(b)** Schematic picture of a lateral Josephson junction based on a strained, undoped HgTe layer. The Dirac states residing at the surfaces of the HgTe layer form the weak link between the Nb electrodes (modified from [18]).

Semiconducting quantum wires under a magnetic field condition are an extraordinary example of a 1D topological superconductor [26, 27]. In the presence of moderate disorder, the conductance of the nanowire is quantized to e^2/h in the single-mode regime, until the chemical potential is not near zero. But these wires had several modes thus challenging to observe the quantization, either a tunnel barrier [28] or a quantum point contact [29] can resolve those issues. Juan *et al*. reported topological superconductivity from 3D TI nanowire by generating a normal superconductor junction [30]. The quantum Hall effect makes the top and bottom regions of the wire insulating when the normal part of the wire is subjected to a perpendicular field. A counterpropagating chiral edge states robust against backscattering due to their spatial separation are created between those regions. This results in a single chiral mode and is an ideal candidate for stabilizing quantized Hall effect [31, 32]. The issue of proximity induced superconductivity of a TI in close contact with an FM has also been investigated with special attention [33]. Yano *et al.* proposed Fe-doped $BiSbTe_2Se$ magnetic topological insulator in order to gain access to proximity-induced surface states [34]. However, the FM/TI bilayer explores many fascinating physics such as creating a gap at the Dirac point, RKKY interaction between two of them, etc. A recent investigation on ferromagnetic metal and doped topological insulator $Bi_{0.8}Sb_{1.2}Te_3$ bilayer suggests proximity induced surface states create a gap on surface states that is directly in contact with FM as well as the surface state that is separated from the FM layer by the TI bulk [35]. A hybrid superconductor-ferromagnet structure was deposited on top of a topological insulator to unveil novel physics behind proximity induced superconductivity and ferromagnetic surface state [36]. In order to achieve topological superconductivity in TI/SC heterostructures, researchers have mostly concentrated on coupling TIs with low-Tc SCs such as $NbSe_2$/ TI heterostructure [37, 38], $Sn-Bi_2Se_3$ interface junctions [17], $Pb-Bi_2Te_3$ Hybrid Structures [39]. The goal of achieving proximity induced superconductivity has been pursued actively also using high Tc materials for practical device applications [40, 41]. High Tc SC material Fe(Te,Se) was used with TI Bi_2Te_3 [42] to achieve large proximity induced gap (~3.5meV), which was a challenging issue for those with low Tc SCs. Lower gap (~ 1meV in case of $NbSe_2$/TI heterostructure) basically limits potential future applications of the detected Majorana modes.

Optoelectronic Device

Optoelectronic devices can convert light into an electric signal or electric signal into light and are made of semiconducting crystalline materials. Application of optoelectronic devices are emitting diode (LED), solar cell, photodiodes, optical fibers, laser diodes, photodetector, and so forth. Importantly, Topological insulators offer intrinsically appealing large surface plasmon resonance and huge bulk refractive index for optical applications. Furthermore, the narrow bandgap of TI improves conductivity and allows for large transparency in the near-infrared region. Various technical applications, such as sensing, imaging, and optical communications, rely on broadband photodetection.

Zhang *et al.* reported wide bandwidth, high-performance photodetection covering from terahertz to infrared region with Bi_2Se_3 thin film [45], the performance of the photodetection increased with reducing thickness of the TI layers. Later in 2015, Zheng group established a near infrared photodetector with Sb_2Te_3 thin film [43]. The schematic of Sb_2Te_3 based photodetector with Sapphire substrate is displayed in Fig. (**4a**). TI Bi_2Se_3 nanostructure [46] with single Dirac cone surface state is produced as a transparent conductor for infrared wavelength, it also possesses low sheet resistance, high durability, more than 70% transparency, good conductivity. It shows better performance compared with ITO (indium tin oxide) films in the context of infrared sensing and imaging applications. High-performance broadband (UV to THz) photodetectors were constructed by Yao group with TI Bi_2Te_3–Si heterostructure [44]. The photodetector exhibited extraordinarily large photosensitivity and detectivity of 7.5×10^5 cm^2 W^{-1} and 2.5×10^{11} cm Hz$^{1/2}$ W^{-1}, respectively. The schematic based on Bi_2Te_3 TI with Pt and Ag electrodes is manifested in Fig. (**4b**). Bi_2Te_3 with the polyethylene terephthalate (PET) substrate leads to a quite big responsivity and detectivity at near infrared regime along with strong light absorption properties [44]. In an exfoliated Bi_2Se_3 thin flake, McIver *et al.* discovered a photocurrent caused by topological helical Dirac fermions [47].

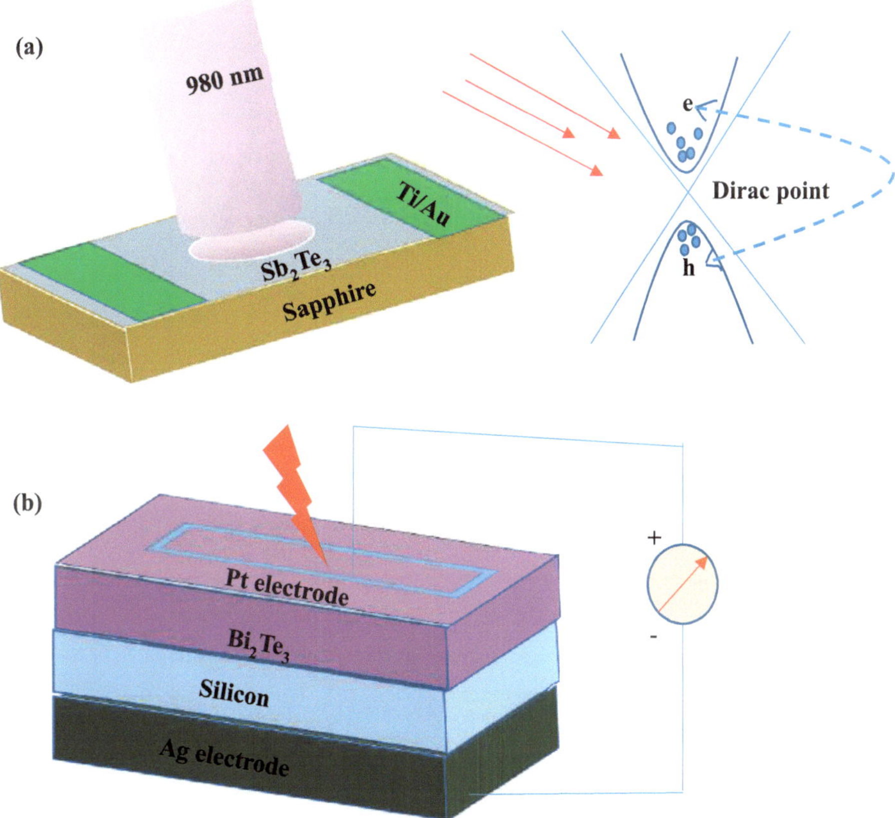

Fig. (4). (a) A near-infrared light photodetector based on Sb$_2$Te$_3$ is shown schematically, Sapphire is used as substrate (modifed from [43]), **(b)** TI Bi$_2$Te$_3$–Si heterojunction photodetector with Pt and Ag electrodes in a two pole structure (modified from [44]).

More surprisingly, the direction of the photocurrent could be changed by reversing the helicity of the light. The topological insulators at the nano scale with higher surface to volume ratios (nanowires and nanoribbons) can be used as a substitute to increase the surface contribution of the photocurrent. The next generation of photodetector-based devices demands robust, durable nanostructured materials due to miniaturization of device dimensions, in this scenario the TIs at the nanoscales are irreplaceable. In 2017, Sharma *et al.* proposed Bi_2Se_3 nanosheet and nanowire [48] to compare photosensivity. They realized that the nanowires are able to produce one order greater photoresponsivity than the nanosheet. The nano confinement effect of the TI surface state is responsible for the rise in photoresponsivity and electron-hole pair production from incident photons. The ability to create high-quality topological insulator-based heterostructures is another advantage of synthetic nanostructures [49] over exfoliated flakes. This is highly useful for improving the effective creation and transmission of photocarriers at the interface, hence increasing the photocurrent while preserving the detecting spectral width.

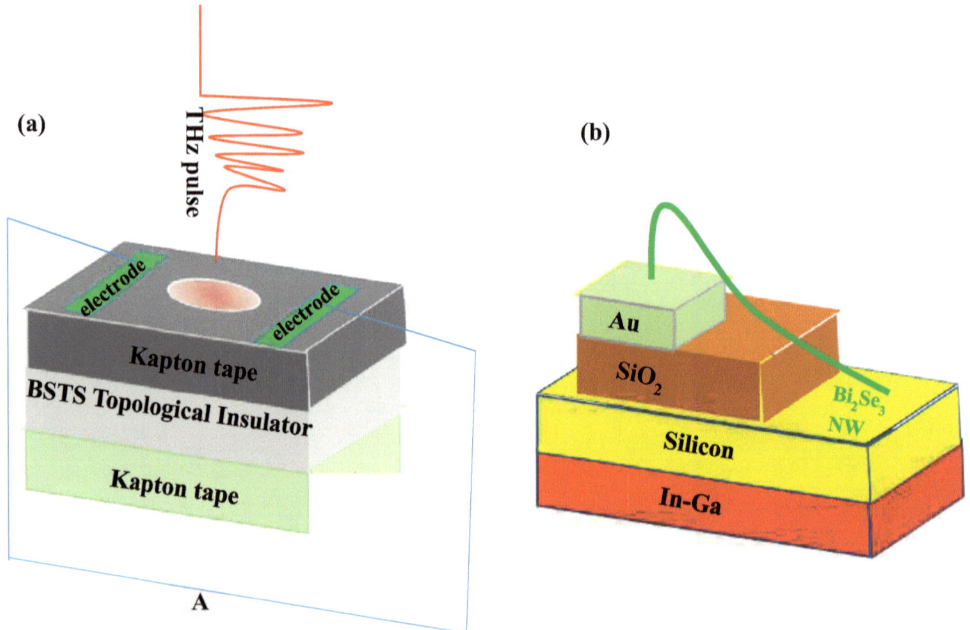

Fig. (5). (a) Topological insulator BSTS based THz modulator, Kapton/BSTS/Kapton sandwich-structure with Si electrode (modified from [50], **(b)** Schematic diagram of the Bi_2Se_3 NW/Si heterostructure photodetector, Au electrode deposited on SiO_2/Si wafer (modified from [51].

Researchers demonostrated electronically-tunable extraordinarily large(THz) modulator comprised of $Bi_{1.5}Sb_{0.5}Te_{1.8}Se_{1.2}$ single crystal [50]. Such a large modulation is associated with thermally-activated carrier absorption in the semiconducting bulk states. The schematic of Kapton tape based TI THz modulator is depicted in Fig. (**5a**).

Moreover, there are many advantages of using Bi_2Se_3/Si heterostructure, it produces a robust Schottky barrier at the interface as well as increases photogenerated carrier life time so that a photoemission could be achieved from Bi_2Se_3 to silicon. The inbuilt potential in such a heterostructure is able to block the recombination of photo produced carriers, resulting in a significant photocurrent in the external circuit. Bi_2Se_3/Si heterostructure [52] proposed by Zhang group had the highest light responsivity and detectivity reported so far. It also possessed a quick response speed ($\sim$10^{-6} order). The photodetector could detect wavelengths ranging from ultraviolet to optical telecommunication wavelengths. Another group in 2016 reported a heterostructure of Bi_2Se_3 nanowire on Si substrate [51] with a responsivity of $\sim$10^3 A W^{-1}. A quick and broadband response property was also achieved by this structure. The schematic of the hetorostructure is illustrated in Fig. (**5b**). Organic materials are also used with TI to access wideband detection (450-3500 nm) and high performance photodetection in a report [53]. The schematic is manifested in Fig. (**6a**).

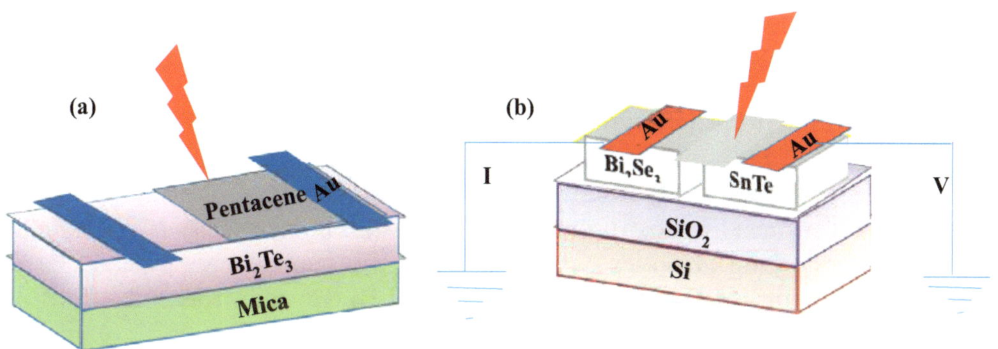

Fig. (6). (a) Schematic of organic heterostructure (modified from [53], **(b)**Basic electrical transport characteristic of the SnTe/Bi_2Se_3 heterojunction device (modified from [54].

A recent research on SnTe/Bi_2Se_3 heterostructure devices enables low-power and low-cost optical detection and a great opportunity for ultrafast memory devices [54]. The schematic of SnTe/Bi_2Se_3 heterostructure is shown in Fig. (**6b**).

Some unconventional structure was also used for achieving greater absorption length for light absorption. External Cu doping was used to balance the intrinsic defect in Bi_2Se_3 [55]. Generally reduced dimensions (nanoplates, nanoribbons) are used to achieve high suface to bulk phenomena, but they sometimes limit large scale applications, the unconventional nanostructured films are useful in such scenarios [56].

Field- effect Transistors

The field-effect transistor (FET), based on nanostructured semiconducting materials, has been explored as one of the most important components for circuit technologies and upcoming nanoelectronic devices. In addition to optoelectronics and spintronics, topological insulator FETs are also very useful for future electronic applications. To efficiently tune the Fermi level and minimise the residual bulk carrier density, field-effect gating can be utilised. For the research of novel topological surface phenomena and future device applications, control over the topological surface states is greatly required [57]. Modulating the chemical potential in TIs was first executed in Bi_2Se_3 with NO_2 molecules on the surface [58]. An FET device with MBE-grown topological insulator thin films $Cr_{0.15}(Bi_{0.1}Sb_{0.9})_{1.85}Te_3$ was developed, the advantage of Sb doping in Bi reduces the intrinsic defects inside the crystal, field effect gating provides fine tuning of E_F over the various regime of the band structure [59]. In the work of Liu group, field-effect transistors (FET) with large ON/OFF ratios were achieved in the ultrathin $(Bi_{1-x}Sb_x)_2Se_3$ using the molecular beam epitaxy (MBE) method [60]. A field effect transistors (FETs) with near-zero cutoff current, sharp turn-on, large On/Off current ratio ($\sim 10^8$) and well satuarated output current were accomplished by the Zhu group [61]. They also managed to separate the surface conduction from the bulk through proper field effect gating. Ambipolar field effect on $(Bi_xSb_{1-x})_2Se_3$ nanoplates facilitates precise control over carrier type and concentration for the TI nanostructures [62, 63]. Ambipolar transport *via* standard dielectric layers gating provides carrier tuning at the surface state of nano Bi_2Se_3 [64]. Hong group utilized zinc oxide-encapsulated Bi_2Se_3 nanoribbons to diminish bulk electron concentration and achieved electron to hole carrier through field effect gating [65]. Yuan *et al.* used ionic-liquid (IL) gating for modulating the electronic states of Bi_2Te_3 [66]. Similar electric-double-layer transistor (EDLT) configurations with ionic-liquid dielectrics was also used in $(Bi_{1-x}Sb_x)_2Te_3$ and $BiSbTeSe_2$ to control the surface carriers efficiently [67, 68]. EDLT with ionic liquid dielectric gating holds promise for modulating the Fermi level in topological insulators as higher carrier density and ambipolar transport can be achieved under such conditions. When compared to conventional field effect transistors with an oxide gate dielectric, this method is more efficient [69, 70]. The suppression of bulk carrier density transport was demonstrated in TI thin films as well as nanoplates, demonstrating the ambipolar field effect [63, 66, 71]. Simultaneous approach to material synthesis at the nanoscale and fine tuning of fermi level through electrostatic gating results in significantly reduced bulk conduction and thus promotes a prominent surface transport [72, 73]. Furthermore, due to the more pronounced contribution of the surface state to the overall conductance in nanostructures, larger ambipolar field effects can be detected in thinner materials or at lower temperatures [63, 74].

p-n Junction Devices

A topological p-n junction (TPNJ) is a key idea for controlling spin and charge transport on a three-dimensional topological insulator surface (3D-TIs). Si is a common p-type material that has been used to make p-n junctions for decades. These junctions have a large output current, a low leakage current, and other impressive features. Si, on the other hand, has a very small band gap, which makes it difficult to align bands with standard n-type oxide semiconductors, which has a wider band gap. As a result, the search for both p and n-type counterparts for junction manufacturing continues, and various TIs with metallic surface states are unique for this purpose. Among them, Bi_2Se_3 is one of the most versatile n-type topological insulators with many predominant features like high electrical conductivity and small band gap (~0.35 eV) close to Si. This avoids the undesired mismatch between the junction components that occurred in Si. The spin momentum locking property facilitates enhanced diode properties by providing separate forward and reverse bias. A unique union of p-Si NW/n-Bi_2Se_3 adopted for p-n junction diode to get an extraordinary application for NIR photodetector [75]. Particularly, $Bi_{2-x}Sb_xTe_{3-y}Se_y$ has very high resistivity, also the energy position of the Dirac point in the band gap is easily tunable while maintaining the insulating bulk and high mobility surface state. Moreover, such BSTS system gives an excellent platform for developing dual-gate configuration for the electric control of spins or p-n junction fabrications that are essential for semiconductor technology. A theoretical proposal on topological p-n junction using $(Bi_{1-x}Sb_x)_2Te_3$ was discovered [76]. Han Tu *et al.* [77] proposed an experimental p-n junction using ultrathin BSTS topological insulator systems.

Memory device and Spintronics

In recent years, spintronics exploiting TIs as strong spin–orbit coupling materials has arisen and progressed rapidly. TIs, unlike typical heavy metals, have very large SOC and nontrivial topological surface states that originate from the bulk band topology, which may be used to manipulate the nearby magnetic materials very efficiently when a charge current is passed through them. The spin-orbit coupling links the charge current to spin current *via* the spin Hall effect (SHE). Magnetic tunnel junctions (MTJ) are used as magnetoresistive random access memory (MRAM). But those MTJ owing to the large current in the `write' process experience some disadvantages like dielectric breakdown at the tunnel barrier. Heavy metals and topological insulators with high spin-orbit coupling (SOC) are one option for resolving this issue, the unique band topology of the surface state spin torques could be created in the MTJ materials [78]. The advantage of using TIs is to prevent charge current from crossing the tunnel barrier during spin torque generation and allow them for a separate 'read' and 'write' current path. Moreover, topological insulators are considered to be good candidates for investigating spin–orbit torque (SOT)-related physics due to their inherent strong SOC. Fan *et al.* [79] Fan *et al.* investigated magnetization switching in a chromium-doped TI bilayer heterostructure generated by huge SOT. In 2016 [80], the same group demonstrated gate controlled spin torque device. They were able to enhance the SOT by 4 times with proper control of gate voltage. TI Bi_2Se_3 thin film with adjacent permalloy ($Ni_{81}Fe_{19}$) was investigated for generating spin-transfer torque (STT) [81]. Other topological insulator-based spin torque effects have also been proposed thoroughly [82, 83]. Furthermore, spin–torque ferromagnetic resonance (ST-FMR) measurements [81, 84], SOT induced magnetization switching in magnetic TI [79, 80] and spin pumping/spin injection experiment [85 - 87] were executed for various spintronic applications.

Ultrafast Photodetectors

Infrared photodetection benefits from Dirac electronic systems, such as graphene, topological insulators, and Dirac semimetals [88]. The bulk structure allows topological semimetals to have a higher density of states, which means they can ensure higher optical absorption and stronger light-matter interaction than atomically thin graphene, allowing for high photodetection sensitivity [89 - 91]. Gapless bulk states [92, 93], strong carrier mobility [93 - 95], and ultrafast transient [89, 96] duration have all been reported in Cd_3As_2, a stable Dirac semimetal with several electrical features appealing for photodetectors [89]. At room temperature, a wideband photodetector with improved performance has been reported on the Cd_3As_2 heterojunction [97]. The ultrabroadband photoresponse of Weyl semimetal TaAs was proven experimentally at room temperature (RT) [98]. Based on its massless Weyl nodes and very significant absorptance across a wide spectrum range, TaAs clearly have the advantages of a broad detection region and ambient operating temperature when compared to graphene and narrow-gap semiconductors. Gao *et al.* [99] reported PLD-grown WTe_2 ultrathin film WSM that exhibits a unique ultrafast response and persists over a broad spectral range, at least covering 1.2 to 2.5 μm, and makes it an efficient fast saturable absorber. Another novel work [100] on T_d-WTe_2 reported photocurrent and responsivity can be as large as 40 μA and 250 A W^{-1} for a 3.8 μm laser at 77 K. Broadband responses from 532 nm to 10.6 μm have been investigated experimentally Td-MoTe$_2$ [101], with a detection range that might be extended to far-infrared and terahertz. Another Weyl semimetal TaIrTe$_4$ [102] outperforms the 3D Dirac semimetal Cd$_3$As$_2$, which offers similar advantages in terms of photosensitivity and broadband response as a photosensing material.

Quantum Computation

Topological insulators exhibit novel physics at the interfaces due to the presence of the proximity effect with different superconductors and, magnets giving rise to applications in topological quantum computation [103 - 108] and topological spintronics field. A superconducting pairing gap leads to a condition that supports Majorana fermions, potentially opening up new avenues for topological quantum computation. When superconducting state is created at the edge or surfaces of TIs *via* the superconducting proximity effect, the spin-momentum locking property allows TIs to host Majorana fermions. Majorana Fermions are particles that are their own anti-particle. Topological insulators possess non spin degeneracy because of the spin helical surface state. As a matter of fact, the intra-band pairing must be odd to become a p-type superconductor that carries Majorana fermions. This means that at the vortex core of the aforementioned persuaded p+ip Topological SCs, a Majorana state can be bound [109]. The zero mode Majorana Fermions also emerge at the edge in the one-dimensional p-wave superconductor [110, 111] but are not experimentally proved. The topological SCs is a current quantum state of matter that, in terms of adiabaticity, varies from the superconducting state formed by the Bose–Einstein condensate of Cooper pairs. Non-Abelian anyons, which are believed to revolutionise quantum computation, are Majorana fermions that obey non-Abelian statistics. Both, the bound and edge states of Majorana have a lot of potential for forming qubits [112 - 114]. This is still true in condensed matter physics, and the superposition of electron and hole excitations can be used to describe the Majorana fermion. A Majorana fermion is defined as a quasi-particle or excitation whose creation and annihilation operators are the same. The appearance of this kind of excitation is one of the most distinguishing feature of a topological SCs, which can be used as a topological qubit in topological quantum computation [115].

CONCLUSON

In this review, we briefly introduced the diversity of topological fields and recent progress from the application point of view. Although the word 'topological insulators' was invented a while ago, the field of topological materials (topological crystalline insulators TCI, WSMs, DSMs, NLSMs etc) has blossomed, and the current theoretical works indicate that ~25% of all the materials of our knowledge may be topological in nature [116]. In the recent years of condensed matter research, topological materials have sparked into multidimensional fields and suggest one of the most fascinating and significant research frontiers these days. Topological proximity effect with dissipation less conduction, field- effect transistors with fast- switching, broadband photodetector for optoelectronic devices, an ultrafast photodetector for innovative photodetection technologies, topological p-n junction diode for electronic devices and qubits for robust quantum computers based on 1D topological superconductors, are crucial for potential device applications. The TI surface state is ideal for producing SOT *via* SHE, the effective control over the SOT by the gate voltage represents energy-efficient gate-controlled nonvolatile spintronic memory and logic devices. TIs are also shown to be very successful at converting spin to charge, implying that they could be used as spin detectors or efficient spin-to-charge converters. In summary, initial theoretical breakthroughs and subsequent experimental verification have led to current advances in the physics of topological insulators. The exotic metallic surfaces of these insulators could lead to the development of new spintronic/electronic devices and the discovery of newly quantum state with rich physics.

CONSENT FOR PUBLICATION

Not applicable.

CONFLICT OF INTEREST

The author declares no conflict of interest, financial or otherwise.

ACKNOWLEDGEMENTS

Declared none.

REFERENCES

[1]   Jauregui, L.A.; Pettes, M.T.; Rokhinson, L.P.; Shi, L.; Chen, Y.P. Gate tunable relativistic mass and Berry's phase in topological insulator nanoribbon field effect devices. *Sci. Rep.,* **2015**, *5*(1), 8452.
[http://dx.doi.org/10.1038/srep08452] [PMID: 25677703]

[2]   Jauregui, L.A.; Pettes, M.T.; Rokhinson, L.P.; Shi, L.; Chen, Y.P. Gate tunable relativistic mass and Berry's phase in topological insulator nanoribbon field effect devices. *Sci. Rep.,* **2015**, *5*(1), 8452.
[http://dx.doi.org/10.1038/srep08452] [PMID: 25677703]

[3]   Deng, Y.; Zhou, X. song; Wei, G. dan; Liu, J.; Nan, C. W.; Zhao, S. jing. Solvothermal Preparation and Characterization of Nanocrystalline Bi2Te3 Powder with Different Morphology. *J. Phys. Chem. Solids,* **2002**, *63*(11), 2119-2121.
[http://dx.doi.org/10.1016/S0022-3697(02)00261-5]

[4]   Yu, H.; Gibbons, P.C.; Buhro, W.E. Bismuth{,} Tellurium{,} and Bismuth Telluride Nanowires. *J. Mater. Chem.,* **2004**, *14*(4), 595-602.
[http://dx.doi.org/10.1039/b312820b]

[5]   Purkayastha, A.; Lupo, F.; Kim, S.; Borca-Tasciuc, T.; Ramanath, G. Low-Temperature, Template-Free Synthesis of Single-Crystal Bismuth Telluride Nanorods. *Adv. Mater.,* **2006**, *18*(4), 496-500.https://doi.org/https://doi.org/10.1002/adma.200501339
       [http://dx.doi.org/10.1002/adma.200501339]

[6]   Xiao, F.; Yoo, B.; Lee, K.H.; Myung, N.V. Synthesis of Bi2Te3 nanotubes by galvanic displacement. *J. Am. Chem. Soc.,* **2007**, *129*(33), 10068-10069.
       [http://dx.doi.org/10.1021/ja073032w] [PMID: 17655307]

[7]   Keuleyan, S.; Lhuillier, E.; Guyot-Sionnest, P. Synthesis of colloidal HgTe quantum dots for narrow mid-IR emission and detection. *J. Am. Chem. Soc.,* **2011**, *133*(41), 16422-16424.
       [http://dx.doi.org/10.1021/ja2079509] [PMID: 21942339]

[8]   Jauregui, L.A.; Kayyalha, M.; Kazakov, A.; Miotkowski, I.; Rokhinson, L.P.; Chen, Y.P. Gate-Tunable Supercurrent and Multiple Andreev Reflections in a Superconductor-Topological Insulator Nanoribbon-Superconductor Hybrid Device. *Appl. Phys. Lett.,* **2018**, *112*(9), 93105.
       [http://dx.doi.org/10.1063/1.5008746]

[9]   Ando, Y. Topological Insulator Materials. *J. Phys. Soc. Jpn.,* **2013**, *82*(10), 102001.
       [http://dx.doi.org/10.7566/JPSJ.82.102001]

[10]   Weng, H.; Dai, X.; Fang, Z. Topological semimetals predicted from first-principles calculations. *J. Phys. Condens. Matter,* **2016**, *28*(30), 303001.
       [http://dx.doi.org/10.1088/0953-8984/28/30/303001] [PMID: 27269048]

[11]   Galletti, L.; Charpentier, S.; Iavarone, M.; Lucignano, P.; Massarotti, D.; Arpaia, R.; Suzuki, Y.; Kadowaki, K.; Bauch, T.; Tagliacozzo, A.; Tafuri, F.; Lombardi, F. Influence of Topological Edge States on the Properties of Al / Bi 2 Se 3 / Al Hybrid Josephson Devices. *Phys. Rev. B Condens. Matter Mater. Phys.,* **2014**, *89*(13), 134512.
       [http://dx.doi.org/10.1103/PhysRevB.89.134512]

[12]   Potter, A.C.; Fu, L. Anomalous Supercurrent from Majorana States in Topological Insulator Josephson Junctions. *Phys. Rev. B Condens. Matter Mater. Phys.,* **2013**, *88*(12), 121109.
       [http://dx.doi.org/10.1103/PhysRevB.88.121109]

[13]   Sacépé, B.; Oostinga, J.B.; Li, J.; Ubaldini, A.; Couto, N.J.G.; Giannini, E.; Morpurgo, A.F. Gate-tuned normal and superconducting transport at the surface of a topological insulator. *Nat. Commun.,* **2011**, *2*(1), 575.
       [http://dx.doi.org/10.1038/ncomms1586] [PMID: 22146394]

[14]   Williams, J.R.; Bestwick, A.J.; Gallagher, P.; Hong, S.S.; Cui, Y.; Bleich, A.S.; Analytis, J.G.; Fisher, I.R.; Goldhaber-Gordon, D. Unconventional Josephson effect in hybrid superconductor-topological insulator devices. *Phys. Rev. Lett.,* **2012**, *109*(5), 056803.
       [http://dx.doi.org/10.1103/PhysRevLett.109.056803] [PMID: 23006196]

[15]   Veldhorst, M.; Snelder, M.; Hoek, M.; Gang, T.; Guduru, V.K.; Wang, X.L.; Zeitler, U.; van der Wiel, W.G.; Golubov, A.A.; Hilgenkamp, H.; Brinkman, A. Josephson supercurrent through a topological insulator surface state. *Nat. Mater.,* **2012**, *11*(5), 417-421.
       [http://dx.doi.org/10.1038/nmat3255] [PMID: 22344327]

[16]   Wang, J.; Chang, C-Z.; Li, H.; He, K.; Zhang, D.; Singh, M.; Ma, X-C.; Samarth, N.; Xie, M.; Xue, Q-K.; Chan, M.H.W. Interplay between Topological Insulators and Superconductors. *Phys. Rev. B Condens. Matter Mater. Phys.,* **2012**, *85*(4), 045415.
       [http://dx.doi.org/10.1103/PhysRevB.85.045415]

[17]   Yang, F.; Ding, Y.; Qu, F.; Shen, J.; Chen, J.; Wei, Z.; Ji, Z.; Liu, G.; Fan, J.; Yang, C.; Xiang, T.; Lu, L. Proximity Effect at Superconducting Sn-Bi 2 Se 3 Interface. *Phys. Rev. B Condens. Matter Mater. Phys.,* **2012**, *85*(10), 104508.
       [http://dx.doi.org/10.1103/PhysRevB.85.104508]

[18]   Oostinga, J.B.; Maier, L.; Schüffelgen, P.; Knott, D.; Ames, C.; Brüne, C.; Tkachov, G.; Buhmann, H.;

Molenkamp, L.W. Josephson Supercurrent through the Topological Surface States of Strained Bulk HgTe. *Phys. Rev. X,* **2013**, *3*(2), 021007.
[http://dx.doi.org/10.1103/PhysRevX.3.021007]

[19]	Wiedenmann, J.; Bocquillon, E.; Deacon, R.S.; Hartinger, S.; Herrmann, O.; Klapwijk, T.M.; Maier, L.; Ames, C.; Brüne, C.; Gould, C.; Oiwa, A.; Ishibashi, K.; Tarucha, S.; Buhmann, H.; Molenkamp, L.W. 4π-periodic Josephson supercurrent in HgTe-based topological Josephson junctions. *Nat. Commun.,* **2016**, *7*(1), 10303.
[http://dx.doi.org/10.1038/ncomms10303] [PMID: 26792013]

[20]	Flötotto, D.; Ota, Y.; Bai, Y.; Zhang, C.; Okazaki, K.; Tsuzuki, A.; Hashimoto, T.; Eckstein, J. N.; Shin, S.; Chiang, T.-C. *Superconducting Pairing of Topological Surface States in Bismuth Selenide Films on Niobium,* **2018**.
[http://dx.doi.org/10.1126/sciadv.aar7214]

[21]	Sun, H-H.; Wang, M-X.; Zhu, F.; Wang, G-Y.; Ma, H-Y.; Xu, Z-A.; Liao, Q.; Lu, Y.; Gao, C-L.; Li, Y-Y.; Liu, C.; Qian, D.; Guan, D.; Jia, J-F. Coexistence of Topological Edge State and Superconductivity in Bismuth Ultrathin Film. *Nano Lett.,* **2017**, *17*(5), 3035-3039.
[http://dx.doi.org/10.1021/acs.nanolett.7b00365] [PMID: 28415840]

[22]	Xu, S-Y.; Alidoust, N.; Belopolski, I.; Richardella, A.; Liu, C.; Neupane, M.; Bian, G.; Huang, S-H.; Sankar, R.; Fang, C.; Dellabetta, B.; Dai, W.; Li, Q.; Gilbert, M.J.; Chou, F.; Samarth, N.; Hasan, M.Z. Momentum-Space Imaging of Cooper Pairing in a Half-Dirac-Gas Topological Superconductor. *Nat. Phys.,* **2014**, *10*(12), 943-950.
[http://dx.doi.org/10.1038/nphys3139]

[23]	Nadj-Perge, S.; Drozdov, I. K.; Li, J.; Chen, H.; Jeon, S.; Seo, J.; MacDonald, A. H.; Bernevig, B. A.; Yazdani, A. Observation of Majorana Fermions in Ferromagnetic Atomic Chains on a Superconductor. *Science (80-.),* **2014**, *346*(6209)
[http://dx.doi.org/10.1126/science.1259327]

[24]	Trang, C.X.; Shimamura, N.; Nakayama, K.; Souma, S.; Sugawara, K.; Watanabe, I.; Yamauchi, K.; Oguchi, T.; Segawa, K.; Takahashi, T.; Ando, Y.; Sato, T. Conversion of a conventional superconductor into a topological superconductor by topological proximity effect. *Nat. Commun.,* **2020**, *11*(1), 159.
[http://dx.doi.org/10.1038/s41467-019-13946-0] [PMID: 31919356]

[25]	Park, K.; Csire, G.; Ujfalussy, B. Proximity Effect in a Superconductor–Topological Insulator Heterostructure Based on First Principles. *Phys. Rev. B,* **2020**, *102*(13), 134504.
[http://dx.doi.org/10.1103/PhysRevB.102.134504]

[26]	Lutchyn, R.M.; Sau, J.D.; Das Sarma, S. Majorana fermions and a topological phase transition in semiconductor-superconductor heterostructures. *Phys. Rev. Lett.,* **2010**, *105*(7), 077001.
[http://dx.doi.org/10.1103/PhysRevLett.105.077001] [PMID: 20868069]

[27]	Oreg, Y.; Refael, G.; von Oppen, F. Helical liquids and Majorana bound states in quantum wires. *Phys. Rev. Lett.,* **2010**, *105*(17), 177002.
[http://dx.doi.org/10.1103/PhysRevLett.105.177002] [PMID: 21231073]

[28]	Das, A.; Ronen, Y.; Most, Y.; Oreg, Y.; Heiblum, M.; Shtrikman, H. Zero-Bias Peaks and Splitting in an Al–InAs Nanowire Topological Superconductor as a Signature of Majorana Fermions. *Nat. Phys.,* **2012**, *8*(12), 887-895.
[http://dx.doi.org/10.1038/nphys2479]

[29]	Wimmer, M.; Akhmerov, A.R.; Dahlhaus, J.P.; Beenakker, C.W.J. Quantum Point Contact as a Probe of a Topological Superconductor. *New J. Phys.,* **2011**, *13*(5), 053016.
[http://dx.doi.org/10.1088/1367-2630/13/5/053016]

[30]	de Juan, F.; Ilan, R.; Bardarson, J.H. Robust transport signatures of topological superconductivity in topological insulator nanowires. *Phys. Rev. Lett.,* **2014**, *113*(10), 107003.
[http://dx.doi.org/10.1103/PhysRevLett.113.107003] [PMID: 25238379]

[31]   Lu, D-H. Surface states of topological insulators: the Dirac fermion in curved two-dimensional spaces. *Phys. Rev. Lett.,* **2009**, *103*(19), 196804.
[http://dx.doi.org/10.1103/PhysRevLett.103.196804] [PMID: 20365943]

[32]   Brey, L.; Fertig, H.A. Electronic States of Wires and Slabs of Topological Insulators: Quantum Hall Effects and Edge Transport. *Phys. Rev. B Condens. Matter Mater. Phys.,* **2014**, *89*(8), 085305.
[http://dx.doi.org/10.1103/PhysRevB.89.085305]

[33]   Hugdal, H.G.; Rex, S.; Nogueira, F.S.; Sudbø, A. Magnon-Induced Superconductivity in a Topological Insulator Coupled to Ferromagnetic and Antiferromagnetic Insulators. *Phys. Rev. B,* **2018**, *97*(19), 195438.
[http://dx.doi.org/10.1103/PhysRevB.97.195438]

[34]   Yano, R.; Hirose, H. T.; Tsumura, K.; Yamamoto, S.; Koyanagi, M.; Kanou, M.; Kashiwaya, H.; Sasagawa, T.; Kashiwaya, S. Proximity-Induced Superconducting States of Magnetically Doped 3D Topological Insulators with High Bulk Insulation.
[http://dx.doi.org/10.3390/condmat4010009]

[35]   Jarach, Y.; Koren, G.; Lindner, N.H.; Kanigel, A. Carrier Density and Thickness-Dependent Proximity Effect in Doped-Topological-Insulator–Metallic-Ferromagnet Bilayers. *Phys. Rev. B,* **2021**, *103*(20), 205407.
[http://dx.doi.org/10.1103/PhysRevB.103.205407]

[36]   Linder, J.; Tanaka, Y.; Yokoyama, T.; Sudbø, A.; Nagaosa, N. Interplay between Superconductivity and Ferromagnetism on a Topological Insulator. *Phys. Rev. B Condens. Matter Mater. Phys.,* **2010**, *81*(18), 184525.
[http://dx.doi.org/10.1103/PhysRevB.81.184525]

[37]   Xu, J-P.; Wang, M-X.; Liu, Z.L.; Ge, J-F.; Yang, X.; Liu, C.; Xu, Z.A.; Guan, D.; Gao, C.L.; Qian, D.; Liu, Y.; Wang, Q-H.; Zhang, F-C.; Xue, Q-K.; Jia, J-F. Experimental detection of a Majorana mode in the core of a magnetic vortex inside a topological insulator-superconductor heterostructure. *Phys. Rev. Lett.,* **2015**, *114*(1), 017001.
[http://dx.doi.org/10.1103/PhysRevLett.114.017001] [PMID: 25615497]

[38]   Sun, H-H.; Zhang, K-W.; Hu, L-H.; Li, C.; Wang, G-Y.; Ma, H-Y.; Xu, Z-A.; Gao, C-L.; Guan, D-D.; Li, Y-Y.; Liu, C.; Qian, D.; Zhou, Y.; Fu, L.; Li, S-C.; Zhang, F-C.; Jia, J-F. Majorana Zero Mode Detected with Spin Selective Andreev Reflection in the Vortex of a Topological Superconductor. *Phys. Rev. Lett.,* **2016**, *116*(25), 257003.
[http://dx.doi.org/10.1103/PhysRevLett.116.257003] [PMID: 27391745]

[39]   Qu, F.; Yang, F.; Shen, J.; Ding, Y.; Chen, J.; Ji, Z.; Liu, G.; Fan, J.; Jing, X.; Yang, C.; Lu, L. Strong superconducting proximity effect in pb-bi(2)te(3) hybrid structures. *Sci. Rep.,* **2012**, *2*(1), 339.
[http://dx.doi.org/10.1038/srep00339] [PMID: 22468226]

[40]   Wang, E.; Ding, H.; Fedorov, A.V.; Yao, W.; Li, Z.; Lv, Y-F.; Zhao, K.; Zhang, L-G.; Xu, Z.; Schneeloch, J.; Zhong, R.; Ji, S-H.; Wang, L.; He, K.; Ma, X.; Gu, G.; Yao, H.; Xue, Q-K.; Chen, X.; Zhou, S. Fully Gapped Topological Surface States in Bi_2Se_3 Films Induced by a D-Wave High-Temperature Superconductor. *Nat. Phys.,* **2013**, *9*(10), 621-625.
[http://dx.doi.org/10.1038/nphys2744]

[41]   Yilmaz, T.; Pletikosić, I.; Weber, A.P.; Sadowski, J.T.; Gu, G.D.; Caruso, A.N.; Sinkovic, B.; Valla, T. Absence of a proximity effect for a thin-films of a Bi2Se3 topological insulator grown on top of a Bi2Sr2CaCu2O(8+δ) cuprate superconductor. *Phys. Rev. Lett.,* **2014**, *113*(6), 067003.
[http://dx.doi.org/10.1103/PhysRevLett.113.067003] [PMID: 25148345]

[42]   Zhao, H.; Rachmilowitz, B.; Ren, Z.; Han, R.; Schneeloch, J.; Zhong, R.; Gu, G.; Wang, Z.; Zeljkovic, I. Superconducting Proximity Effect in a Topological Insulator Using Fe(Te, Se). *Phys. Rev. B,* **2018**, *97*(22), 224504.
[http://dx.doi.org/10.1103/PhysRevB.97.224504]

[43]   Zheng, K.; Luo, L-B.; Zhang, T-F.; Liu, Y-H.; Yu, Y-Q.; Lu, R.; Qiu, H-L.; Li, Z-J.; Andrew Huang,

J.C. Optoelectronic Characteristics of a near Infrared Light Photodetector Based on a Topological Insulator Sb2Te3 Film. *J. Mater. Chem. C Mater. Opt. Electron. Devices,* **2015**, *3*(35), 9154-9160.
[http://dx.doi.org/10.1039/C5TC01772F]

[44] Yao, J.; Shao, J.; Wang, Y.; Zhao, Z.; Yang, G. Ultra-broadband and high response of the Bi_2Te_3-Si heterojunction and its application as a photodetector at room temperature in harsh working environments. *Nanoscale,* **2015**, *7*(29), 12535-12541.
[http://dx.doi.org/10.1039/C5NR02953H] [PMID: 26138000]

[45] Zhang, X.; Wang, J.; Zhang, S-C. Topological Insulators for High-Performance Terahertz to Infrared Applications. *Phys. Rev. B Condens. Matter Mater. Phys.,* **2010**, *82*(24), 245107.
[http://dx.doi.org/10.1103/PhysRevB.82.245107]

[46] Peng, H.; Dang, W.; Cao, J.; Chen, Y.; Wu, D.; Zheng, W.; Li, H.; Shen, Z-X.; Liu, Z. Topological insulator nanostructures for near-infrared transparent flexible electrodes. *Nat. Chem.,* **2012**, *4*(4), 281-286.
[http://dx.doi.org/10.1038/nchem.1277] [PMID: 22437712]

[47] McIver, J.W.; Hsieh, D.; Steinberg, H.; Jarillo-Herrero, P.; Gedik, N. Control over topological insulator photocurrents with light polarization. *Nat. Nanotechnol.,* **2011**, *7*(2), 96-100.
[http://dx.doi.org/10.1038/nnano.2011.214] [PMID: 22138862]

[48] Sharma, A.; Srivastava, A.K.; Senguttuvan, T.D.; Husale, S. Robust broad spectral photodetection (UV-NIR) and ultra high responsivity investigated in nanosheets and nanowires of Bi_2Te_3 under harsh nano-milling conditions. *Sci. Rep.,* **2017**, *7*(1), 17911.
[http://dx.doi.org/10.1038/s41598-017-18166-4] [PMID: 29263434]

[49] Fei, F.; Wei, Z.; Wang, Q.; Lu, P.; Wang, S.; Qin, Y.; Pan, D.; Zhao, B.; Wang, X.; Sun, J.; Wang, X.; Wang, P.; Wan, J.; Zhou, J.; Han, M.; Song, F.; Wang, B.; Wang, G. Solvothermal Synthesis of Lateral Heterojunction Sb2Te3/Bi2Te3 Nanoplates. *Nano Lett.,* **2015**, *15*(9), 5905-5911.
[http://dx.doi.org/10.1021/acs.nanolett.5b01987] [PMID: 26305696]

[50] Wang, X.B.; Cheng, L.; Wu, Y.; Zhu, D.P.; Wang, L.; Zhu, J.X.; Yang, H.; Chia, E.E.M. Topological-insulator-based terahertz modulator. *Sci. Rep.,* **2017**, *7*(1), 13486.
[http://dx.doi.org/10.1038/s41598-017-13701-9] [PMID: 29044164]

[51] Liu, C.; Zhang, H.; Sun, Z.; Ding, K.; Mao, J.; Shao, Z.; Jie, J. Topological Insulator Bi2Se3 Nanowire/Si Heterostructure Photodetectors with Ultrahigh Responsivity and Broadband Response. *J. Mater. Chem. C Mater. Opt. Electron. Devices,* **2016**, *4*(24), 5648-5655.
[http://dx.doi.org/10.1039/C6TC01083K]

[52] Zhang, H.; Zhang, X.; Liu, C.; Lee, S-T.; Jie, J. High-Responsivity, High-Detectivity, Ultrafast Topological Insulator Bi2Se3/Silicon Heterostructure Broadband Photodetectors. *ACS Nano,* **2016**, *10*(5), 5113-5122.
[http://dx.doi.org/10.1021/acsnano.6b00272] [PMID: 27116332]

[53] Yang, M.; Wang, J.; Zhao, Y.; He, L.; Ji, C.; Liu, X.; Zhou, H.; Wu, Z.; Wang, X.; Jiang, Y. **2018**.
[http://dx.doi.org/10.1021/acsnano.8b08056]

[54] Zhang, H.; Song, Z.; Li, D.; Xu, Y.; Li, J.; Bai, C.; Man, B. Near-Infrared Photodetection Based on Topological Insulator P-N Heterojunction of SnTe/Bi2Se3. *Appl. Surf. Sci.,* **2020**, *509*, 145290.
[http://dx.doi.org/10.1016/j.apsusc.2020.145290]

[55] Li, M.; Wang, Z.; Gao, X.P.A.; Zhang, Z. Vertically Oriented Topological Insulator Bi 2 Se 3 Nanoplates on Silicon for Broadband Photodetection. *J. Phys. Chem. C,* **2020**, *124*(18), 10135-10142.
[http://dx.doi.org/10.1021/acs.jpcc.0c01978]

[56] Li, M.; Wang, Z.; Yang, L.; Pan, D.; Li, D.; Gao, X.P.A.; Zhang, Z. Growth and quantum transport properties of vertical Bi_2Se_3 nanoplate films on Si substrates. *Nanotechnology,* **2018**, *29*(31), 315706.
[http://dx.doi.org/10.1088/1361-6528/aac457] [PMID: 29757160]

[57] Xiu, F.; He, L.; Wang, Y.; Cheng, L.; Chang, L-T.; Lang, M.; Huang, G.; Kou, X.; Zhou, Y.; Jiang,

X.; Chen, Z.; Zou, J.; Shailos, A.; Wang, K.L. Manipulating surface states in topological insulator nanoribbons. *Nat. Nanotechnol.,* **2011**, *6*(4), 216-221.
[http://dx.doi.org/10.1038/nnano.2011.19] [PMID: 21317891]

[58]   Hsieh, D.; Xia, Y.; Qian, D.; Wray, L.; Dil, J.H.; Meier, F.; Osterwalder, J.; Patthey, L.; Checkelsky, J.G.; Ong, N.P.; Fedorov, A.V.; Lin, H.; Bansil, A.; Grauer, D.; Hor, Y.S.; Cava, R.J.; Hasan, M.Z. A tunable topological insulator in the spin helical Dirac transport regime. *Nature,* **2009**, *460*(7259), 1101-1105.
[http://dx.doi.org/10.1038/nature08234] [PMID: 19620959]

[59]   Zhang, Z.; Feng, X.; Guo, M.; Li, K.; Zhang, J.; Ou, Y.; Feng, Y.; Wang, L.; Chen, X.; He, K.; Ma, X.; Xue, Q.; Wang, Y. Electrically tuned magnetic order and magnetoresistance in a topological insulator. *Nat. Commun.,* **2014**, *5*(1), 4915.
[http://dx.doi.org/10.1038/ncomms5915] [PMID: 25222696]

[60]   Liu, Y-H.; Chong, C-W.; FanChiang, C.M.; Huang, J.C.; Han, H.C.; Li, Z.; Qiu, H.; Li, Y.C.; Liu, C.P. Ultrathin $(Bi_{1-x}Sb_x)_2Se_3$ Field Effect Transistor with Large ON/OFF Ratio. *ACS Appl. Mater. Interfaces,* **2017**, *9*(14), 12859-12864.
[http://dx.doi.org/10.1021/acsami.7b00541] [PMID: 28318226]

[61]   Zhu, H.; Zhao, E.; Richter, C.A.; Li, Q. Topological Insulator Bi2Se3 Nanowire Field Effect Transistors. *ECS Trans.,* **2014**, *64*(17), 51-59.
[http://dx.doi.org/10.1149/06417.0051ecst]

[62]   Kong, D.; Koski, K.J.; Cha, J.J.; Hong, S.S.; Cui, Y. Ambipolar field effect in Sb-doped Bi2Se3 nanoplates by solvothermal synthesis. *Nano Lett.,* **2013**, *13*(2), 632-636.
[http://dx.doi.org/10.1021/nl304212u] [PMID: 23323715]

[63]   Kong, D.; Chen, Y.; Cha, J. J.; Zhang, Q.; Analytis, J. G.; Lai, K.; Liu, Z.; Hong, S. S.; Koski, K. J.; Mo, S.-K.; Hussain, Z.; Fisher, I. R.; Shen, Z.-X.; Cui, Y. **2011**.
[http://dx.doi.org/10.1038/nnano.2011.172]

[64]   Steinberg, H.; Gardner, D.R.; Lee, Y.S.; Jarillo-Herrero, P. Surface state transport and ambipolar electric field effect in Bi_2Se_3 nanodevices. *Nano Lett.,* **2010**, *10*(12), 5032-5036.
[http://dx.doi.org/10.1021/nl1032183] [PMID: 21038914]

[65]   Sae, Hong.; seung Cha, J. J; Kong, D.; Cui, Y. Ultra-Low Carrier Concentration and Surface-Dominant Transport in Antimony-Doped Bi 2 Se 3 Topological Insulator Nanoribbons. **2012**.
[http://dx.doi.org/10.1038/ncomms1771]

[66]   Yuan, H.; Liu, H.; Shimotani, H.; Guo, H.; Chen, M.; Xue, Q.; Iwasa, Y. Liquid-gated ambipolar transport in ultrathin films of a topological insulator Bi2Te3. *Nano Lett.,* **2011**, *11*(7), 2601-2605.
[http://dx.doi.org/10.1021/nl201561u] [PMID: 21696167]

[67]   Shimizu, S.; Yoshimi, R.; Hatano, T.; Takahashi, K.S.; Tsukazaki, A.; Kawasaki, M.; Iwasa, Y.; Tokura, Y. Gate Control of Surface Transport in MBE-Grown Topological Insulator (Bi 1 − x Sb x) 2 Te 3 Thin Films. *Phys. Rev. B Condens. Matter Mater. Phys.,* **2012**, *86*(4), 045319.
[http://dx.doi.org/10.1103/PhysRevB.86.045319]

[68]   Segawa, K.; Ren, Z.; Sasaki, S.; Tsuda, T.; Kuwabata, S.; Ando, Y. Ambipolar Transport in Bulk Crystals of a Topological Insulator by Gating with Ionic Liquid. *Phys. Rev. B Condens. Matter Mater. Phys.,* **2012**, *86*(7), 075306.
[http://dx.doi.org/10.1103/PhysRevB.86.075306]

[69]   Chen, J.; Qin, H.J.; Yang, F.; Liu, J.; Guan, T.; Qu, F.M.; Zhang, G.H.; Shi, J.R.; Xie, X.C.; Yang, C.L.; Wu, K.H.; Li, Y.Q.; Lu, L. Gate-voltage control of chemical potential and weak antilocalization in Bi_2Se_3. *Phys. Rev. Lett.,* **2010**, *105*(17), 176602.
[http://dx.doi.org/10.1103/PhysRevLett.105.176602] [PMID: 21231064]

[70]   Checkelsky, J.G.; Hor, Y.S.; Cava, R.J.; Ong, N.P. Bulk band gap and surface state conduction observed in voltage-tuned crystals of the topological insulator Bi2Se3. *Phys. Rev. Lett.,* **2011**, *106*(19), 196801.

[http://dx.doi.org/10.1103/PhysRevLett.106.196801] [PMID: 21668185]

[71]	Liu, H.; Liu, S.; Yi, Y.; He, H.; Wang, J. Shubnikov–de Haas Oscillations in n and p Type $Bi_2 Se_3$ Flakes. *2D Mater,* **2015**, *2*(4), 045002.
	[http://dx.doi.org/10.1088/2053-1583/2/4/045002]

[72]	Bansal, N.; Koirala, N.; Brahlek, M.; Han, M-G.; Zhu, Y.; Cao, Y.; Waugh, J.; Dessau, D.S.; Oh, S. Robust Topological Surface States of $Bi_2 Se_3$ Thin Films on Amorphous SiO_2/Si Substrate and a Large Ambipolar Gating Effect. *Appl. Phys. Lett.,* **2014**, *104*(24), 241606.
	[http://dx.doi.org/10.1063/1.4884348]

[73]	Steinberg, H.; Laloë, J-B.; Fatemi, V.; Moodera, J.S.; Jarillo-Herrero, P. Electrically Tunable Surface-to-Bulk Coherent Coupling in Topological Insulator Thin Films. *Phys. Rev. B Condens. Matter Mater. Phys.,* **2011**, *84*(23), 233101.
	[http://dx.doi.org/10.1103/PhysRevB.84.233101]

[74]	Wang, Z.; Qiu, R.L.J.; Lee, C.H.; Zhang, Z.; Gao, X.P.A. Ambipolar surface conduction in ternary topological insulator $Bi_2(Te_1\text{-xSex})_3$ nanoribbons. *ACS Nano,* **2013**, *7*(3), 2126-2131.
	[http://dx.doi.org/10.1021/nn304684b] [PMID: 23441571]

[75]	Das, B.; Das, N.S.; Sarkar, S.; Chatterjee, B.K.; Chattopadhyay, K.K. Topological Insulator Bi_2Se_3/Si-Nanowire-Based p-n Junction Diode for High-Performance Near-Infrared Photodetector. *ACS Appl. Mater. Interfaces,* **2017**, *9*(27), 22788-22798.
	[http://dx.doi.org/10.1021/acsami.7b00759] [PMID: 28621513]

[76]	Wang, J.; Chen, X.; Zhu, B.F.; Zhang, S.C. Topological P-n Junction. *Phys. Rev. B Condens. Matter Mater. Phys.,* **2012**, *85*(23), 1-4.
	[http://dx.doi.org/10.1103/PhysRevB.85.235131]

[77]	Tu, N.H.; Tanabe, Y.; Satake, Y.; Huynh, K.K.; Tanigaki, K. In-plane topological p-n junction in the three-dimensional topological insulator $Bi_{2-x}Sb_xTe_{3-y}Se_y$. *Nat. Commun.,* **2016**, *7*(1), 13763.
	[http://dx.doi.org/10.1038/ncomms13763] [PMID: 27934857]

[78]	Kent, A.D.; Worledge, D.C. A new spin on magnetic memories. *Nat. Nanotechnol.,* **2015**, *10*(3), 187-191.
	[http://dx.doi.org/10.1038/nnano.2015.24] [PMID: 25740126]

[79]	Fan, Y.; Upadhyaya, P.; Kou, X.; Lang, M.; Takei, S.; Wang, Z.; Tang, J.; He, L.; Chang, L.-T.; Montazeri, M.; Yu, G.; Jiang, W.; Nie, T.; Schwartz, R. N.; Tserkovnyak, Y.; Wang, K. L. *Magnetization Switching through Giant Spin-Orbit Torque in a Magnetically Doped Topological Insulator Heterostructure.,* **2014**.
	[http://dx.doi.org/10.1038/nmat3973]

[80]	Fan, Y.; Kou, X.; Upadhyaya, P.; Shao, Q.; Pan, L.; Lang, M.; Che, X.; Tang, J.; Montazeri, M.; Murata, K.; Chang, L-T.; Akyol, M.; Yu, G.; Nie, T.; Wong, K.L.; Liu, J.; Wang, Y.; Tserkovnyak, Y.; Wang, K.L. Electric-field control of spin-orbit torque in a magnetically doped topological insulator. *Nat. Nanotechnol.,* **2016**, *11*(4), 352-359.
	[http://dx.doi.org/10.1038/nnano.2015.294] [PMID: 26727198]

[81]	Mellnik, A.R.; Lee, J.S.; Richardella, A.; Grab, J.L.; Mintun, P.J.; Fischer, M.H.; Vaezi, A.; Manchon, A.; Kim, E.A.; Samarth, N.; Ralph, D.C. Spin-transfer torque generated by a topological insulator. *Nature,* **2014**, *511*(7510), 449-451.
	[http://dx.doi.org/10.1038/nature13534] [PMID: 25056062]

[82]	Mahfouzi, F.; Nagaosa, N.; Nikolić, B.K. Spin-orbit coupling induced spin-transfer torque and current polarization in topological-insulator/ferromagnet vertical heterostructures. *Phys. Rev. Lett.,* **2012**, *109*(16), 166602.
	[http://dx.doi.org/10.1103/PhysRevLett.109.166602] [PMID: 23215105]

[83]	Yokoyama, T. Current-Induced Magnetization Reversal on the Surface of a Topological Insulator. *Phys. Rev. B Condens. Matter Mater. Phys.,* **2011**, *84*(11), 113407.
	[http://dx.doi.org/10.1103/PhysRevB.84.113407]

[84]   Wang, Y.; Deorani, P.; Banerjee, K.; Koirala, N.; Brahlek, M.; Oh, S.; Yang, H. Topological Surface
        States Originated Spin-Orbit Torques in Bi(2)Se(3). *Phys. Rev. Lett.,* **2015**, *114*(25), 257202.
        [http://dx.doi.org/10.1103/PhysRevLett.114.257202] [PMID: 26197141]

[85]   Deorani, P.; Son, J.; Banerjee, K.; Koirala, N.; Brahlek, M.; Oh, S.; Yang, H. Observation of Inverse
        Spin Hall Effect in Bismuth Selenide. *Phys. Rev. B Condens. Matter Mater. Phys.,* **2014**, *90*(9),
        094403.
        [http://dx.doi.org/10.1103/PhysRevB.90.094403]

[86]   Shiomi, Y.; Nomura, K.; Kajiwara, Y.; Eto, K.; Novak, M.; Segawa, K.; Ando, Y.; Saitoh, E. Spin-
        electricity conversion induced by spin injection into topological insulators. *Phys. Rev. Lett.,* **2014**,
        113(19), 196601.
        [http://dx.doi.org/10.1103/PhysRevLett.113.196601] [PMID: 25415913]

[87]   Wang, H.; Kally, J.; Lee, J.S.; Liu, T.; Chang, H.; Hickey, D.R.; Mkhoyan, K.A.; Wu, M.; Richardella,
        A.; Samarth, N. Surface-State-Dominated Spin-Charge Current Conversion in Topological-Insulato-
        -Ferromagnetic-Insulator Heterostructures. *Phys. Rev. Lett.,* **2016**, *117*(7), 076601.
        [http://dx.doi.org/10.1103/PhysRevLett.117.076601] [PMID: 27563980]

[88]   Wang, A-Q.; Ye, X-G.; Yu, D-P.; Liao, Z-M. Topological Semimetal Nanostructures: From Properties
        to Topotronics. *ACS Nano,* **2020**, *14*(4), 3755-3778.
        [http://dx.doi.org/10.1021/acsnano.9b07990] [PMID: 32286783]

[89]   Wang, Q.; Li, C-Z.; Ge, S.; Li, J-G.; Lu, W.; Lai, J.; Liu, X.; Ma, J.; Yu, D-P.; Liao, Z-M.; Sun, D.
        Ultrafast Broadband Photodetectors Based on Three-Dimensional Dirac Semimetal Cd_3As_2. *Nano
        Lett.,* **2017**, *17*(2), 834-841.
        [http://dx.doi.org/10.1021/acs.nanolett.6b04084] [PMID: 28099030]

[90]   Britnell, L.; Ribeiro, R. M.; Eckmann, A.; Jalil, R.; Belle, B. D.; Mishchenko, A. Strong Light-Matter
        Interactions in Heterostructures of Atomically Thin Films. *Science (80-.),* **2013**, *340*(6138), 1311-
        1314.
        [http://dx.doi.org/10.1126/science.1235547]

[91]   Novoselov, K. S.; Mishchenko, A.; Carvalho, A. 2D Materials and van Der Waals Heterostructures.
        Science (80-.), **2016**, *353*(6298), aac9439.
        [http://dx.doi.org/10.1126/science.aac9439]

[92]   Wang, Z.; Weng, H.; Wu, Q.; Dai, X.; Fang, Z. Three-Dimensional Dirac Semimetal and Quantum
        Transport in Cd 3 As 2. *Phys. Rev. B Condens. Matter Mater. Phys.,* **2013**, *88*(12), 125427.
        [http://dx.doi.org/10.1103/PhysRevB.88.125427]

[93]   Neupane, M.; Xu, S-Y.; Sankar, R.; Alidoust, N.; Bian, G.; Liu, C.; Belopolski, I.; Chang, T-R.; Jeng,
        H-T.; Lin, H.; Bansil, A.; Chou, F.; Hasan, M.Z. Observation of a three-dimensional topological Dirac
        semimetal phase in high-mobility Cd3As2. *Nat. Commun.,* **2014**, *5*(1), 3786.
        [http://dx.doi.org/10.1038/ncomms4786] [PMID: 24807399]

[94]   Jeon, S.; Zhou, B.B.; Gyenis, A.; Feldman, B.E.; Kimchi, I.; Potter, A.C.; Gibson, Q.D.; Cava, R.J.;
        Vishwanath, A.; Yazdani, A. Landau quantization and quasiparticle interference in the three-
        dimensional Dirac semimetal Cd_3As_2. *Nat. Mater.,* **2014**, *13*(9), 851-856.
        [http://dx.doi.org/10.1038/nmat4023] [PMID: 24974888]

[95]   Liang, T.; Gibson, Q.; Ali, M.N.; Liu, M.; Cava, R.J.; Ong, N.P. Ultrahigh mobility and giant
        magnetoresistance in the Dirac semimetal Cd3As2. *Nat. Mater.,* **2015**, *14*(3), 280-284.
        [http://dx.doi.org/10.1038/nmat4143] [PMID: 25419815]

[96]   Lu, W.; Ge, S.; Liu, X.; Lu, H.; Li, C.; Lai, J.; Zhao, C.; Liao, Z.; Jia, S.; Sun, D. Ultrafast Relaxation
        Dynamics of Photoexcited Dirac Fermions in the Three-Dimensional Dirac Semimetal $Cd_3 As_2$. *Phys.
        Rev. B,* **2017**, *95*(2), 024303.
        [http://dx.doi.org/10.1103/PhysRevB.95.024303]

[97]   Yang, M.; Wang, J.; Han, J.; Ling, J.; Ji, C.; Kong, X.; Liu, X.; Huang, Z.; Gou, J.; Liu, Z.; Xiu, F.;

Jiang, Y. Enhanced Performance of Wideband Room Temperature Photodetector Based on Cd_3As_2 Thin Film/Pentacene Heterojunction. *ACS Photonics,* **2018**, *5*(8), 3438-3445.
[http://dx.doi.org/10.1021/acsphotonics.8b00727]

[98]   Chi, S.; Li, Z.; Xie, Y.; Zhao, Y.; Wang, Z.; Li, L.; Yu, H.; Wang, G.; Weng, H.; Zhang, H.; Wang, J. A Wide-Range Photosensitive Weyl Semimetal Single Crystal-TaAs. *Adv. Mater.,* **2018**, *30*(43), e1801372.https://doi.org/https://doi.org/10.1002/adma.201801372
[http://dx.doi.org/10.1002/adma.201801372] [PMID: 30260577]

[99]   Gao, W.; Huang, L.; Xu, J.; Chen, Y.; Zhu, C.; Nie, Z.; Li, Y.; Wang, X.; Xie, Z.; Zhu, S.; Xu, J.; Wan, X.; Zhang, C.; Xu, Y.; Shi, Y.; Wang, F. Broadband Photocarrier Dynamics and Nonlinear Absorption of PLD-Grown WTe2 Semimetal Films. *Appl. Phys. Lett.,* **2018**, *112*(17), 171112.
[http://dx.doi.org/10.1063/1.5024777]

[100]   Zhou, W.; Chen, J.; Gao, H.; Hu, T.; Ruan, S.; Stroppa, A.; Ren, W. Anomalous and Polarization-Sensitive Photoresponse of T_d -WTe$_2$ from Visible to Infrared Light. *Adv. Mater.,* **2019**, *31*(5), e1804629.https://doi.org/https://doi.org/10.1002/adma.201804629
[http://dx.doi.org/10.1002/adma.201804629] [PMID: 30516849]

[101]   Lai, J.; Liu, X.; Ma, J.; Wang, Q.; Zhang, K.; Ren, X.; Liu, Y.; Gu, Q.; Zhuo, X.; Lu, W.; Wu, Y.; Li, Y.; Feng, J.; Zhou, S.; Chen, J-H.; Sun, D. Anisotropic Broadband Photoresponse of Layered Type-II Weyl Semimetal MoTe$_2$. *Adv. Mater.,* **2018**, *30*(22), e1707152.https://doi.org/https://doi.org/10.1002/adma.201707152
[http://dx.doi.org/10.1002/adma.201707152] [PMID: 29665162]

[102]   Lai, J.; Liu, Y.; Ma, J.; Zhuo, X.; Peng, Y.; Lu, W.; Liu, Z.; Chen, J.; Sun, D. Broadband Anisotropic Photoresponse of the "Hydrogen Atom" Version Type-II Weyl Semimetal Candidate TaIrTe$_4$. *ACS Nano,* **2018**, *12*(4), 4055-4061.
[http://dx.doi.org/10.1021/acsnano.8b01897] [PMID: 29621401]

[103]   Lian, B.; Sun, X-Q.; Vaezi, A.; Qi, X-L.; Zhang, S-C. Topological quantum computation based on chiral Majorana fermions. *Proc. Natl. Acad. Sci. USA,* **2018**, *115*(43), 10938-10942.
[http://dx.doi.org/10.1073/pnas.1810003115] [PMID: 30297431]

[104]   Ivanov, D.A. Non-Abelian statistics of half-quantum vortices in p-wave superconductors. *Phys. Rev. Lett.,* **2001**, *86*(2), 268-271.
[http://dx.doi.org/10.1103/PhysRevLett.86.268] [PMID: 11177808]

[105]   Aasen, D.; Hell, M.; Mishmash, R.V.; Higginbotham, A.; Danon, J.; Leijnse, M.; Jespersen, T.S.; Folk, J.A.; Marcus, C.M.; Flensberg, K.; Alicea, J. Milestones Toward Majorana-Based Quantum Computing. *Phys. Rev. X,* **2016**, *6*(3), 031016.
[http://dx.doi.org/10.1103/PhysRevX.6.031016]

[106]   Karzig, T.; Knapp, C.; Lutchyn, R.M.; Bonderson, P.; Hastings, M.B.; Nayak, C.; Alicea, J.; Flensberg, K.; Plugge, S.; Oreg, Y.; Marcus, C.M.; Freedman, M.H. Scalable Designs for Quasiparticle-Poisoning-Protected Topological Quantum Computation with Majorana Zero Modes. *Phys. Rev. B,* **2017**, *95*(23), 235305.
[http://dx.doi.org/10.1103/PhysRevB.95.235305]

[107]   Alicea, J. New directions in the pursuit of Majorana fermions in solid state systems. *Rep. Prog. Phys.,* **2012**, *75*(7), 076501.
[http://dx.doi.org/10.1088/0034-4885/75/7/076501] [PMID: 22790778]

[108]   Alicea, J.; Oreg, Y.; Refael, G.; von Oppen, F.; Fisher, M.P.A. Non-Abelian Statistics and Topological Quantum Information Processing in 1D Wire Networks. *Nat. Phys.,* **2011**, *7*(5), 412-417.
[http://dx.doi.org/10.1038/nphys1915]

[109]   Read, N.; Green, D. Paired States of Fermions in Two Dimensions with Breaking of Parity and Time-Reversal Symmetries and the Fractional Quantum Hall Effect. *Phys. Rev. B Condens. Matter,* **2000**, *61*(15), 10267-10297.
[http://dx.doi.org/10.1103/PhysRevB.61.10267]

[110]   Kitaev, A.Y. Unpaired Majorana Fermions in Quantum Wires. *Phys. Uspekhi,* **2001**, *44*(10S), 131-136.
[http://dx.doi.org/10.1070/1063-7869/44/10S/S29]

[111]   Kitaev, A.Y. Fault-Tolerant Quantum Computation by Anyons. *Ann. Phys.,* **2003**, *303*(1), 2-30.https://doi.org/https://doi.org/10.1016/S0003-4916(02)00018-0
[http://dx.doi.org/10.1016/S0003-4916(02)00018-0]

[112]   Teo, J.C.Y.; Kane, C.L. Majorana fermions and non-Abelian statistics in three dimensions. *Phys. Rev. Lett.,* **2010**, *104*(4), 046401.
[http://dx.doi.org/10.1103/PhysRevLett.104.046401] [PMID: 20366722]

[113]   Fu, L.; Kane, C.L. Superconducting proximity effect and majorana fermions at the surface of a topological insulator. *Phys. Rev. Lett.,* **2008**, *100*(9), 096407.
[http://dx.doi.org/10.1103/PhysRevLett.100.096407] [PMID: 18352737]

[114]   Chiu, C-K.; Teo, J.C.Y.; Schnyder, A.P.; Ryu, S. Classification of Topological Quantum Matter with Symmetries. *Rev. Mod. Phys.,* **2016**, *88*(3), 035005.
[http://dx.doi.org/10.1103/RevModPhys.88.035005]

[115]   He, M.; Sun, H.; He, Q.L. Topological Insulator: Spintronics and Quantum Computations. *Front. Phys.,* **2019**, *14*(4), 43401.
[http://dx.doi.org/10.1007/s11467-019-0893-4]

[116]   Gibney, E. Thousands of exotic 'topological' materials discovered through sweeping search. *Nature,* **2018**, *560*(7717), 151-152.
[http://dx.doi.org/10.1038/d41586-018-05913-4] [PMID: 30089928]

CHAPTER 11

The Advancement in Research and Technology with New Kinds of Hollow Structures

Sakshi Sharma[1,*], A. K. Shrivastav[1], Anjali Oudhia[2] and Mohan L Verma[3]

[1] Department of Physics, National Institute of Technology, Raipur, India

[2] Department of Physics, Government Nagarjuna P.G. Science College, Raipur, India

[3] Department of Applied Physics, FET-SSGI Shri Shankaracharya Technical Campus, Junwani, Bhilai, India

Abstract: Hollow structures are one of the most highlighted topics of research in nanotechnology. These hollow structures can be in the form of nanospheres, nanocages, nanorods, nano boxes, *etc.* They can be single-layered or multi-layered, with different kinds of doping. All these variations in hollow structures open up various fields of research, from biomedicines to optoelectronics. With the discovery of hollow structures like carbon buckyball, nanotubes, *etc.*, several application-based -research was carried out, both theoretically as well as experimentally. Modifications are observed in the properties of a material when formed in a hollow shape like better conductivity, trapping capacity, catalytic effect, *etc.* These properties were the highlight of the studies. This field is still under investigation, and there is a lot of scope for new possibilities in the future. This chapter covers the basic information about different kinds of hollow structures like carbon buckyball, variations in their properties, along with recent developments and their applications. This chapter also includes detailed research about buckyball structures of ZnO, ZnS, and Al-doped ZnO using simulations with their comparative study and future applications.

Keywords: Al-doped ZnO, Buckyball, Hollow structures, Nanocages, ZnO, ZnS.

INTRODUCTION

Hollow structures came into the limelight due to their wide range of applications. These structures have a huge number of applications because of their configurable pore size, high value of the specific surface area, and large surface-to-volume ratio [1]. For example, in the green and sustainable energy conversion system with better efficiency and minimum cost, hollow nanostructures have proven to be excellent electrocatalysts [2]. In the field of biomedicines, spherical hollow

---

[*] **Corresponding author Sakshi Sharma**: Department of Physics, National Institute of Technology, Raipur, India; Tel:8770190792; E-mail: 9423sharmasakshi@gmail.com

Dibya Prakash Rai (Ed.)

structures are at the top. Due to their desired shape, they can be used as a drug delivery tool or in biosensors and bioimaging devices [3]. They also have several other applications in the areas of optoelectronic devices, water treatment, armoury, energy storage devices, sensors, *etc.*

Hollow structures can be in various shapes like spherical cages, square boxes, cylindrical rods, *etc.* Each kind of hollow structure has its own advantages and application. For example, nano boxes are used in the fields like fuel cells, batteries, *etc.*, while nanowires are majorly used in Field-Effect Transistors, photocatalysis, *etc* [2].

Hollow structures of different elements have different properties of their own, which imply specifications in their applications. Among hollow structures of different elements present in nature, carbon is one of its kind, with a huge range of applications. Carbon nanomaterials have some very interesting and important qualities like good mechanical and chemical stability, a large value of electrical conductivity, *etc* [4]. These properties grab the attention of researchers to explore more. Different kinds of hollow structures made up of carbon atoms are known as Fullerenes. It can be a hollow-sphere or a hollow-tube, *etc.* The study of fullerene is very vast due to the presence of a huge range of possible structures and isomers. Two different carbon hollow structures, carbon nanotube and carbon buckyball, are shown below in Fig. (**1**).

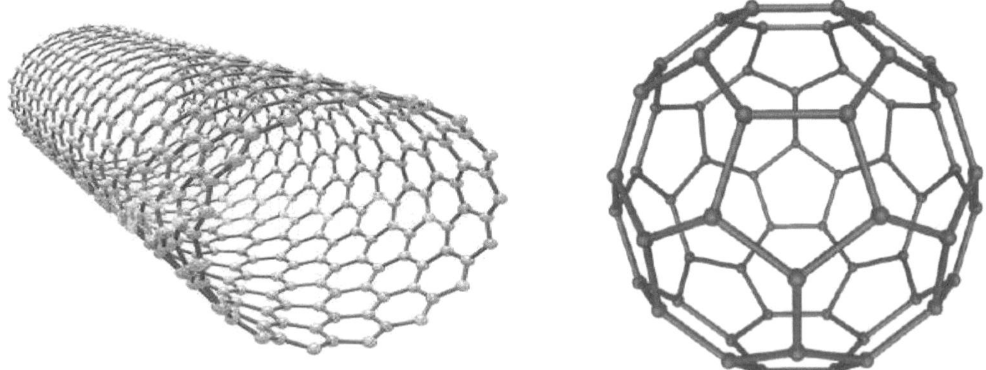

Fig. (1). (a) Hollow cylindrical structure of carbon nanotube [Photos by: PIXTA "Carbon nanotubes molecular 3d structure", Eugene Sergeev/ PIXTA, Copyright: PIXTA]. (b) Hollow spherical structure of carbon buckyball [Photos created by Michael Ströck (mstroeck), "A 3D model of a C-60 molecule", on 6th of February, 2006, by using iMol for Mac OS X and Photoshop CS2].

Among different kinds of Fullerenes, Buckminsterfullerene with 60 carbon atoms (Buckyball) is one of a special kind. This structure has mesmerized researchers from around the world since its discovery [4, 5]. The carbon buckyball structure

(C_{60}) is made up of carbon atoms arranged in six-fold rings, where few are replaced by some five-fold rings in the hollow-cage-like structure. There are 20 hexagons and 12 pentagons in this structure, where no two pentagons share a vertex [6]. The Swiss mathematician Leonhard Euler showed that a geodesic structure must have 12 pentagons to form a spheroid, where the number of hexagons may vary. This outcome in the 18th century led to some specifications in the fullerene structures. The fullerene structures are represented by 'C_n' where $n \geq 20$, with 12 pentagonal faces, 3 connected vertices, and $3n/2$ number of edges. With this formula, it could be verified that C_{60} has 20 hexagonal faces while C_{70} has 25 [4]. The Isolated Pentagon Rule (IPR) is followed for the formation of stable fullerenes. This rule states that *only when the pentagons at the surface are separated, a non-reactive and stable fullerene is formed.* There are several reported works on buckyball structures with different numbers of carbon atoms. As we know, the buckminsterfullerene structure consists of 60 carbon atoms. Thus, C_{60} with only one isomer is the smallest stable fullerene structure. Other than that, C_{20}, C_{26}, C_{28}, C_{70}, C_{74}, C_{90}, *etc.*, are also present. The difference in their stability and symmetry are discussed in various publications [7]. Fullerenes like C_{20} or C_{36} have formed already, but due to their high reactivity, the study is possible only under some special environment [8].

Other advances in the field of fullerene include variations like fullerene dendrimers, fullerene-substituted acetylenic polymers, endohedral metallo-fullerenes, *etc.* These variations have applications like tumour treatment, chromatographic purification, *etc* [9].

There are some previously reported works on the buckyball structures with not only carbon atoms but others as well. In some cases, one of the atoms of carbon is replaced by another atom, and in other cases, more than one is replaced. Later, a whole bunch of buckyballs with different elemental atoms were formed. For example, Boron Buckyball structure. Before Boron, some other elements were also visualized as buckyballs like Uranium, Silicon, Molybdenum disulfide, Gold, Tin, *etc* [10]. However, these structures were not that close to the resemblance of Carbon Buckyball. The Boron buckyball structure with 80 boron atoms resembles the most to C_{60}. Theoretical discovery of B_{80} was done in 2007, and after seven years of this discovery, experimental evidence of its existence came to light. This particular structure showed its application in hydrogen storage systems. Buckyball structures of multi-walled Boron Nitride were also reported earlier [11].

The hollow microspheres of Triple-shelled $ZnO/ZnFe_2O_4$ were recently explored due to their sensing capacity of acetone vapour. The unique heterojunction of ZnO and $ZnFe_2O_4$ with high specific surface area proved to be superior sensors for acetone vapour [12].

This led researchers to further investigate more about hollow structures and to try out finding other buckyball-like structures with different combinations of atoms. Most cage-like structures are not that stable and require some special environment for their existence. So, doping is done to increase the stability of such cage structures just like the doping of Silicon with Zr or Ti atoms to form a stable structure [13]. Some other kinds of doped cage-like structures include endohedral doped Ga, Al, In, Sn, Ge, and Pb clusters; cages of Ag, Au, and Cu doped with transition metal atoms; the cage-like structures of MgO and ZnO [14].

To the best of our knowledge, we have introduced the different kinds of hollow structures and especially focused on hollow spheres. We have also discussed the variations in the spherical cages other than carbon buckyball, their properties and applications. Further, we would like to introduce some of our research works to further examine and explain a new kind of hollow structure. We have previously investigated ZnO, ZnS and Al-doped ZnO buckyball structures theoretically [15 - 17]. Our studies showed some very useful results that are included in this chapter.

NOVEL MATERIALS SIMULATED IN OUR LAB AND THEIR COMPARATIVE STUDY

ZnO Buckyball (ZnO BB) and ZnS Buckyball (ZnS BB) Structures

The II-VI group materials have been studied for application-based research for more than a century. The large bandgap and high electron mobility are the characteristics of compounds in this group which increases their applicability over a larger area [18, 19]. Even though several research works have been conducted in these areas for years, every time researchers find some breakthroughs to carry on.

The structural variations in the material led to some specific changes in the properties of that material and hence increased their applications. There are previously reported works that provide pieces of evidence to justify this statement [20]. The structural variations include modifications in the shapes and sizes. Some known structures are quantum dots, nanoflowers, nanowires, nanorods, *etc.*

When we move from bulk to nano range, properties like the bandgap and conductivity also change. ZnO and ZnS at the bulk state have a broad direct bandgap and high exciton binding energy. As we move from bulk to nano region, the quantum confinement effect occurs, and the bandgap increases. Quantum confinement means the changes in the properties of a material when its size is reduced to a level comparable with that of the wavelength of an electron [21]. These materials have numerous applications in the fields like photovoltaics, optoelectronic devices, gas sensors, biosensors, *etc* [22, 23].

As mentioned in the introduction section, when we replace one of the carbon atoms in the carbon-buckyball structure with any other elements' atom, properties get modified. This led to the conclusion that one or all carbon atoms could be replaced in the buckyball structure and a new kind of buckyball can be formed, just like Boron- Buckyball structure (B_{80}) [11]. This led us to study the ZnO BB and ZnS BB structures in the first place [15]. We also examined the interaction between Zig-Zag Graphene with ZnO BB structure and the effect of Aluminum(Al) doping in ZnO BB structure in some of our previously reported works [16, 17]. Both of these studies showed some incredible results like Al-ZnO BB showed better electrical conductivity, while the Graphene-ZnO BB compound showed better optical conductivity and reduction in reflectivity. These advancements in the cage-like structure of ZnO could be used as an inspiration to do further studies with other compounds from the same group.

The structural, electronic, transport, magnetic, and optical properties are defined by the behaviour of valence electrons of the element. The interaction between the electron and the nuclei can be explained and examined by using quantum mechanical laws. Thus a Density Functional Theory (DFT)-based study is used to interpret the intrinsic properties of the material. Here we have used the approximations to solve the many-body problem by using DFT [24]. The DFT-based study is performed by using software named "Spanish Initiative for Electronic Simulations with Thousands of Atoms (SIESTA)" in this work. This method gives a very widespread plan to execute calculations in a very fast and accurate manner [25]. SIESTA uses the standard Kohn-Sham self-consistent DFT method in the local density approximation (LDA) or generalized gradient approximation (GGA). It is also capable of describing the van der Waals interactions. As we know, the basis sets are the main feature to calculate the Hamiltonian and overlap matrices in operations, and the atomic orbitals with finite support are used as a basis set in SIESTA, which allows limitless angular momenta, multiple-zeta, off-site orbitals, and polarization. It estimates the density and electron-wave functions onto a real-space grid for calculating the Hartree and exchange-correlation potentials along with its matrix elements [26]. This section of the chapter includes the study of the structural and electronic behaviour of ZnO BB and ZnS BB.

Stability and Structural Behaviour

The structural analysis provides an outlook on how the material looks, what are its bond lengths and angles, cell volume, packing fraction, *etc*. With the structural properties, we can easily define the physical behaviour of the material. All these parameters are very useful to grab a deep knowledge about a particular system.

For example, variations in the bond length lead to changes in the elastic properties of the material like stress or strain [27]. Also, a stable structure is formed only when the surface energy of the structure is reduced to its minimum [28].

Under the structural analysis of ZnO BB and ZnS BB, we optimized the structures and studied the system by using parameters like Fermi energy, volumes, bond lengths, *etc*. As exhibited in Figs. (**2a** and **2b**), ZnO BB and ZnS BB both look like a ball.

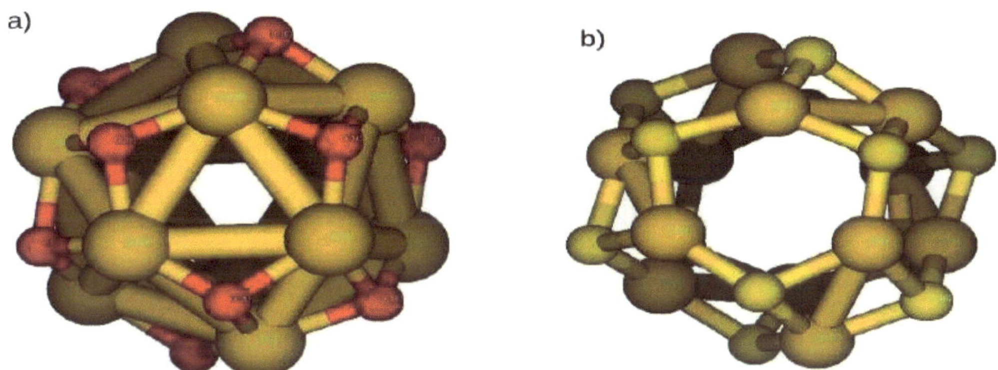

Fig. (2). Structural representation of **a)** ZnO BB **b)** ZnS BB.

Table **1** shows the input and calculated parameters for both structures. As reported in Table 1, the Fermi level of the ZnO BB structure (-5.16 eV) is greater than that of the ZnS BB structure (-4.63 eV). The formula used for the calculation of Atomic Packing Fraction (APF) and Atomic Density (AD) is APF=Volume taken by atoms/Cell Volume and AD=Atomic mass/Cell Volume, respectively [29].

Table 1. Calculated parameters of ZnO BB and ZnS BB structures using SIESTA software.

SN	Structure	Fermi energy (eV)	Cell Volume ($A^{\circ 3}$)	Atomic Volume ($A^{\circ 3}$)	Packing Fraction (%)	Atomic Density (Kg/ m^3)
1.	ZnO BB	-5.16	395.78	134.57	34.01	4.13 x 10^3
2.	ZnS BB	-4.63	570.99	173.99	30.47	3.43 x 10^3

Figs. (**2a** and **2b**) show the optimized ZnO BB and ZnS BB structures by a ball and stick model. Here both ZnO BB and ZnS BB have a spherical structure. The structural representation consists of an atom's symbol along with its number, where the number is just a numerical representation for the nth number of atoms in the system, for a better visual understanding.

Among all the comparing figures, the one with the most symmetric structure is the most stable one [30]. As per our results, both ZnO BB and ZnS BB are symmetric and hence stable structures.Tables **2** and **3** give the details about the obtained values of bond length and bond angle of both systems.

Table 2. Values of bond lengths for ZnO BB and ZnS BB structures.

SN	Structure	Bond Between Atoms	Bond Length(Å)
1.	ZnO BB	Zn-Zn	3.18
		Zn-O	1.92
2.	ZnS BB	Zn-Zn	3.20
		Zn-S	2.35

Table 3. Input and calculated parameters of ZnO BB and Al-ZnO BB structures.

SN	Structure	Fermi Energy (eV)	Cell Volume (A$^{\circ 3}$)	Atomic Volume (A$^{\circ 3}$)	Packing Fraction (%)	Atomic Density (Kg/ m^3)
1.	ZnO BB	-5.16	395.78	134.57	34.01	4.13 x 10^3
2.	Al-ZnO BB	-4.18	419.26	132.45	31.59	3.75 x 10^3

The electronegativity in an atom is its tendency to attract electrons in a shared covalent bond. As observed here, the bond length of Zn-S in ZnS BB is greater than Zn-O in ZnO BB, as sulfur is less electronegative than oxygen [31, 32]. Also, the value of cell volume in ZnS BB is around 570.99 A$^{\circ 3}$, which is greater than that of the value of cell volume in ZnO BB, which is around 395.78 A$^{\circ 3}$. This might be due to the difference between the bond lengths of ZnO BB and ZnS BB. As per studies, the change in cell volume attributes to the changes in the energy density of the system.

Electronic Properties

The electronic properties of a material describe the state of electrons in that system. We can define the DOS of a system as the number of states available to get occupied by the electrons, per interval of energy at each energy level. Higher DOS value at a particular energy level signifies that more states are available to get occupied and vice-versa for lower DOS value [33]. The forbidden energy gap between the HOMO (highest occupied molecular orbital) and LUMO (lowest unoccupied molecular orbital) show the bandgap of the material, which explains the conductivity of that particular material [34]. The charge transfer properties in any material is depending on the energy bandgap of that molecule. Hence, the study of the DOS is one of the most important parameters for understanding the electronic behaviour of any material [35, 36].

The expression for DOS in 3-dimensions for the carriers in the conduction band(CB) or the valence band(VB) under the approximation for the band energies are given below in equation 1.

$$\rho_c(E) = \frac{\sqrt{2}m_c^{3/2}}{n^2\hbar^3}\sqrt{E-E_g'} \quad \text{and} \quad \rho_v(E) = \frac{\sqrt{2}m_v^{3/2}}{n^2\hbar^3}\sqrt{-E} \tag{1}$$

Where, ρ_{pC} (E) and ρ_{pv} (E)= Density of States for CB and VB, respectively; m_C and m_V = mass of carriers in CB and VB, respectively; E= Energy value ; Eg = Fermi energy [37].

The Density of States (DOS)

Here, the DOS for ZnO BB and ZnS BB is shown in Figs. (**3a** and **3b**), respectively. The ZnO BB structure shows the energy band gap value of around 1.52 eV, which is a noticeably smaller value than the regular structured ZnO (3.37 eV) [16, 38]. Similarly, in the case of the ZnS BB, the bandgap is reduced to approximately 1.91 eV, which is smaller than the regular structured ZnS (3.45 eV) [39].

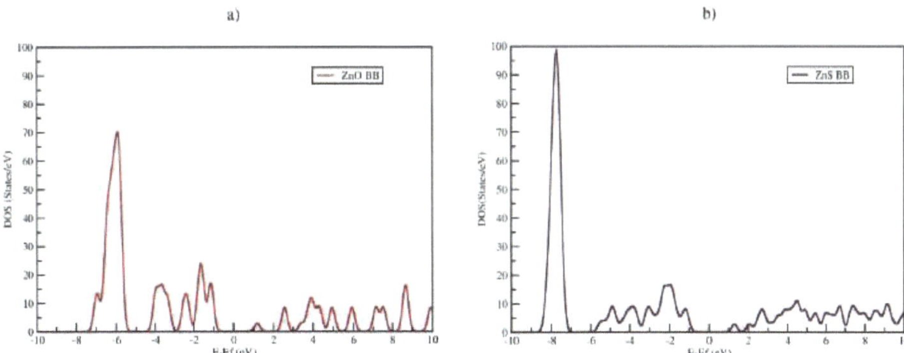

Fig. (3). The Density of States of a) ZnO BB structure b) ZnS BB structure.

The decrease in the value of the bandgap can be explained by the fact that, as defect levels increase in the valence band region, conductivity increases. Also, we can observe that the DOS plot of ZnS BB comprises more discrete peaks as compared to ZnO BB, which indicates more levels of quantum confinement effect in ZnS BB [15, 16].

The Projected Density of States (PDOS)

Basically, the PDOS plot explains the contribution of different orbitals and atoms to the DOS of that system. Here, Fig. (**4a**) shows the PDOS of ZnO BB, and Fig. (**4b**) shows the PDOS of ZnS BB.

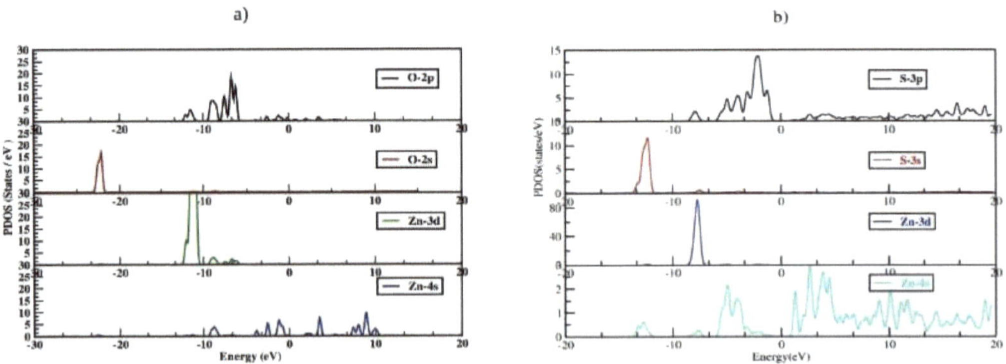

Fig. (4). The Projected Density of states of **a)** ZnO BB structure **b)** ZnS BB structure.

In Fig. (**4a**), the PDOS plot of ZnO BB shows that in the VB region, mainly Zn-4s, Zn-3d, and O-2p orbitals are contributing between the energy range of 0 to -20 eV. Whereas, O- 2s orbital shows its maximum control in the energy range of -20 to -30 eV, with a peak centred around -23 eV. Similarly, in the CB region, at the bottom of the CB, Zn-4s orbital is dominating with small peaks of O-2p. The value of the bandgap in ZnO BB is influenced by both Zn-4s and O-2p orbitals [15, 40].

In Fig. (**4b**), the PDOS plot of ZnS BB shows that in the VB region between the energy range of 0 to -10 eV, S-3p, Zn-3d and Zn-4s are dominating. Whereas, in the energy range of -10 eV to -20 eV, S-3s orbital is dominating. Similarly, in the case of CB, S-3p and Zn-4s are major contributors [15, 40].

ZnO Buckyball (ZnO BB) and Aluminum-doped ZnO Buckyball (Al-ZnO BB) Structures

As the results obtained from the buckyball design of ZnO were good in terms of bandgap engineering and its applications, we further extended our study by looking for some good dopants for the structure. There are various papers available for metal-doped ZnO. We wanted to look for a cost-effective, non-hazardous, efficient, and abundant material as a dopant with proper applications. So we found papers with Al as a suitable doping material with ZnO. The combination of Al-doped ZnO showed an extensive variety of applications in the field of LEDs, glasses, flat-panel displays, solar cells, *etc* [41]. These applications are possible because of the property of Al-doped ZnO to generate Transparent Conducting Oxides (TCOs). Oxides like Indium Tin Oxides (ITOs) and Fluorine doped Tin Oxides (FTOs) are popular for creating TCOs. With the introduction of Al-doped ZnO, this industry got new material for TCOs, as it was convinced to be a better replacement for ITOs and FTOs. The reason behind this is the non-toxic nature, high transparency, cost-effectiveness, low resistivity, and better conductivity with Al [42]. Simple ZnO shows less stability and lower electrical conductivity. When Al is added as a dopant in ZnO, these limitations get diminished with higher transparency [43]. Not only this, after doping there are some significant changes in the bandgap value, which was published in our previous work [17]. The detailed study of the structure and its outcomes are discussed further in this chapter.

Stability and Structural Behaviour

As discussed earlier in the case of ZnO BB, all the studies included in Al-ZnO BB are the same. The structure of Al-ZnO BB is formed by replacing one of the Zn atoms from the original structure with an Al atom, which is shown in Fig. (**5**), given below.

a)
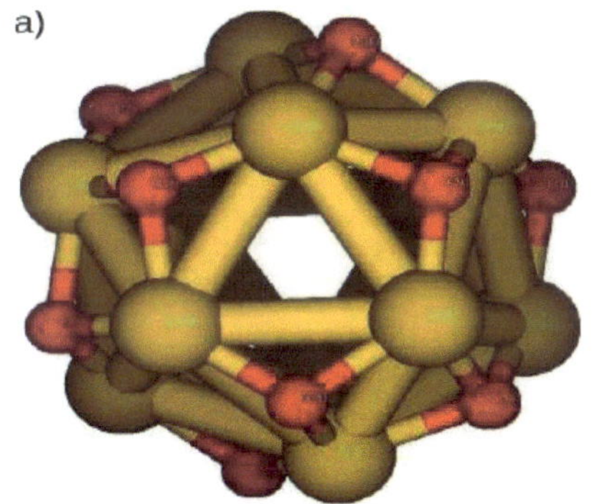
b)
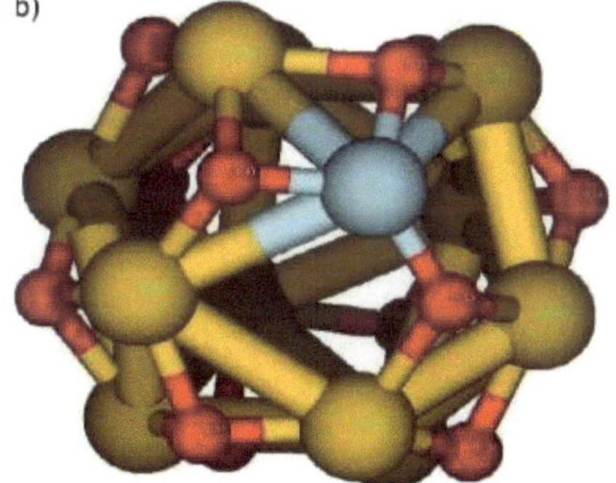

Fig. (5). Structural representation of **a)** ZnO BB **b)** Al-ZnO BB.

In the above Table **4**, it shows the average of all the bond lengths present in the structure. As observed here, when ZnO BB was doped with Al, the system was restructured in such a way that a stable structure was obtained with minimum surface energy [28]. With the replacement of a Zn atom with an Al atom, the Al-Zn bonds and Al-O bonds (marked as a-b and c-d) showed some reduction in their values as compared to Zn-Zn and Zn-O bonds in the same place for simple ZnO BB. Such a reduction in the value of bond length might be due to the size difference between Al and Zn atoms. Al is smaller than the Zn atom. However, other Zn-Zn bonds and Zn-O bonds (marked as e-f and g-h) present in Al-doped ZnO has shown an increase in their values as compared to undoped ZnO BB. This explains the overall increase in the cell volume of Al-ZnO BB. This is just a basic analysis of the outcomes from our structure and there is a lot more to study in detail for a better understanding.

Table 4. Values of bond lengths for ZnO BB and Al-ZnO BB structures.

SN	Atom's Description	The Bond Length in ZnO BB (A°)	The Bond Length in Al-ZnO BB (A°)
1.	Replaced atoms	Zn (a)-Zn(b) = 3.05	Al (a)-Zn(b) = 3.02
		Zn(c) -O (d) = 1.99	Al(c)-O(d) = 1.78
2.	Non-replaced atoms	Zn (e)- Zn(f) = 3.18	Zn (e)- Zn(f) = 4.23
		Zn(g) - O(h) = 1.98	Zn(g) - O(h) = 1.99

Electronic Properties

As we have already explained the study of the electronic properties of ZnO BB in the above section, it is not repeated here. A detailed explanation for DOS and PDOS of Al-ZnO BB is given in this section along with its comparison with undoped ZnO BB.

The Density of States (DOS)

Figs. (**6a** and **6b**) show the DOS of ZnO BB and Al-ZnO BB, respectively. The ZnO BB structure shows the energy band gap value of around 1.52 eV, which is a noticeably smaller value than the regular structured ZnO (3.37 eV) [16, 17, 44].

After doping, we observed in the DOS plot that the value of the bandgap is increased. Also, as compared with undoped ZnO BB, the Al-ZnO BB structure showed some discrete peaks. The Al-ZnO BB structure had a bandgap value of around 2.95 eV. This result was analogous to the experimental results, which states that the value of the bandgap increases after doping ZnO with Al due to the fact that the average crystallite size of ZnO is decreased by the addition of Al in it and it will go on decreasing with an increasing Al concentration [17, 27, 44].

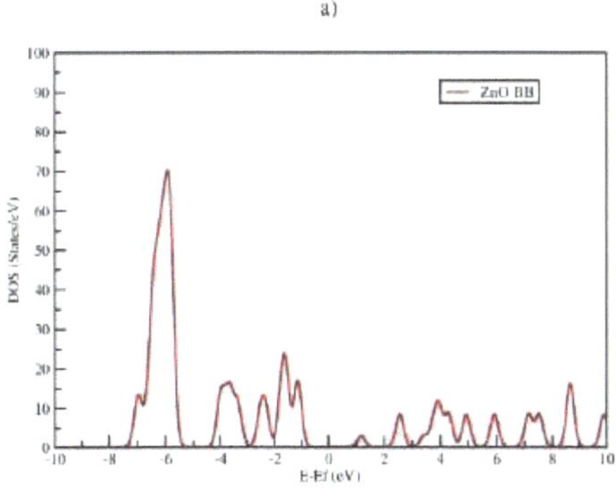 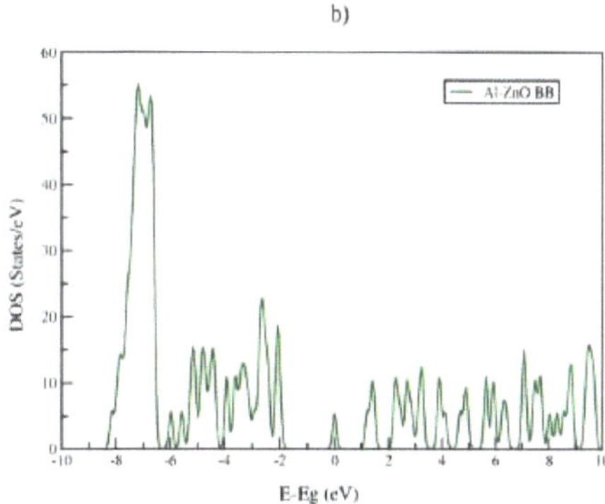

Fig. (6). The Density of States of **a)** ZnO BB structure **b)** Al-ZnO BB structure [17].

Also, it is astonishing to point out the fact that the n-type conductivity increases even after the bandgap value has been increased. This was explained earlier in a research paper that there might be no new local energy levels generated due to Al atoms; instead, the electrons hop directly to the conduction band [43]. The formation of discrete peaks in the DOS plot and an increase in the value of bandgap after doping justifies the quantum confinement effect.

The Projected Density of States (PDOS)

We obtained the PDOS plot after doping ZnO BB with an Al atom. As we can see in Fig. (**7a**), the O-2s orbital shows a peak at the VB around -23 eV. The O-2p and Zn-3d orbitals have a major contribution under the range of 0 to -15 eV in the VB region. These results resembled undoped ZnO BB. The addition of the Al atom can be seen by the contribution of its orbitals from the above graph. The Al-3p orbital has major peaks in the CB. So we can sum up the overall contributions in such a way that Zn-4s, Zn-3d, O-2s, and O-2p orbitals are the major contributors in the VB while Zn-4s and Al-3p are major contributors in the CB [17, 43]. Al-3s orbital showed some minor peaks in both regions. The donor levels in the CB of Al-ZnO BB are due to the presence of Al atoms in the system, which hybridizes with O-2p orbital and increases the conductivity [45].

The study of the cage-like structure of ZnO, ZnS, and Al-ZnO BB is a very unique and promising area of research. A DFT-based study was used to obtain the details regarding the structural and electronic properties of undoped/doped-ZnO BB and ZnS BB. It is verified by the DOS plots that the value of the bandgap of ZnO BB (1.52 eV) and ZnS BB (1.91 eV) is smaller than that of the existing values from the literature that justify the bandgap engineering through structural variations. Also, the value of the bandgap increases after Al doping in ZnO BB

(3.25 eV). The most symmetric structures are also the most stable ones, and here all the structures are stable. These are very useful results and there is further scope for a more detailed study of electronic, mechanical, optical, and magnetic properties of doped/undoped ZnO BB and ZnS BB structures. These structures will have a broad range of applications like sensors, drug delivery tools, catalysts, and optoelectronic devices.

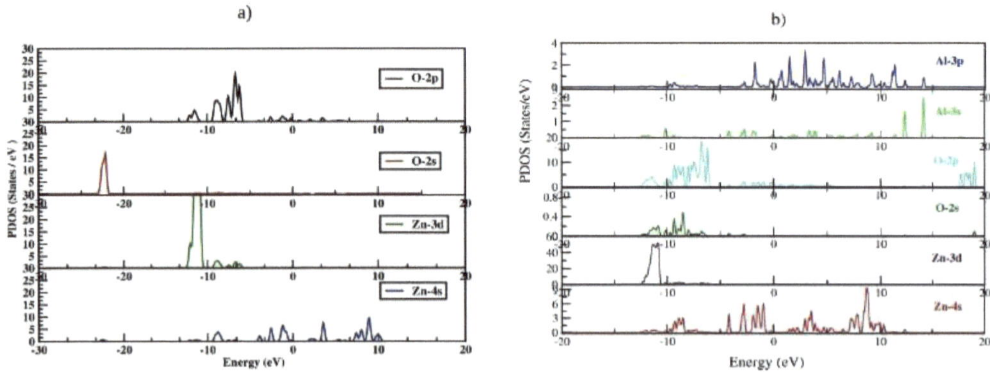

Fig. (7). The Projected Density of states of **a)** ZnO BB structure **b)** Al-ZnO BB structure [17].

APPLICATIONS OF HOLLOW STRUCTURES

Biomedical

Carbon buckyball has shown some of the most important medical applications. For example, it can be used in cancer treatment. One of the derivatives of C-60 fullerene DF1, called dendro fullerene, has shown good water solubility and is used in anti-cancer and anti-tumour treatments. Overall we can say that the high mechanical strength, high specific surface area, and unusual optical qualities are useful for their applications in drug and gene delivery, bioimaging and sensors, *etc* [3]. Not only this, they have a wide range of other applications like antioxidants, antiviral activity, photosensitizers, orthopaedic treatments, *etc*. Being a spherical hollow structure with non-toxic nature, ZnO BB might have such applications as well. There is a lot of scope for further research in this field.

Nanoreactors

The hollow nanostructures enclosing catalytic material inside them have been proven to be excellent nanoreactors. These reactors are helpful to make the reaction mechanism fast and scout the nanocrystal growth. For example, nanoreactors are made up of hollow structures of silica or cobalt nanoparticles [46].

Ferromagnetism

It has been reported that C_{60} compounds like tetrakis (dimethylamino) ethylene-fullerene (60) [TDAE-C60] and 3-aminophenyl-methano-fullerene (60) - cobaltocene has shown ferromagnetic behaviour below 17 K and 19 K, respectively. Also, palladium fullerides (C_{60} Pdn) and hydrogenated fullerenes ($C_{60}H_{36}$) might show ferromagnetism. Also with further study, it was observed that in TDAE-C_{60} with a change in the orientation of C_{60} molecules its magnetic state gets affected [47].

Solar Cells and other Optoelectronic Devices

Carbon buckyballs are proven to be good electron acceptors and thus, they have a broad area of applications in optoelectronic devices like solar cells. In organic solar cells, derivatives of fullerenes have proven to be a good active material due to their built-in bandgaps [48]. Exceeding efficiency has been reported with carbon buckyballs. Not only this, ZnO BB and its derivatives like Al-ZnO BB and composite of Graphene and ZnO BB have shown some significant results in electronic and optical properties that could benefit the solar cell industry [15, 16].

Energy Storage Devices

Hollow structures of different materials have proven to be excellent storage systems for energy, due to their trapping capacity and high reactivity. $FeCO_3$ in the form of Rambutan-like hollow microspheres were examined as anode materials for lithium batteries and they showed ideal conditions for contact with electrolyte with a low diffusion path of Li+ ions. Also, multi-shelled $Co_3O_4@Co_3V_2O_8$ hollow nano boxes and the triple-shelled $Co_3O_4@Co_3V_2O_8$ nano boxes showed controlled, high storage and better cycling capacity in Li-batteries [2].

Hydrogen Gas Storage Systems

It has been reported that the buckyball structures are immensely useful for hydrogen storage implementation. The C_{60} molecule can be hydrogenated and then dehydrogenated in a very quick and easy process, just because of the presence of C = C bond which breaks to form C-C and C-H bonds and further can break C-H bond without harming the C-C bond when heated up, due to the difference between their bond strength. Other than carbon atoms, a buckyball structure with boron atoms and with transition metals like titanium and scandium has shown some very significant results for hydrogen storage [49, 50]. Also, the Komatsu group had reported a new development in buckyball structure with atoms of sulfur used in it for trapping hydrogen molecules once they enter the cage by filling its gap at a temperature around 340 °C within 2 hours [50, 51]. Not only hydrogen but helium atoms can also be trapped in a few instances. In the case of endohedral fullerenes, metals can be trapped inside the cage of fullerenes.

Hardening of Materials

The mechanical strength of the carbon buckyball and some other fullerenes are extremely high. They might replace metals in various places in the future. Not only this, due to their resemblance with diamond structure, they can further replace diamond in various electronic devices making it more efficient and cost-effective [52].

Water Purifications

Recently, Fe_3O_4@polydopamine (PDA)-Ag hollow microspheres which looked like an urchin were investigated for their adsorption and catalytic performance to remove organic dyes like rhodamine B and methylene blue, from water [53]. Hierarchical superparamagnetic Fe_3O_4 hollow nanospheres had shown simple and fast elimination of a toxin named "Microcystin-LR" from water. Also, some carbon nitride hollow tubes showed high photocatalytic degradation of phenol [54, 55].

Applications of ZnO Buckyball Structure in Particular

The cage-like structure of ZnO has various known applications. For example, a ZnO cage structure can be used in photocatalytic activity as they show a higher value of photocatalytic degradation because of the larger number of oxygen vacancies on their surface as a result of its unique structure [56]. With its cage-like structure, non-toxic nature, and biocompatibility it can be used in biomedicines as well [57]. Due to the decrease in the value of the bandgap of the ZnO cage-like structure, better conductivity is observed here. This can be a useful result for applications in optoelectronic devices [15 - 17, 48].

Applications of Al-doped ZnO Buckyball Structure

There are various research works that are carried out particularly in the field of Aluminum metal doping in ZnO [41 - 43]. Those works showed some specific results like improvement in conductivity, transparency, and also biomedical applications of this nano composite [42, 43]. Also, there is a reported work that suggests that Al doping in ZnO induces chances of room temperature ferromagnetism in ZnO [58]. Earlier there were reports regarding ZnO doped with magnetic ions like Fe, Co, and Mn which showed magnetic properties. But, it was not clear by the studies if the induced magnetic properties are due to structural variations in ZnO or due to dopant magnetic ions [58]. So, later non-magnetic ions such as Ag or Al were introduced as a dopant in ZnO, which showed room-temperature ferromagnetism [58, 59]. Thus, there is a scope of study of the magnetic properties of Al-doped ZnO BB. This study is under progress by our group, and we hope to get some interesting results in the future.

There are many more kinds of hollow structures with numerous applications. But it was out of scope for this chapter to include them all. These are just a few to explain their major contributions.

CONCLUDING REMARKS

Hollow structures have a wide range of applications due to their configurable pore size, the high value of the specific surface area, and the large surface-to-volume ratio. But hollow structures are mostly unstable and they require some special conditions and doping to become stable. The fullerene materials had a huge impact on the field of research. Carbon buckyball structure being one of its kind showed some exceptional qualities which improvised several industries from electronics to biomedicine. With an increase in the number of applications, further investigations about more kinds of buckyball or cage-like structures of new elements are considered. Modified buckyball structures of Boron, Sulphur, Tin, Silver, *etc* came to the limelight. Further, this led to an investigation of a cage-like structure of ZnO. With the help of this idea and some previously reported works, we studied the buckyball structure of ZnO and later doped it with an aluminium atom. We also studied the ZnS buckyball structure and compared it with ZnO. These studies gave some important results like modifications in the band gaps that can be useful for optoelectronic devices. There are a huge number of applications of all kinds of buckyball structures, and it was not possible to include all of them in this chapter, so we mentioned a few from solar cells, sensors, biomedical treatments as well as devices. There is a need to study these structures more closely and widely as there is a lot more to be explored in the future.

CONSENT FOR PUBLICATION

Not applicable.

CONFLICT OF INTEREST

The author declares no conflict of interest, financial or otherwise.

ACKNOWLEDGEMENTS

The authors want to thank the management of Shri Shankaracharya Technical Campus-SSGI for granting permission to access SIESTA software using their computer lab. They also want to acknowledge the Department of Physics, NIT Raipur for their kind support.

REFERENCES

[1]   Liu, F.; Cheng, Y.; Tan, J.; Li, J.; Cheng, H.; Hu, H.; Du, C.; Zhao, S.; Yan, Y.; Liu, M. Carbon Nanomaterials With Hollow Structures: A Mini-Review. *Front Chem.,* **2021**, *9*, 668336.
[http://dx.doi.org/10.3389/fchem.2021.668336] [PMID: 33859976]

[2]   Soares, S.F.; Fernandes, T.; Daniel-da-Silva, A.L.; Trindade, T. The controlled synthesis of complex hollow nanostructures and prospective applications[†]. *Proc.- Royal Soc., Math. Phys. Eng. Sci.,* **2019**, *475*(2224), 20180677.
[http://dx.doi.org/10.1098/rspa.2018.0677] [PMID: 31105450]

[3]   Mohajeri, M.; Behnam, B.; Sahebkar, A. Biomedical applications of carbon nanomaterials: Drug and gene delivery potentials. *J. Cell. Physiol.,* **2018**, *234*(1), 298-319.
[http://dx.doi.org/10.1002/jcp.26899] [PMID: 30078182]

[4]   Dinadayalane, T.C.; Leszczynski, J. Remarkable diversity of carbon-carbon bonds: structures and properties of fullerenes, carbon nanotubes, and graphene. *Struct. Chem.,* **2010**, *21*(6), 1155-1169.
[http://dx.doi.org/10.1007/s11224-010-9670-2]

[5]   Kroto, H.W.; Heath, J.R.; O'Brien, S.C.; Curl, R.F.; Smalley, R.E. C60: Buckminsterfullerene. *Nature,* **1985**, *318*(6042), 162-163.
[http://dx.doi.org/10.1038/318162a0]

[6]   Shanbogh, P.P.; Sundaram, N.G. Fullerenes Revisited: Materials Chemistry and Applications of C_{60} Molecules. *Resonance,* **2015**, *20*(2), 123-135.
[http://dx.doi.org/10.1007/s12045-015-0160-0]

[7]   Manolopoulos, D.E.; Fowler, P.W. Structural proposals for endohedral metal-fullerene complexes. *Chem. Phys. Lett.,* **1991**, *187*(1-2), 1-7.
[http://dx.doi.org/10.1016/0009-2614(91)90475-O]

[8]   Knupfer, M. Electronic properties of carbon nanostructures. *Surf. Sci. Rep.,* **2001**, *42*(1-2), 1-74.
[http://dx.doi.org/10.1016/S0167-5729(00)00012-1]

[9]   Edelmann, F.T. Filled Buckyballs: Recent Developments from the Endohedral Metallofullerenes of Lanthanides. *Angew. Chem. Int. Ed. Engl.,* **1995**, *34*(9), 981-985.
[http://dx.doi.org/10.1002/anie.199509811]

[10]   *First "Buckyball" Molecules Created from Boron.,* https://www.scientificamerican.com/article/first-buckyball-molecules-created-from-boron/

[11]   Gonzalez Szwacki, N.; Sadrzadeh, A.; Yakobson, B.I. B80 fullerene: an Ab initio prediction of geometry, stability, and electronic structure. *Phys. Rev. Lett.,* **2007**, *98*(16), 166804.
[http://dx.doi.org/10.1103/PhysRevLett.98.166804] [PMID: 17501448]

[12]   Song, X-Z.; Qiao, L.; Sun, K-M.; Tan, Z.; Ma, W.; Kang, X-L.; Sun, F-F.; Huang, T.; Wang, X-F. Triple-shelled ZnO/ZnFe2O4 heterojunction hollow microspheres derived from Prussian Blue analog as high-performance acetone sensors. *Sens. Actuators B Chem.,* **2017**, *256*, 374-382.

[http://dx.doi.org/10.1016/j.snb.2017.10.081]

[13]   Zhao, J.; Du, Q.; Zhou, S.; Kumar, V. Endohedrally Doped Cage Clusters. *Chem. Rev.,* **2020**, *120*(17), 9021-9163.
[http://dx.doi.org/10.1021/acs.chemrev.9b00651] [PMID: 32865417]

[14]   Peyghan, A.A.; Noei, M. Electronic Response of Nano-sized Cages of ZnO and MgO to Presence of Nitric Oxide. *Chin. J. Chem. Phys.,* **2013**, *26*(2), 2.
[http://dx.doi.org/10.1063/1674-0068/26/02/231-236]

[15]   Sharma, S.; Shrivastav, A.K.; Oudhia, A.; Verma, M.L. A First principle study of structural and electronic properties of ZnO and ZnS Buckyball structures. IOP conf. Ser. *Mater. Sci. Eng.,* **2020**, *798*, 012032.

[16]   Sharma, S.; Shrivastav, A.K.; Oudhia, A.; Verma, M.L. Ab Initio Investigation on Interaction of Zig-Zag Graphene Nanoribbon and ZnO Buckyball. *J. Inst. Eng. India Ser. E,* In Press
[http://dx.doi.org/10.1007/s40034-021-00204-6]

[17]   Sharma, S.; Shrivastav, A.K.; Oudhia, A.; Verma, M.L. A first-principle density functional study of metal-doped ZnO buckyball structure for electronic applications. **2021**.
[http://dx.doi.org/10.1088/1757-899X/1120/1/012014]

[18]   Lee, J.H.; Ko, K.H.; Park, B.O. Electrical and Optical Properties of ZnO Transparent Conducting Films by the Sol-Gel Method. *J. Cryst. Growth,* **2003**, *247*(1-2), 119-125.
[http://dx.doi.org/10.1016/S0022-0248(02)01907-3]

[19]   Chen, S.; Liu, Y.; Shao, C.; Mu, R.; Lu, Y.; Zhang, J.; Shen, D.; Fan, X. Structural and optical properties of uniform ZnO nanosheets. *Adv. Mater.,* **2005**, *7*(5), 586-590.
[http://dx.doi.org/10.1002/adma.200401263]

[20]   Xu, X.; Zhuang, J.; Wang, X. SnO_2 quantum dots and quantum wires: controllable synthesis, self-assembled 2D architectures, and gas-sensing properties. *J. Am. Chem. Soc.,* **2008**, *130*(37), 12527-12535.
[http://dx.doi.org/10.1021/ja8040527] [PMID: 18715007]

[21]   Neikov, O.D.; Yefimov, N.A. *Handbook of Non-Ferrous Metal Powders,* 2nd ed; Elsevier: Amsterdam, **2019**.

[22]   Zhang, L.; Yin, L.; Wang, C.; Lun, N.; Qi, Y.; Xiang, D. Origin of Visible Photoluminescence of ZnO Quantum Dots: Defect-Dependent and Size-Dependent. *J. Phys. Chem. C,* **2010**, *114*(21), 9651-9658.
[http://dx.doi.org/10.1021/jp101324a]

[23]   Taori, V.; Shrivastava, A.K; Sharma, S.; Shukla, N.; Oudhia, A. Growth and characterizations of chemically deposited ZnO: Ag NWs 1. *International Journal of Luminescence and applications,* https://www.researchgate.net/publication/334746877_Growth_and_characterizations_of_chemically_d eposited_ZnOAg_NWs_1?_sg=bKd5vP5VsPPliVLADKBzKZ8KEaVyFdRFw5rqyNiSwNY5u1mEpx o2SWDOk6KeqETTctnBOFVLb73QtIKPeeMm4nYGnjKoGFaFt_dI9AN2.0Fx0xo3WpKBcep5zjXZ RKGYCsOiU0GTLTMsk8dLzBS3ch_begzu_I8bGOO8vlocz5ul6gTsSHgnxfYrZPVOolQ**2019**.

[24]   Verma, H.; Rao, B.K.; Verma, M.L.; Chauhan, J. First-principles study of breaking energy and mechanical strength of Kevlar-29. *Bull. Mater. Sci.,* **2019**, *42*(2), 2.
[http://dx.doi.org/10.1007/s12034-019-1747-y]

[25]   Soler, J.M.; Artacho, E.; Gale, J.D.; García, A.; Junquera, J.; Ordejón, P.; Portal, D.S. The SIESTA method for ab initio order-N materials simulation. *J. Phys. Condens. Matter,* **2002**, *14*(11), 2745-2779.
[http://dx.doi.org/10.1088/0953-8984/14/11/302]

[26]   Siesta web page. *siesta,* https://departments.icmab.es/leem/siesta/About/overview.html

[27]   Hou, X.; Xie, Z.; Li, C.; Li, G.; Chen, Z. Study of Electronic Structure, Thermal Conductivity, Elastic and Optical Properties of α, β, γ-Graphyne. *Materials (Basel),* **2018**, *11*(2), 188.
[http://dx.doi.org/10.3390/ma11020188] [PMID: 29370070]

[28]   Cao, G. *Nanostructures & Nanomaterials,* 1ˢᵗ edition; Imperial College 283 Press: London, **2003**.

[29]   Raton, B.; Moore, Lesley. E; Smart, Elaine., Eds. *Journal of applied crystallography*; Taylor & Francis: Florida, **2006**, Vol. 39, p. 288.

[30]   Obayes, H.R.; Al Obaidy, A.H.; Alwan, G.H.; Al-Amier, A.A. Molecular simulation for novel carbon buckyball materials. *Cogent Chem.,* **2015**, *1*(1), 1.
[http://dx.doi.org/10.1080/23312009.2015.1026638]

[31]   Lindquist, B.A.; Dunning, T.H., Jr The nature of the SO bond of chlorinated sulfur–oxygen compounds. *Theor. Chem. Acc.,* **2014**, *33*(2), 1443.
[http://dx.doi.org/10.1007/s00214-013-1443-8]

[32]   Supatutkul, C.; Pramchu, S.; Jaroenjittichai, A.P.; Laosiritaworn, Y. Electronic properties of two-dimensional zinc oxide in hexagonal, (4,4)-tetragonal, and (4,8)-tetragonal structures by using Hybrid Functional calculation. *IOP Conf. Series: Journal of Physics: Conf. Series,* **2017**.

[33]   Global, I.G.I. *What is Density of States.,* https://www.igi-global.com/dictionary/density--f-states/40458

[34]   Selvaraju, K.; Jothi, M.; Kumaradhas, P. Probing the Effect of Applied Electric Field in Charge Density Distribution and Electrostatic Properties of Au Substituted Saturated Polycyclic Hydrocarbon Molecular Nanowires *via* Quantum Chemical and Charge Density Study. *J. Comput. Theor. Nanosci.,* **2014**, *10*(1), 1-26.
[http://dx.doi.org/10.4086/toc.2014.v010a001]

[35]   Djurovich, P.I.; Mayo, E.I.; Forest, S.R.; Thompson, M.E. Measurement of the lowest unoccupied molecular orbital energies of molecular organic semiconductors. *Org. Electron.,* **2009**, *10*(3), 515-520.
[http://dx.doi.org/10.1016/j.orgel.2008.12.011]

[36]   Rangel, T.; Rignanese, G.M.; Olevano, V. Can molecular projected density of states (PDOS) be systematically used in electronic conductance analysis? *Beilstein J. Nanotechnol.,* **2015**, *6*, 1247-1259.
[http://dx.doi.org/10.3762/bjnano.6.128] [PMID: 26171300]

[37]   Haus, J.W. *Fundamentals and Applications of Nanophotonics,* 1st ed; Elsevier Woodhead Publishing: USA, **2016**.

[38]   Kamarulzaman, N.; Kasim, M.F.; Rusdi, R. Band Gap Narrowing and Widening of ZnO Nanostructures and Doped Materials. *Nanoscale Res. Lett.,* **2015**, *10*(1), 1034.
[http://dx.doi.org/10.1186/s11671-015-1034-9] [PMID: 26319225]

[39]   Sun, Y.; Xiaiea, S.; Meng, X. A first-principles study on hydrogen in ZnS: Structure, stability, and diffusion. *Phys. Lett. A,* **2014**, *379*(5), 487-490.
[http://dx.doi.org/10.1016/j.physleta.2014.12.002]

[40]   Upma, V.; Verma, M.L. M.L. First-Principles Approach to Study the Structural, Electronic and Transport Properties of Dimer Chitosan with Graphene Electrodes. *J. Electron. Mater.,* **2019**, *48*(6), 4007-4016.
[http://dx.doi.org/10.1007/s11664-019-07163-0]

[41]   Lewis, B.G.; Paine, D.C. Applications and Processing of Transparent Conducting Oxides. *MRS Bull.,* **2011**, *25*(8), 22-27.
[http://dx.doi.org/10.1557/mrs2000.147]

[42]   Murugesan, M.; Arjunraj, D.; Mayandi, J.; Venkatachalapathy, V.; Pearce, J.M. Properties of Al-doped zinc oxide and In-doped zinc oxide bilayer transparent conducting oxides for solar cell applications. *Mater. Lett.,* **2018**, *222*, 50-53.
[http://dx.doi.org/10.1016/j.matlet.2018.03.097]

[43]   Maldonado, F.; Stashans, A. Al-doped ZnO: Electronic, electrical and structural properties. *J. Phys. Chem. Solids,* **2010**, *71*(5), 784-787.
[http://dx.doi.org/10.1016/j.jpcs.2010.02.001]

[44]   Mallika, A.N.; Reddy, A.R.; SowriBabu, K.; Reddy, K.V. Synthesis and optical characterization of aluminum-doped ZnO nanoparticles. *Ceram. Int.,* **2014**, *40*(8), 12171-12177.
[http://dx.doi.org/10.1016/j.ceramint.2014.04.057]

[45]   Palacios, P.; Sanchez, K.; Wahnon, P. Ab-initio valence band spectra of Al, In doped ZnO. *Thin Solid Films,* **2009**, *517*(7), 2448-2451.
[http://dx.doi.org/10.1016/j.tsf.2008.11.037]

[46]   Lee, J.; Kim, S.; Lee, I. Functionalization of hollow nanoparticles for nanoreactor applications. *Nano Today,* **2014**, *9*(5), 631-667.
[http://dx.doi.org/10.1016/j.nantod.2014.09.003]

[47]   Makarova, T.L.; Sundqvist, B.; Höhne, R.; Esquinazi, P.; Kopelevich, Y.; Scharff, P.; Davydov, V.A.; Kashevarova, L.S.; Rakhmanina, A.V. Magnetic carbon. *Nature,* **2001**, *413*(6857), 716-718.
[http://dx.doi.org/10.1038/35099527] [PMID: 11607027]

[48]   Zhang, Z.; Wei, L.; Qin, X.; Li, Y. Carbon nanomaterials for photovoltaic process. *Nano Energy,* **2015**, *15*, 490-522.
[http://dx.doi.org/10.1016/j.nanoen.2015.04.003]

[49]   Yildirim, T.; Íñiguez, J.; Ciraci, S. Molecular and dissociative adsorption of multiple hydrogen molecules on transition metal decorated C60. *Phys. Rev. B Condens. Matter Mater. Phys.,* **2005**, *72*(15), 153403-153404.
[http://dx.doi.org/10.1103/PhysRevB.72.153403]

[50]   Komatsu, K.; Murata, M.; Murata, Y. Encapsulation of molecular hydrogen in fullerene C60 by organic synthesis. *Science,* **2005**, *307*(5707), 238-240.
[http://dx.doi.org/10.1126/science.1106185] [PMID: 15653499]

[51]   Mohan, M.; Sharma, V.K.; Kumar, E.A.; Gayathri, V. *Hydrogen storage in carbon materials- A review*; Energy Storage, **2019**, p. 1.

[52]   Modern Synthesis Technology. *Fullerene Uses and Applications – The Comprehensive Overview.,* https://mstnano.com/fullerene-uses/

[53]   Cui, K.; Yan, B.; Xie, Y.; Qian, H.; Wang, X.; Huang, Q.; He, Y.; Jin, S.; Zeng, H. Regenerable urchin-like Fe$_3$O$_4$@PDA-Ag hollow microspheres as catalyst and adsorbent for enhanced removal of organic dyes. *J. Hazard. Mater.,* **2018**, *350*, 66-75.
[http://dx.doi.org/10.1016/j.jhazmat.2018.02.011] [PMID: 29453121]

[54]   Narasimha, G.; Mahesh, A.G.; Manorama, S.V. Citrate Stabilized Hierarchical SPIO Nanostructures: Synthesis and Application Towards Effective Removal of Toxin, Microcystin-LR from Water. *ChemistrySelect,* **2017**, *2*(18), 5226-5233.
[http://dx.doi.org/10.1002/slct.201700664]

[55]   Tian, L.; Li, J.; Liang, F.; Wang, J.; Li, S.; Zhang, H.; Zhang, S. Molten salt synthesis of tetragonal carbon nitride hollow tubes and their application for removal of pollutants from wastewater. *Appl. Catal. B,* **2017**, *225*, 307-313.
[http://dx.doi.org/10.1016/j.apcatb.2017.11.082]

[56]   Yang, S.; Wang, J.; Li, X.; Zhai, H.; Han, D.; Wei, B.; Wang, D.; Yang, J. One-step synthesis of bird cage-like ZnO and other controlled morphologies: Structural, growth mechanism and photocatalytic properties. *Appl. Surf. Sci.,* **2014**, *319*, 211-215.
[http://dx.doi.org/10.1016/j.apsusc.2014.07.165]

[57]   Wang, Z.L. Nanostructures of zinc oxide. *Mater. Today,* **2004**, *7*(6), 26-33.
[http://dx.doi.org/10.1016/S1369-7021(04)00286-X]

[58] Silvearv, F.; Rosa, A.L.; Lebègue, S.; Ahuja, R. An ab-initio study of (Mn,Al) doped ZnO including strong correlation effects. *Physica E,* **2012**, *44*(6), 1095-1097.
[http://dx.doi.org/10.1016/j.physe.2010.12.007]

[59] Ali, N.; A R, V.; Khan, Z.A.; Tarafder, K.; Kumar, A.; Wadhwa, M.K.; Singh, B.; Ghosh, S. Ferromagnetism from non-magnetic ions: Ag-doped ZnO. *Sci. Rep.,* **2019**, *9*(1), 20039.
[http://dx.doi.org/10.1038/s41598-019-56568-8] [PMID: 31882806]

SUBJECT INDEX

A

Absorption 33, 42, 100, 127, 134, 135, 177
  ozone 177
  photon 127
  properties 134
  spectrum 135
  visible-light 177
Acetylacetonate 100
Acid 18, 96, 97, 99, 100, 145, 179, 180
  ascorbic 97, 99
  Bronsfed 18
  carboxylic 96
  Lewis 179
  nitric 100
  nucleic 145, 180
  triflic 18
Action, photocatalytic 177
Activated carbon 92, 95, 96, 97, 100, 101, 103, 105, 182
  commercial 105
Activity, photocatalytic 227
Adsorption 1, 3, 4, 5, 6, 9, 10, 11, 14, 15, 16, 17, 33, 44, 45, 74, 75, 76, 77, 78, 80, 85, 97, 102, 144
  adsorbents post 102
  capacity 9, 97
  energy 1, 3, 4, 6, 15, 16, 17
  of nanoparticles 144
  of nitrogen molecule 80
  process 14, 78
  properties 44, 45
Ag electrodes 196
Agents 103, 139
  chemotherapeutic 103
Age-related macular degeneration 34
Agglomerate 96, 134
Agglomeration 94, 97, 102, 182
Alloys 75, 95, 96, 159, 162, 165, 167, 168
  iron-based 162
  superelastic 165
Amalgamation 92
Ammonia adsorption 5

Analysis 10, 158, 159, 162
  elastic recoil detection 10
  finite element 158, 159, 162
  theoretical 10
Antibacterial 120, 139, 140, 142, 143, 144, 145
  activity 120, 140, 142, 143, 144, 145
  agent 139, 140
Antibiotic drugs 141
Antiferromagnetic coupling 133
Anti-microbial activities 120
Anti-tumour treatments 225
Applications 92, 104, 132, 195, 199, 201
  electronic 199
  multi-faceted 92
  of optoelectronic devices 195
  orthopedic 104
  spintronic 132, 201
Atomic packing fraction (APF) 218

B

Bacteria 67, 139, 140, 141, 145
  drug-resistant 141
  infectious 140, 145
Bacterial growth 121, 139, 140, 141
  inhibiting 121, 139
Bacteriemia 141
Band gap energy 139
Behavior 93, 121, 122, 135
  anti-ferromagnetic 135
Biomedical 4, 93, 104, 145, 120, 122, 174, 225, 228
  agents 120
  applications 4, 93, 104, 145, 174, 228
  treatments 228
Biosensors 214, 216
Boltzmann constant 7
Boron nitride 1, 4, 5, 7, 12, 13, 14, 15, 16, 18
  bilayer 15
  hybrid 16
  nanosheets 5, 7, 13, 15, 18

hydrogen-fueled power 2
Pollutants 2, 92, 102, 180
  harmful 2
  organic 102
  ranging 92
Polymer Electrolyte Membrane (PEM) 82,
    103
  fuel cell (PEMFC) 103
Polymer matrix nanocomposites (PMNC) 94
Pressure 1, 2, 3, 7, 9, 10, 11, 12, 40, 66, 67,
    82, 86, 98
  catalysts 82
  conditions 66, 67
  equilibrium 7
  storage method 3
Processes 6, 8, 34, 54, 58, 68, 69, 76, 77, 79,
    97, 102, 108, 122, 126, 168, 175, 178,
    176, 179
  based catalytic 175
  capacity fade 58
  desorption 6, 8
  eclectic hydrogenation 178
  hybrid 69
  light emission 126
  metal-based 68
  nanotechnological 108
  organic 179
  semiconductor photocatalytic 176
  solid phase extraction 102
Production 62, 66, 68, 74, 77, 81, 82, 83, 122,
    145, 172, 177, 197
  electron-hole pair 197
  of ammonia 66, 68, 74, 77, 82, 83
  waste products 62
Projected density of states (PDOS) 8, 9, 14,
    221, 224, 225
Properties 40, 44, 93, 108, 109, 120, 121, 122,
    124, 125, 129, 132, 133, 135, 136, 137,
    138, 139, 140, 141, 144, 159, 162, 170,
    176, 213, 216, 218
  antibacterial 141, 144
  biological 120
  chemical 108, 122, 125
  desired 40, 170
  dielectric 44
  elastic 162, 218
  electrical-optical 140
  fascination 135
  ferromagnetic 129, 137
  optoelectronic 109

physicochemical 176
  sensing 136
Proteins 67, 145, 180
  cellular 67
Pseudoelasticity 168
Pseudomonas 140
  aeruginosa 140
  fluorescens 140

Q

Quantum 190, 191, 193, 202
  computation 190, 202
  computing 191, 193
Quantum dots (QDs) 124, 175, 192, 216
  semiconductors 124

R

Raman 98, 108, 118
  analysis 98
  spectra of binary solutions of water 118
  spectroscopy 108, 118
Reaction 55, 62, 67, 68, 70, 74, 76, 78, 83, 84,
    85, 86, 97, 132, 172, 176, 179, 181
  cathodic 84
  chemical 62, 70, 84, 132
  electrochemical catalytic reduction 67
  electrochemical reduction 68
  organic 179
  oxidation 55, 181
Reactive oxygen species (ROS) 142, 144, 145
Reduction 58, 66, 93, 96, 98, 121, 129, 134,
    173, 177, 217, 223
  photocatalytic 173
  sustainable 66
  reactions 98, 177
  treatments 96
Resistance 93, 120, 139, 141
  antibiotic 120
  microbial 139
Resources, non-sustainable energy 170
Retinal 33, 34, 35, 38, 42, 48
  ganglion cells (RGC) 34, 38
  implant applications 34, 42, 48
  implant system 33
  prosthetic projects 33
  tissues, damaged 35
Retinitis pigmentosa 34